Mordecai Cubitt Cooke

Mordecai Cubitt Cooke

Victorian Naturalist, Mycologist, Teacher & Eccentric

Mary P English

© Biopress Ltd 1987

All rights reserved. No part of this publication may be reproduced, stored in a retrieval system, or transmitted, in any form or by any means, electronic, mechanical, photocopying, recording or otherwise, without the prior permission of the copyright owner.

ISBN 0-948 737-02-6

PUBLISHED BY:

Biopress Ltd.
'The Orchard'
Clanage Road
Bristol
BS3 2JX
England

Publication of this book has been assisted by a grant from the British Mycological Society

British Library Cataloguing in Publication Data

English, Mary
   Mordecai Cubitt Cooke : Victorian Eccentric,
   educationalist, naturalist and mycologist.
   1. Cooke, Mordecai Cubitt 2. Mycologists—
   Great Britain—Biography
   I. Title
589.2'092'4   QK600.45.C6/
   ISBN 0-948737-02-6

Printed in Great Britain by Henry Ling Ltd., at the Dorset Press, Dorchester, Dorset.

To the memory of
  my mother
    GLADYS NELLIE née CUBITT
  and father
    MARCUS CLAUDE ENGLISH

Contents

Acknowledgements	vii
1. Sources	1
2. Cookes and Cubitts. 1825–1850	7
3. Holy Trinity School, Lambeth. 1851–1860	23
4. Out of School-Hours. 1851–1860	39
5. Interregnum. 1860–1862	53
6. The India Museum. Family Matters. 1862–1880	73
7. Lovers of Nature. 1862–1866	94
8. The Quekett Microscopical Club. 1865–1880	119
9. The Making of a Mycologist. 1862–1870	149
10. 'A maze of Correspondence'. 1868–1880	171
11. Royal Botanic Gardens, Kew. 1880–1884	204
12. Years of Disaster. 1885–1892	233
13. 'Forays Amongst the Funguses'. 1880–1889	250
14. Missionary for Mycology. 1890–1892	269
15. Troubled Retirement. 1893–1898	279

16. 'Nature Never Did Betray the Heart
    that Loved Her'. (W. Wordsworth)
    1899–1914   298

    Postscript   325

    Post-Postscript   329

    Appendix A. Chronological Table of the Life & Work
    of Mordecai Cubitt Cooke.   332

    Appendix B. List of Works by M. C. Cooke.   337

    References & Bibliography   342

    Index   350

Acknowledgements

This biography was made possible only by the enthusiastic collaboration, endless patience, skill and knowledge of mycological friends and colleagues, of my family, of Mordecai Cooke's relatives and their friends, and of librarians, archivists and historians all over Britain and abroad. My most grateful thanks are extended to them all. The specific contributions of some are described in Chapter 1 and elsewhere in the text: here I would like to offer more general thanks.

Dr. G. C. Ainsworth, with his unrivalled knowledge of mycology and its history, was generous with ideas and encouragement throughout, and his unique *Introduction to the History of Mycology* was indispensible. Dr. M. F. Madelin (University of Bristol) read and criticised an early draft of the mycological chapters, and he and Dr. P. M. Stockdale (Commonwealth Mycological Institute) were always ready to discuss my problems and give advice.

For information about the Cooke and Cubitt families, and about Cooke's early life in Norfolk, I am deeply indebted to my distant cousin, Mr. D. Cubitt, who also prepared the family trees and checked relevant parts of the text, and to Mr. F. Sayer (Norwich Central Library). Between them they not only disinterred much fascinating family detail, but discovered some of Cooke's living relatives.

As background for the Natural history sections of the book I made extensive use of Mr. D. E. Allen's superb study, *The Naturalist in Britain*, and Mr. Allen was kind enough to read critically an early draft of these sections. Much material was obtained from the library and archives of the Quekett Microscopical Club, from whose officers, especially Mr. D. A. Chalkley, Mr. A. L. F. Barron, Mr. D. Perrott and the late Mr. E. P. Herlihy, I received unstinted help.

During my lengthy researches at the Royal Botanic Gardens, Kew, Mr. P. S. Green, Deputy Director (now retired), was a constant source of help and guidance, as were the mycologists of the Herbarium and the archivist, Miss E. Smith. The Botany Library and Archives of the British Museum (Natural History) were the site of many of Cooke's letters and drawings and of other original sources, and I am most grateful to Mrs. J. Diment and her predecessors who guided me to these. Miss J. A. Moore (BM(NH)) and Prof. F. E. Round (Bristol University) commented on Cooke's phycology and Mr. K. Hyatt on his entomology.

Professor D. P. Rogers (University of Illinois, U.S.A) took endless pains to locate sources and check facts concerning Cooke's connections with American mycologists and institutions.

Dr. S. Buczacki (Stratford-upon-Avon) was a mine of information concerning the Rev. M. J. Berkeley and his times. Dr. R. Watling (Royal Botanic Gardens, Edinburgh) and Dr. M. Noble delved into Cooke's work for Scottish mycology, and Dr. J. Walker (Rydalmere, New South Wales) and Mr. R. Hilton (University of Western Australia) did the same for his Australian contributions. Miss A. V. Brooks drew my attention to the existence of Cooke's microscope at the Royal Horticultural Society's gardens at Wisley.

Mr. D. Mellor guided me through the maze of the Victorian education system. The discovery of details of Cooke's employment at the India Museum was due to Mr. R. Desmond (India Office Library). Mr. D. Muspratt was tireless in uncovering the connections of Mordecai and Ebenezer Cooke with the Working Men's College. Mr. E. A. Ellis (lately of the Castle Museum, Norwich) provided information and photographs of the Cooke family, and my sister, Miss S. I. F. English, was responsible for tracing Ebenezer's descendents.

My most grateful thanks are extended to the Librarians, Archivists, Curators and Historians of the following institutions, and to others whom space does not allow me to name, who went out of their way to aid my researches: Islington Central Library (Especially Mr. E. A. Willets), St. Pancras Library (especially Mr. W. Maidment), Marylebone Public Library, Lambeth Central Library, Leatherhead Branch Library, Stockton-on-Tees Central Library, Carlisle Public Library, Department of Education and Science Library, Imperial College of Science and Technology Library and Archive (especially Mrs. J. Pingree), the National Society, the Baptist Union, the Society for the Propagation of Christian Knowledge, the Royal Society of Arts, the Primrose League, the Linnean Society, the Royal Horticultural Society's Library, the Passmore Edwards Museum, Stratford, the Tolson Museum, Huddersfield (especially Mr. Aubrook), Messrs Wheldon and Wesley, Biological Booksellers, St. Lawrence University, New York, Yale University, Acadamy of Natural Sciences, Philadelphia, Farlow Reference Library, Harvard University, Instituto di Botanico dell'Universita Padova, Italy. The Medical Illustration Department of Bristol University was of great assistance with the illustrations.

To the late Miss L. A. Cooke I owe almost all the more personal details about her father that remain to us, and much more besides. His granddaughters, Mrs. D. Wynn (Kent), Mrs. E. M. Girard and Mrs. M. Dunkinson (Canada), Mrs. M. Coward and Mrs. B. Hobson (Australia) and Mrs. A. Evans (New Zealand) all supplied information essential to the piecing together of Cooke's family life. His great nieces the late Miss E. Cooke and Mrs. H. Geary of Norwich and Miss K. Evans (London), and great nephew, Mr. W. H. Barrett, and a more distant relative, Mrs. M. Drury (Herts.), were very helpful, and Miss Leila Cooke's devoted friends Miss A. H. Bulley and Dr. D. Cotton gave me much much help and encouragement.

Other friends, colleagues and fellow members of the British Mycological society whom, unfortunately, space does not allow me to mention by name, gave endless support, information and gifts of copies of Cooke's books, of relics and photographs. To them all I am immensely grateful.

I am grateful to Mrs. D. Wynne, holder of the copyright in the correspondence of M. C. Cooke, for permission to publish extracts from some letters.

For permission to use illustrations I am indebted to the following copyright holders or owners of originals copied:

American Phytopathological Society, frontispiece.
Mr. E. A. Ellis, Figs. 2/2, 6/4.
Mr. A. L. E. Barron, Fig. 2/3.
Mr. W. H. Barrett, Fig. 2/4.
British Library, Fig. 3/1 (pressmark 10825 e 25).
Quekett Microscopical Club, Figs. 3/2, 3/6, 5/2, 5/4, 7/2, 9/12, 10/7.
Department of Education and Science Library, Fig. 3/5
Working Men's College, Fig. 4/1.
Museum of London, Figs. 5/3, 6/1.
Illustrated London News, Figs. 5/7, 5/8, 5/9, 6/2
Mrs. D. Wynne, Fig. 5/10.
Copyright of the Trustees © R. B. G. Kew, 1985. Reproduced with permission, Figs. 6/5, 10/5, 10/6, 11/1, 11/2, 11/3, 11/4, 11/11, 13/3, 16/1, 16/7.
Mrs. B. Hobson, Figs. 6/6, 12/1, 15/2, 15/3.
Gardeners' Chronicle, Fig. 7/1.
University College, London, Fig. 7/7.
Leatherhead and District Local History Society, Fig. 8/2.
British Mycological Society, Figs. 8/3, 8/6, 10/13, 13/1, 13/2, 13/4, 13/6, 14/1, 15/7, 15/8, 16/4.
Hereford and Worcester County Libraries, Fig. 8/4
Woolhope Naturalists' Field Club, Figs. 8/5, 13/5.
Dr. S. Buczacki, Figs. 9/1, 9/2.
Constable Publishers, Figs. 9/4, 16/5.
Royal Horticultural Society, Lindley Library, Figs. 9/8, 9/9, 13/7.
Mr. E. P. Herlihy, Figs. 9/10, 16/6.
British Museum (Natural History), Botany Library, Fig. 9/11.
Farlow Reference Library and Herbarium, U.S.A., Figs. 10/1, 10/9, 11/7, 11/9, 11/10.
The Essex Institute, U.S.A., Fig. 10/2.
University of Padua, Italy, Fig. 10/10.
Vineland Historical and Antiquarian Society, U.S.A., Fig. 10/11.
Mrs. K. Evans, Fig. 12/2.

Mrs. M. Coward, Fig. 12/3.
Passmore Edwards Museum, Fig. 14/2.
Basil Blackwell, Oxford and Mrs. E. Dobbs, Fig. 14/4.
Messrs. Wheldon and Wesley, Fig. 15/1.

List of Illustrations

Figure.		Page.
Frontispiece	Mordecai Cubitt Cooke, 1825–1914.	
2.1	Neatishead Baptist Chapel.	9
2.2	Mordecai Cooke, 1799–1869.	11
2.3	Horning Post Office, birthplace of M. C. Cooke.	12
2.4	Cottage where M. C. Cooke attended day school.	16
2.5	St. Alphege's Church, Greenwich.	21
3.1	The Rev. James Gillman, vicar of Holy Trinity, Lambeth.	25
3.2	Holy Trinity School, Lambeth, not later than 1930.	27
3.3	Holy Trinity School, Lambeth, in 1975.	27
3.4	Holy Trinity School, Lambeth, in 1975. Score marks on the bricks.	27
3.5	Engraving by M. C. Cooke illustrating his article in *The School and the Teacher*.	34
3.6	Handbill announcing M. C. Cooke's evening classes.	37
4.1	Ebenezer Cooke, 1837–1913.	41
4.2	M. C. Cooke's illustration of *Amanita virosa*.	44
5.1	No. 8, Winchester Street, Pimlico.	54
5.2	Dr. Edwin Lankester, 1814–1874.	57
5.3	No. 192, Piccadilly, in 1840.	59
5.4	Robert Hardwicke, 1822–1875	60
5.5	Page from *A Manual of Structural Botany*.	61
5.6	The honey fungus, an illustration from *British Fungi*.	63
5.7	The South Kensington building for the Great International Exhibition of 1862.	68
5.8	The New Zealand Pavilion at the 1862 International Exhibition.	68
5.9	Kentish Town in 1862.	69
5.10	Annie Elizabeth Thornton Biggs, 1844–1920	70
6.1	Fife House, Whitehall Yard, in 1805.	74
6.2	The India Museum, Fife House, in 1861.	75

6.3	No. 146, Junction Road, in 1974.	81
6.4	Harry Linnaeus Cooke, 1862–1873.	87
6.5	M. C. Cooke aged 48.	87
6.6	Annie (Cooke) Cubitt, née Biggs, aged about 40.	92
7.1	Worthington George Smith, 1835–1917.	99
7.2	Frontispiece for the Proceedings of the Society of Amateur Botanists.	102
7.3	Illustration by W. G. Smith from *Science Gossip*.	105
7.4	Ebenezer Cooke's frontispiece for *Our Reptiles*.	106
7.5	Illustration by Ebenezer Cooke for *Our Reptiles*.	107
7.6	Title page of *Hardwicke's Science Gossip*, Vol. 2.	108
7.7	General Library of University College, London.	117
8.1	Illustration from *A Fern Book for Everybody*.	123
8.2	The Swan Hotel, Leatherhead.	126
8.3	En route to a Woolhope excursion.	137
8.4	Dr. Henry Graves Bull	137
8.5	The Announcement of the first fungus foray.	138
8.6	The Woolhope Club setting out on a fungus foray in the rain.	139
8.7	T. H. Huxley's understanding of the relationship between Torula (yeast), Penicillium and Bacterium.	142
8.8	Lichens.	144
9.1	The Rev. Miles Joseph Berkeley, 1803–1889.	151
9.2	Christopher Edmund Broome, 1812–1886.	153
9.3	Description of the genus Valsa, from Cooke's *Handbook of British Fungi*.	156
9.4	Original drawing by M. C. Cooke of *Sphaerella inaequalis*.	156
9.5	Illustration by J. E. Sowerby of infections by smut fungi.	160
9.6	A volume of Cooke's Exsiccati.	161
9.7	Pages from Cooke's Exsiccati.	161
9.8	Cooke's microscope and accessories.	165
9.9	Note accompanying Cooke's microscope to the Royal Horticultural Society.	166
9.10	M. C. Cooke, aged about 85 and almost blind, with his microscope.	166
9.11	Part of a letter from M. C. Cooke to C. E. Broome.	166
9.12	Cooke's erroneous drawing of the sporing head of a species of *Aspergillus*.	168

10.1	Henry W. Ravenel.	172
10.2	The Rev. E. C. Bolles.	174
10.3	Title page of *The Handbook of British Fungi*, Vol. 2	176
10.4	Frontispiece of *The Handbook of British Fungi*, Vol. 2.	176
10.5	Title page of *Grevillea*, Vol. I.	180
10.6	Illustration from an article by W. G. Smith in *Grevillea*.	180
10.7	Undated portrait of M. C. Cooke.	183
10.8	Illustration from *Mycographia*.	189
10.9	Charles H. Peck.	191
10.10	P. A. Saccardo.	191
10.11	Job B. Ellis.	191
10.12	Illustration by M. J. Berkeley of a section of a potato leaf invaded by the blight fungus.	197
10.13	Microscopists examining the potato blight fungus.	197
10.14	Illustration by J. Rostafinski from Myxomycetes of Great Britain.	199
11.1	J. D. Hooker, Director of the Royal Botanic Gardens, Kew, 1865–1885.	207
11.2	W. Thiselton-Dyer, Director of the Royal Botanic Gardens, Kew, 1885–1905.	207
11.3	Museum No. 1, Royal Botanic Gardens, Kew.	212
11.4	Hunter House, Royal Botanic Gardens, Kew.	213
11.5	*Coprinus comatus*, the lawyer's wig or shaggy cap fungus.	215
11.6	Plate 585 of the *Illustrations of British Fungi*.	216
11.7	W. G. Farlow.	224
11.8	Illustration from *British Algae*.	228
11.9.	Part of a letter from M. C. Cooke to W. G. Farlow.	229
11.10	Undated postcard from M. C. Cooke to W. G. Farlow.	230
11.11	Baron von Mueller.	231
12.1	Ernest Frederick Cooke, 1868–1885.	234
12.2	Ebenezer Cooke, 1837–1913.	236
12.3	M. C. Cooke's Primrose League Medal with Bars.	240
12.4	Illustration from *British Desmids*.	242
12.5	Illustration from *Vegetable Wasps and Plant Worms*.	247
13.1	The French mycologist, M. Le Cornu.	259
13.2	Setting out on a foray.	259
13.3	Foray sketches by M. C. Cooke.	260
13.4	Collecting bracket fungi.	260

13.5	Menu for the Woolhope Club Dinner, 1877.	262
13.6	The Rev. M. J. Berkeley presiding at a Woolhope Club Dinner.	263
13.7	Dr. Henry Bull as seen by Cooke.	265
13.8	Truffle hunting at a Woolhope foray.	267
14.1	"Steady old fogies" examining an earth star.	271
14.2	An Essex Field Club foray, 1910.	272
14.3	The parasol mushroom, *Lepiota procera*.	274
14.4	A giant puff-ball, *Lycoperdon giganteum*.	275
14.5	M. C. Cooke's Christmas card to Miss Rose.	278
15.1	Wheldon's bookshop at 58, Great Queen Street.	282
15.2	Leila Annie Cooke, 1882-1977.	283
15.3	Herbert Stephen (Bertie) Cooke, 1879–1927.	283
15.4	Coloured illustration from *Rambles among the Wild-Flowers*.	286
15.5	Drawing from *Rambles among the Wild-Flowers*.	287
15.6	*Amanita phalloides*, the death-cap toadstool, from *Edible and Poisonous Mushrooms*.	288
15.7	Founders of the British Mycological Society.	290
15.8	The British Mycological Society's Autumn Foray, 1912.	291
15.9	Alfred Clarke of the Yorkshire Naturalists' Union.	292
16.1	The Linnean Society's Gold Medal presented to M. C. Cooke.	301
16.2	Plate from *Fungoid Pests of Cultivated Plants*.	304
16.3	Illustration from *Fungiod Pests of Cultivated Plants*.	305
16.4	M. C. Cooke aged 78.	306
16.5	The life cycle of the black stem rust fungus of wheat.	308
16.6	M. C. Cooke in old age.	316
16.7	Part of the first page of Cooke's autobiographical notes.	319
16.8	M. C. Cooke's gravestone at Finchley (now Islington) cemetery.	322
16.9	The grave in 1974	323
	Family Tree 1	8
	Family Tree 2	84
	Census, 1881	90

Foreword

During the nineteenth century British mycology was dominated in turn by two men–the Rev. Miles Joseph Berkeley and his successor Mordecai Cubitt Cooke–who both gained international reputations for their studies of fungi. Berkeley, after a conventional public school and university education, took holy orders and spent an uneventful life as a country clergyman, supplementing his stipend by keeping a school for boys, and by journalism. Cooke, the son of a village shopkeeper, Mordecai Cooke, and his wife, Mary Cubitt, was of non-conformist Norfolk stock. He received little formal education, was largely self-taught, and the details of his life-long struggle to earn a living in a diversity of occupations, which overlay a complex domestic situation, and his missionary zeal in promoting the study of natural history surely qualify him for a place among the minor English eccentrics.

Today Cooke's name is familiar to taxonomic mycologists throughout the world and his *Handbook of British Fungi* of 1871, if seldom consulted, is still the latest British fungus flora covering all groups, while the eight volumes of his *Illustrations of British Fungi (Hymenomycetes)* with their 1200 coloured plates are prized by both agaricologists and bibliophiles. His life-style and eccentricities, however, were forgotten or unknown until Dr. Mary English, herself a professional mycologist and a distant relative of Cooke (through her mother who was a Norfolk Cubitt), after prolonged and painstaking investigatory and detective work, established all that is now likely to be known about Cooke and wrote this fascinating account of her findings.

After an apprenticeship in the wholesale drapery trade, Cooke, at the age of 21, moved to London which was to be his permanent home, where among other occupations, he became an elementary school teacher, curator of the India Museum, and finally the first cryptogamic botanist at the Royal Botanic Gardens, Kew. It was a chance introduction to edible fungi that initiated his dominating interest in fungi on which he published more than 300 books and articles; and as an incidental result was honoured by an American university. He was a clubbable man with an urge to educate others and, living at the time when local natural history societies were in their heyday, his greatest pleasure was probably the active part he played in many of their meetings. The current interest in fungi shown by the Woolhope Club in Herefordshire, the Yorkshire Naturalists, and the Essex Field Club must in part be attributed to Cooke's influence. He was also a founder of the Queckett Microscopical Club (which in 1965 celebrated its centenary) and the British Mycological Society. Dr. English has set Cooke's life firmly in his times and this biography includes much of interest not only to mycologists and field naturalists but also to a wide range of readers.

Geoffrey C. Ainsworth
Formerly Director of the
Commonwealth Mycological
Institute, Kew.

Chapter One

Sources

The subject of this chronicle, Mordecai Cubitt Cooke, was an East Anglian by birth, was raised in the Calvinist tradition, and his education ceased at the age of 14. At one time or another he worked as an apprentice to a wholesale draper, a solicitor's copying clerk, usher in a Birmingham school, master of an Anglican 'National School', a museum assistant, economic botanist and civil servant, and the first cryptogamic botanist to be appointed to the Royal Botanic Gardens, Kew. In addition he was publisher, printer, inveterate versifyer and after-dinner speaker, ballad singer, flautist, and no mean artist. His home life was tumultuous. All his preoccupations and talents were secondary to his intense activity as a self-taught naturalist and protagonist for Nature, and his consuming interest in fungi, on which he published some 350 books and articles (Appendix B). He founded natural history societies, was a driving force in others, and used every means at his disposal to inculate a love of nature into the lay public. He became internationally recognised as one of the two leading British mycologists of his day, making his mark abroad particularly in the U.S.A. and in Australia. He was honoured by organisations as diverse as Yale Univesity, the Hackney Microscopical Society, the Linnean Society and the Primrose League, but all his life he remained an eccentric outsider among the gentleman scientists of the age.

These are the bare facts of the career of a man whose inner self impelled him to a life of ceaseless activity. But concerning one entire aspect of this life, his domestic arrangements, only the outline is known, and his thoughts and opinions on other than scientific matters are largely unrecorded. The fewer and the less varied the sources of information available to a biographer–the more that is left to one's imagination–the more suspect must be the truth and fairness of the picture one builds up about one's subject and the more that picture will be distorted by one's own preconceptions. I have been very conscious of this while writing about Mordecai Cooke.

When he died he left to his youngest daughter, Miss Leila Cooke, with whom he had lived for many years,

> ... all the first editions of his books, the poems he had written [as a young man]; ... his autobiography, medals from foreign societies and the parchments of his degrees; a volume of original letters from the time of Linnaeus; two unpublished books with original illustrations; two microscopes and about 200 microscope slides, including some done by Sowerby; old daguerrotypes of his father and mother, old photographs of himself taken in his early years, all of which I treasured; his mother's watercolour drawings of plants ...

If he left any other personal papers, which seems doubtful, they would either have been with those treasured possessions or, more likely, would have been destroyed by Miss Cooke on her father's death. But sadly, even the relics were lost, for sometime before the Second World War Miss Cooke went to live in Kenya and:

> when I left England they were all packed in Ernest's old sea-chest [see Chap. 12] and stored at the Army and Navy Stores, where they were completely destroyed by one of Hitler's incendiary bombs [in 1941].

In fact, one of Cooke's grand-daughters told me that her aunt had asked her to take the chest to her home, but that there had been some confusion about the storage fee and the Army and Navy refused to release it: in the meantime the depot was bombed and Mordecai Cooke's personal effects were lost forever.

While virtually no letters to him have survived, there are considerable collections in various parts of the world of some of the vast quantity of letters he wrote to fellow mycologists; but almost all are confined to strictly mycological or financial matters with an occasional allusion to the health or well-being of the recipient. He mentions his own health if it affects his work, but apart from this, comments on his personal affairs, his family, or current events are rare indeed. Every now and then there is a forceful comment on the misdemeanours of some unfortunate mycological colleague, and sometimes his feelings of liking, respect, distrust or annoyance stand out clearly, but little else of a personal nature can be gleaned from these letters.

It is possible to gain some impression of the impact on him of the revolution in scientific thought that was taking place in his time through his popular articles for teachers and amateur naturalists, but considering that he lived in an era of discovery in the biological sciences that was quite as exciting as our own–the period of the final demise of the theory of spontaneous generation, of the birth of the cell theory and of the germ theory of disease, and of the momentous publication of Darwin's *Origin of Species*–it is surprising how little it is possible to discover of Cooke's reaction to any but the last of these.

Three documents are of fundamental importance in determining the events of Cooke's life and career:- *The Cookes of Horning* (1913), a manuscript volume written by his younger brother, William Cubitt Cooke (1827–1916), of which there is a photocopy in the Local History Department of Norwich Public Library; Cooke's own autobiographical notes, deposited at the Royal Botanic Gardens, Kew; and Dr. John Ramsbottom's (Fig. 15.8) marvellously detailed obituary in the *Transactions of the British Mycological Society*.

Cooke's own notes were written when he was 87, two years before his death, and cover 25 foolscap pages in his own handwriting. Some of his descriptions and comments up to the time when he was about 40 are of considerable interest, but later notes mainly concern his mycological activities and his numerous books, and by the 1980s they consist of little more than a list of his publications. Though there are some errors in dates, his accuracy on the whole is remarkable considering his advanced age when he was writing. Throughout, he makes no mention whatsoever of his marriage, his domestic life or his family.

William Cubitt Cooke wrote *The Cookes of Horning* at a time when he and Mordecai were spending many hours together reminiscing about their childhood in Norfolk. He dedicated it to Owen H. Cooke, the son of his younger brother, Josiah (1832–1889). (Chap. 6).* It consists of histories of the Cooke and Cubitt families, with the lore and legend of the former, followed by details of his and Mordecai's grandparents, parents, brothers and sisters, and all their families. Most importantly, and I am sure unintentionally, it provides the clue that led me to the unravelling of the complex skein of Mordecai's family affairs.

Dr. Ramsbottom's obituary of Cooke is a masterpiece of wide-ranging information recorded in a concise but lively style. He was obviously fascinated by his subject and had made a detailed study of the archives available to him as an employee of the British Museum (Natural History) as well as of many other sources, and he also received considerable help from Miss Leila Cooke

---

*The original manuscript from which the Norwich Library photocopy was presumably taken eventually passed into the hands of Owen Cooke's great neice, Mrs. M. Drury. She first contacted me after the completion of this biography and kindly presented me with a typed copy of the manuscript.

While writing the biography, I learned of another manuscript, apparently similar to *The Cookes of Horning*, in the possession of William Cooke's grandson, Cuthbert Cubitt Keet (see p. 14), but I was unable to see it. After the completion of the biography a copy of Keet's manuscript, originally given by Keet to Miss Leila Cooke, was given to me by Miss Cooke's niece, Mrs. Jane Devitt, a granddaughter of another of William's brothers, Ebenezer Chap. 4). It is entitled *Liber de Cookeis Horningi*, and dated 1911. The contents differ in some respects from those of the 1913 manuscript, but the differences do not affect the story of Mordecai Cubitt Cooke.

herself before she left for Kenya. It is testimony to the exhaustive nature of Dr. Ramsbottom's researches that, time and time again in my much more detailed study, I have made what at first I believed to be some new discovery only to find, on re-reading the Obituary, that its author had most certainly made it before me. Unfortunately I was never able to discuss Cooke with Dr. Ramsbottom, for I only became interested in the former shortly before Dr. Ramsbottom's death.

It did, however, seem possible that Leila Cooke might still be alive, and her last address in Kenya was known to the mycologists at Kew Gardens; but a letter to her was returned unopened. No further progress in finding her was possible until, by great good fortune, a friend of her's, Miss A. H. Bulley (see Postscript), wrote to the Commonwealth Mycological Institute at Kew concerning a matter involving Miss Cooke. The Director of the Institute, and Dr. P. M. Stockdale of his staff, told me of this, and as a result I was able to discover that Miss Cooke, now a very old lady, was in hospital in Nairobi where, in 1976, I visited her. Though she was somewhat forgetful about the present her memory for the past was remarkable, and she thoroughly enjoyed the opportunity of talking at length about the father whom she adored, accepting the presence of my tape recorder without a qualm. Inevitably, though, her memories of him were all as an old man, for he was 57 when she was born. She knew nothing, except from the family legends, of his early life, or even of the family during the childhood of her older brothers and sisters, and she would allow no hint of the traumatic events which by then I knew had taken place at that time, and of which she was undoubtedly aware, to pass my lips, let alone her's. She told me about nieces in England, Canada, Australia and New Zealand with whom she had been in touch until fairly recently but, confined as she was to her hospital bed with very few personal possessions around her, she was unable to give me their addresses or even to show me the family photographs she had brought from England. After her death in 1977 her solicitor was kind enough to give me the addresses of two nieces (Cooke's granddaughters) and I have since either met or corresponded with 6 of the 7 of them. They have been able to tell me their childhood memories of their much-loved grandfather, and also to give me valuable additional information about the family, most of which they had obtained from their aunt in earlier days.

In 1951 Miss Cooke paid a brief visit to England in the course of which she visited the Norfolk and Norwich Museum, the Royal Botanic Gardens, Kew, and the Quekett Microscopical Club (of which her father was a co-founder). With Mr. E. A. Ellis, of Norwich, she left some family photographs; at Kew and with the Quekett Club she deposited more photographs; and with all three she left copies of some biographical notes she had written about her father. Inevitably the notes are not entirely reliable, for parts consist of the retelling of tales told to her many years earlier, and are often idealised. But other parts describe her own memories of her father and of their home, and these are fascinating and of unique value.

The leads given by the three documents described, by Miss Cooke, and by her nieces, provided directly or indirectly most of the information available about Cooke. Some of these leads consisted simply of names or of bare facts, and much research was needed to follow them to their conclusions, but without them whole facets of Cooke's activities, and many of his friends and acquaintances, would have remained unknown. In particular his early publications in the pedagogic press would have been lost, and the complex story of his family affairs would have remained hidden. Two valuable sources of information, however, came to light independantly, as a result of a plea for help in the columns of the *Bulletin of the British Mycological Society*. These were a collection of Cooke's letters and other manuscripts at the Tolson Museum, Huddersfield, and his own microscope, which he sold to the Royal Horticultural Society at the end of his life and which is now kept at that Society's Wisley centre.

In undertaking this biography I have had two special advantages. First, as a mycologist myself, I have been able to enlist the enthusiastic help of professional friends and colleagues in Britain and around the world. Secondly I, like Cooke, have Cubitt blood in my veins, and through members of my family have been able to tap sources which might have been inaccessible to others for information about Cooke's antecedants and early life.

Cubitt is not an uncommon name in Norfolk, Cooke's native county. His mother, Mary Cubitt, belonged to a line which had for many generations been based at Neatishead, a village near Norwich. My mother, Gladys Nellie Cubitt, came from another line originating at Bacton, nearer the coast. It seems likely that these two lines have diverged from a common stock, but owing to the loss of certain vital parish registers it has not been possible to prove this. Nevertheless, it was my uncle, George Eaton Stannard Cubitt (p. 206), whose dedicated labours produced the family tree of the Bacton Cubitts, who first told me something of the Neatishead branch. On his death his work was continued by David Cubitt, my distant cousin, who has been able to provide me with much information about the Cooke family as well as the Neatishead Cubitts, and who also discovered that the descendants of one of Mordecai Cooke's brothers still live near Norwich.

From all these sources it has been possible to build up a reasonably comprehensive account of the facts of Cooke's life; a fair impression of his scientific thought, of his contribution to mycology and to the movement for the popularisation of the study of nature can also be pieced together. But seldom is there any direct reference to his opinions on such everyday matters as marriage, fatherhood, politics, religion or morals. One can deduce from the factual evidence that, with advancing years, he moved to the right politically, but it is only a deduction, the matter is never discussed by himself or others. Similarly one can guess at the change in his religious views with the advent of the Darwin era; but no indication survives as to when and how he recanted from the faith of his childhood, whether it was a traumatic episode, or whether he simply drifted into unbelief. Such lacunae, combined with his complete silence (and that of all other sources) about individual members of his complicated household and his relationships with them, result in there being little direct evidence which could throw light on his character or personality.

Any indirect conclusions one is tempted to draw from the purely factual material available often turns out to be invalid on further investigation. For instance, a considerable amount of space in his letters to other mycologists is devoted to money matters, suggesting that he was somewhat mercenary and close-fisted; but such conclusions are probably unjust, for one must remember that he employed no clerk or accountant and therefore had himself to organise the finances and solicit the subscriptions for the several periodical publications which he produced–hence the frequent requests for payment. Again, that he quarrelled with a number of people including some old friends, is certainly true and may give the impression that he was cantankerous and difficult to get on with. But conspicuous events such as quarrels tend to be remembered and remarked upon, while lasting friendships are taken for granted and, especially in the absence of surviving correspondence from those friends, leave no trace for the biographer. A fair judgement cannot be reached when only limited fragments of the evidence survive.

As to family relationships, without a knowledge of which it is impossible to discover the true nature of any man, Cooke himself, his contemporaries and his descendants, have drawn an almost impenetrable veil over these. Only his brother William has raised it very slightly, just enough to allow the bare facts of the story to be disentangled. But how the events came about, the personalities and attitudes of those concerned, the stresses within the family and, above all, the effect of what must at times have been almost intolerable pressures on Cooke himself (or was he able simply to bury himself in his work and shut his mind to such pressures?) will never be known.

Despite the grave shortcomings of the evidence I have been unable to resist the temptation, from time to time, of putting forward my own conclusions about some aspects of Cooke's character and personality; but as, for the reasons just given, these conclusions must be unfairly moulded by my own temperament and outlook, I have tried to let them obtrude as little as possible into my story. Where they are in evidence their inevitable bias must be borne in mind by the reader.

Chapter Two

Cookes and Cubitts From the beginning to 1850

Early in the nineteenth century Neatishead (pronounced Neat's'ed) was a village of over 600 souls lying at a cross-roads in the rich, flat Norfolk countryside about 12 miles north-east of Norwich. Prominent in the community for several generations had been a family by the name of Cubitt, a common enough name in east Norfolk. A John Cubitt, who lived at Neatishead in the seventeenth century, had had a son William who, because there was one son of this name in every succeeding generation, I shall refer to as William I (Tree 1). His son, William II (1724–1794), a substantial farmer who owned some of his land and rented more, married twice, fathering at least 14 children, of whom 3 are of interest to us. William III (1759–1814) was a son of his first wife, Hannah, and he became the village schoolmaster; Benjamin and James were the sons of William II's second wife, Mary Bayes, and thus were William III's step-brothers. Benjamin became the owner of Holly Grove and its 122 acre estate, in the parish of Neatishead and about a mile south of the village, of another large house which he let to a relative, and of some cottages in the village; James (1781–1821) farmed nearby. In the next generation James's son, William Quincey (1818–1872) farmed at Ivy Farm, close to Holly Grove.

William Cubitt III was no ordinary village schoolmaster, struggling to live on the pennies charged to those children whose parents could afford any education at all for them. He inherited land and money from his father, and not only did he build his own small house in the village street, but also the schoolroom adjacent to it, with its stepped gable end, outhouses, garden and meadow. All this must have provided a delightful home for William's wife, Martha, and their family. Until it was rebuilt in 1976, there was a cottage opposite the school-house with the name 'William Cubitt' cut into the stone lintel: possibly it was put up by one of the Williams for a relative, for no William is known to have lived there.

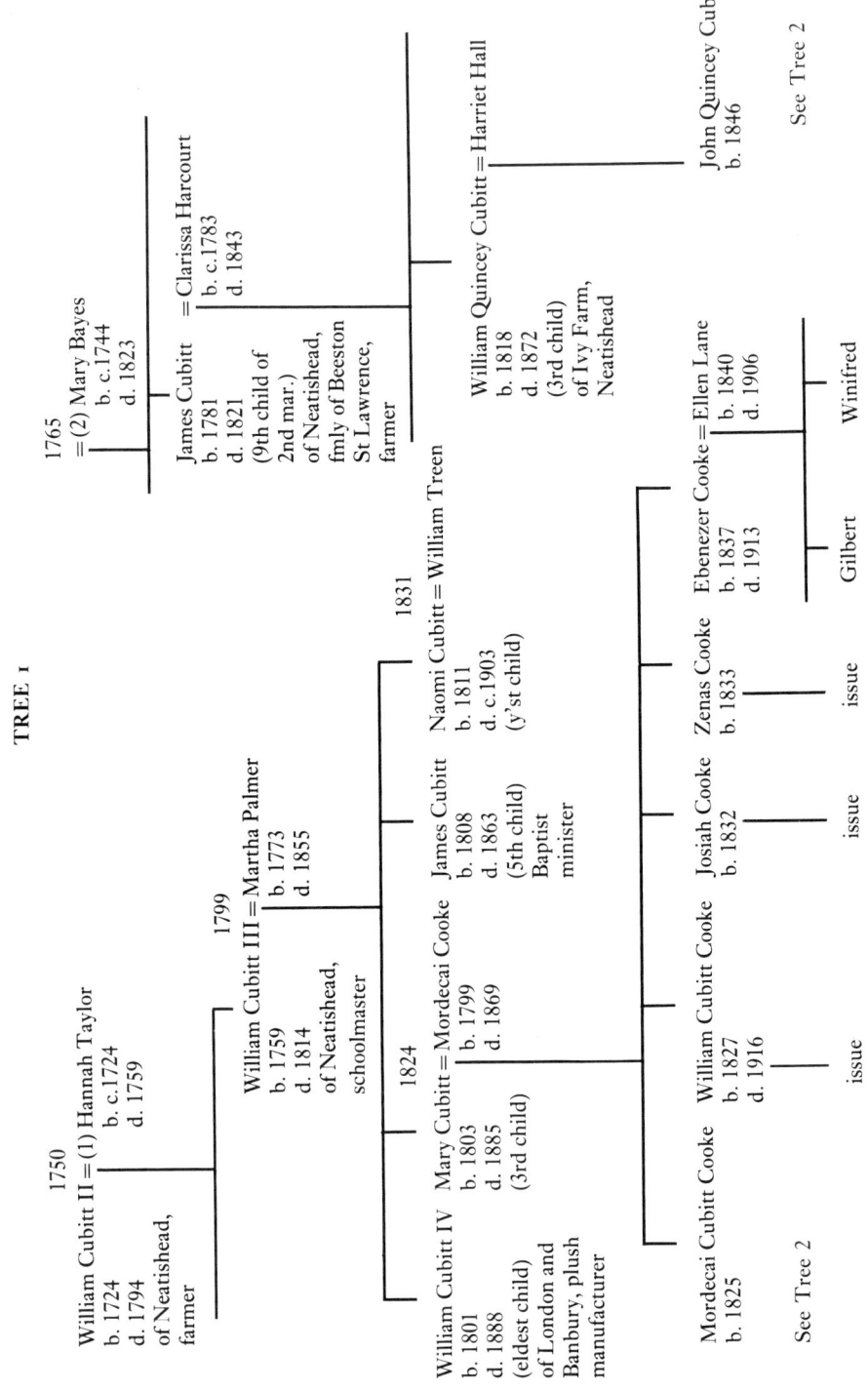

COOKES AND CUBITTS. 1825–1850

Religious dissent of all kinds is deeply rooted in Norfolk, and though the Cubitts originally belonged to the Established Church, a number of them intermarried with dissenters during the eighteenth century, becoming devout and influential Baptists of the more austere kind. William III himself was deeply religious and had a great yearning to spread his gospel among his neighbours, to which end he opened his schoolroom as a preaching station as early as 1798. He wrote later:

> As I hope it pleased the Lord God, who dwells in the heavens above, and rules over the inhabitants of the earth, to choose me from eternity, and in time, to call me by his divine grace and sanctify me by his Holy Spirit, and hearby incline my heart to obey his divine ordinances of this Church Militant; finding the means of grace comfortable to my soul, I opened my schoolroom for the instruction of the ignorant and careless, and began to read the Scriptures, gave exhortations etc. By so doing my neighbours assembled; and after some time had passed thus I invited gospel ministers to preach once a month on Friday evenings. After they had done so for a time preaching commenced on the Sabbath; the Lord in infinite mercy called many to attend, blessed this word to some, and those he inclined openly to espouse his cause in this village [1].

The Lord's Supper was administered for the first time in the schoolroom in 1810, but this was not enough for William. At his instigation and through his enthusiasm the foundations of a Meeting-house to serve the district were laid in the same year, about $1\frac{1}{2}$ miles south of the village but within the parish boundaries. William laid the first brick. The Chapel still stands, (Fig. 2.1), in the middle of fields and halfway to the village of Horning, another stronghold of the Baptist faith. Though there has been no permanent preacher at the Chapel since 1925 the place is well cared for. Many residents of Neatishead and of Horning lie in its tiny burial ground and are listed in its registers, Cookes and Cubitts being prominent among them. There is a plaque to its founder, William Cubitt.

Fig. 2.1 Neatishead Baptist Chapel in 1974.

William and Martha had 7 children of whom two died in infancy. The eldest, William IV, was born in 1801, to be followed by Mary in 1803, James in 1808, Martha in 1809 and Naomi in 1811. Three years later, when the children were still very young, their father died, leaving Martha to bring up her family on her own. Under the terms of her husband's will she was permitted to continue living in the school-house during her lifetime, but it is probable that after a while she moved to Point House, near the Chapel, for her descendants believe that her daughter Mary was married from there.

In the adjoining parish of Horning, the village itself lies on the north bank of a sharp bend in the River Bure, close to Ranworth and Hoveton Broads and about 11 miles north-east of Norwich. The land on the south bank of the river consists of low-lying, marshy scrub, unfit for building. Today Horning is a flourishing tourist centre for the Broads, with holiday homes and retirement bungalows straggling along the high northern bank of the river and clustered in the flat area in the bend, where the village itself lies between the bank and the low but steep escarpment. Boats ply up and down the river ferrying holiday-makers between the village and the Broads; the Swan Hotel (built in 1897) and the adjoining car-park are busy.

Prosperity, though, is not a new experience for Horning, for at the beginning of the nineteenth century, before holidays for all or tourism were dreamt of, it was still a thriving, bustling place, and the reason, then as now, was its easy access by water. Wool, corn, timber and other agricultural products could be moved by water far more cheaply and conveniently than by horse and cart.

At the time Clement Cooke (1759–1845) and his wife Ann lived in the main street of Horning. Clement had been orphaned early and the relatives under whose care he was placed took little interest in his education and failed to give him a proper start in his career, though he may have been assisted later by private means inherited from his parents. He became a wherryman on the river and apparently owned his own craft, for he eventually went into partnership with a timber merchant from Wroxham. Wherries, the flat-bottomed barges that were peculiarly suited to carrying goods on the broads and the slow Norfolk rivers, were ideal for transporting timber. They had a single, large sail, the mast for which could be lowered when passing under the shallow bridges so common in the district. There was a cabin for the crew of two, who had to be very powerful men indeed, for on windless days wherries must be poled along.

Horning's prospects were now improving still further and it was fast becoming a centre for manufacturing, so after a while Clement was able to take a share in the local flour mills and in a factory where the flour was made into ship's biscuits. Later still he had an interest in a factory making bombazine, a cloth much in demand at the time, being used for the mourning that was worn for months after the death of a member of the family, as well as for the garb of religious orders. It had a wool weft and a silk or cotton warp, and its manufacture became such an essential part of Horning's prosperity that when it went out of fashion the village lost its industrial base. The factory stood on the river on the site of the present car-park, ideally placed to receive its raw materials and despatch its finished products by water.

By the time Clement married he must have been reasonably well off, for he bought several adjoining houses in the main street of Horning and settled into one next to the present post office. Ann, his wife, was small, slender and gentle. They had eight children the youngest of whom, Mordecai (1799–1862) (Fig. 2.2), was delicate from birth and subject to violent nose-bleeds which prostrated him for weeks, though he did not let this interfere with an energetic life. He started work in the bombazine factory, but when the trade began to fail he bought the premises (now the Post Office) next his father's house, and the cottages in the rear, and set up as a general shop-keeper 'selling everything that a villager would require; grocery, ironmongery, earthenware, etc., etc.' (Aut).

Like the Neatishead Cubitts, the Cookes adhered rigidly to a strict form of the Baptist faith, and this seems to have left its mark on Mordecai's character, for one of his sons, William, describes his father thus:

> He was a man of considerable natural ability and a rigid Calvinist. His general weak health and frequent depressions from haemorrhage, combined with the austere form of religion which he professed, reacted on his character and gave it a moody, melancholy quality, but when in good health he was humorous, jocular, and was held in much esteem by the merchants with whom he traded. (CH)

Radical political views often went hand in hand with religious dissent, and Mordecai Cooke was no exception. Only a very limited part of the population

Fig. 2.2   Mordecai Cooke, 1799–1869, father of M. C. Cooke.

was eligible to vote at the beginning of the nineteenth century, for not only was the roll confined to males, but there was a property qualification which disenfranchised most of the working class and farm labourers. In Horning this resulted in a total voting strength in the 1835 election of 19, and as the ballot was not secret we know that only one of these, Mordecai Cooke, voted for the Whig candidate, the other 18 supporting the Conservative. The pattern was repeated in 1841 when, with 21 voters on the roll, two, including Mordecai Cooke, voted for the Whig-Radicals, the remainder again voting Conservative. Like other property owners, Cooke had to pay tithes to the vicar, one shilling in 1841, something which, as a Baptist, he must have resented bitterly.

Their common religious faith and their attendance for worship at Neatishead Chapel, half way between the two villages, would have ensured that the Cookes and Cubitts knew each other well. So it is no surprise to find that on Wednesday August 15th, 1824, Mordecai, youngest son of Clement Cooke of Horning, married Mary, eldest daughter of William Cubitt III of Neatishead. Mordecai was 25 and Mary 21. That the wedding took place at St. Peter Parmentergate Church, Norwich, and not in Neatishead Chapel, is because at that time only Church of England marriages were recognised in law. The couple settled into the living quarters of the shop at Horning. It is a small, 2-storied house, opening directly on to the village street. In Cooke's day there was an uninterrupted view across the street and a narrow strip of low-lying land to the river, but today there is a cottage on this land and its garden hedge and trees obscure the view. The front of the house was quite plain, with a window on either side of the porchless front door, three smaller windows up-

Fig. 2.3   Horning Post Office, birthplace of M. C. Cooke, about 1960.

stairs, and a low-pitched, tiled roof; but at some time since a squared bay has been built out from one of the front windows the better to display the wares of the shop (Fig. 2.3). The house has a large back garden with a pump under which, tradition says, all the family used to wash.

On July 12th, 1825, Mary's first child was born and was named Mordecai Cubitt, but as his parents were Baptists he would not be christened until he was old enough to make the decision for himself. As he grew older the young Mordecai was to resent his first name strongly, complaining that because of it he was taunted for being a Jew*. In fact, the name was a traditional one in the Cooke family, one male member of each generation being supposed to hold it. Clement Cooke's brother had been the Mordecai of his generation, and the name can be traced back to at least the seventeenth century. It was said that the tradition had begun in the fifteenth century when a Cooke had married a Jewess who had called her two children Mordecai and Rebecca. It was also said that if the name were not passed on, the Cooke family itself would die out, a prophecy which was to come true, for none of the younger Mordecai's sons were called after him, and he has left no grandsons, only granddaughters.

Whilst discussing family names the question of the possible relationship between the Cookes of Horning and the Cokes, Earls of Leicester, may also be considered. Holkham Hall, near Wells-next-the-Sea on the north coast of Norfolk, is the ancestral home of the Cokes, and in the past, one of the apparently indiscriminate spellings of the name Cooke has been Coke. This has inevitably led to confusion, and there is a persistant belief in part of the Cooke family that they are related to the Cokes. However, William Cooke devotes three pages of his family history to refuting this belief. I think he should have the last word.

Two years separated Mordecai and his younger brother William, and they were very close in childhood. The next infant born to their parents died as a baby, and the fourth was a girl who only lived till she was sixteen; so there was a gap of five years between William and his next brother, Josiah, born in 1832, to be followed by Zenas, Naomi, Ebenezer and Rhoda at two-yearly intervals. Their mother, as might be expected of a daughter of William Cubitt III, was both well educated and as dedicated a Baptist as was her father. To those of her descendants whom I have met she seems to be a much more substantial figure in their family lore than does her husband, and the little information that remains about her suggests that she was a strong personality who might

---

*After going to press, my attention was drawn by a member of the American branch of the Cooke family, Mrs. Frances Cooke Chan, to a letter from Mordecai's younger brother, William, (quoted in Stubbs, Dr. and Mrs. William Carter, *Descendants of Mordecai Cooke of 'Mordecai's Mount' Gloucester Co. Va.*, (1923), New Orleans, of which there is a copy in the Genealogy Room, Library of Congress, Washington, U.S.A.) stating that, because of Mordecai's dislike of his first name, he was always known in the family as 'M.C.'

well have been the dominant partner in the marriage. Certainly the name Cubitt persisted in an extraordinary way in the Cooke family, for not only did one of Mordecai's sons, Willie, add it to his surname, becoming known as Cubitt Cooke, but in William Cooke's family, where there were no sons, the name passed to his grandson through his daughter (Note, p. 3).

The little that we know about Mary Cooke comes to us through her grand-daughter, Leila, who though she was only three when her grandmother died and thus had no memories of her own, has recorded some of her father Mordecai's recollections of his mother, to whom he was very devoted. The family grew many of its own fruits and vegetables and Mary, who was an excellent cook, made jams and preserves, baked her own bread, salted down beef and cured bacon, the children helping in many of these tasks. It may be an indication of her prowess in the kitchen that her burnished copper milk saucepan and her preserving pan are still the treasured possessions of different branches of the Cooke family. Not content with her household duties, sometime about 1845 Mary opened a Post Office in her husband's shop, an enterprise which has flourished ever since. She was also much respected as the village herbalist, and her calf-bound copy of Culpepper's Herbal with its woodcut illustrations, inscribed 'M. Cooke' on the fly-leaf, was one of Mordecai's treasured possessions. At that time a brisk trade, for which no licence or qualification were required, was still being carried on in herbal as well as in patent medicines. Of course, as a herbalist, Mary knew all the local wild flowers, and she also did beautiful water-colour paintings of them. It was doubtless from her that two of her sons, Mordecai and Ebenezer, inherited their gift for drawing and their love of nature.

Leila Cooke tells one story that is almost unbelievable to us to-day. When 'smallpox came to Horning . . . all the children were put together in one room so that they would all get it and she [Mary] could nurse them all together and not have it drag on for months.' Could smallpox ever have been treated as casually as that? Leila says that her father was badly marked as a young man and carried a few pits all his life.

No detailed account remains of Mordecai's childhood at Horning; all that is known is to be found in brief snatches scattered through his books and papers. That he had an abiding love or Norfolk which never left him throughout his long life away from the county is clear. He recollects that that was the time when the tinder-box and rush-light were beginning to be replaced by 'Lucifers' (matches), at the then enormous price of $2\frac{1}{2}$ d each. They caused great astonishment in rural districts where it was considered that Lucifer himself did indeed have something to do with their magic. Mordecai and William could wander at will in the exciting Norfolk countryside, and both boys seem to have been fascinated by every aspect of the natural world. They 'bird's-nested, caught butterflies and worried reptiles with all the pertinacity of youth' [3] as well as collecting wild flowers in the lanes with their mother and learning their local names. The only article Mordecai ever wrote on birds was on the bearded tit he had studied in the Norfolk marshes, where he notes that it is known as the reed pheasant [4]. About ladybirds he tells us:

> In Norfolk they are known as Bishop Barnabee, and the children hold them in their open hands when caught, chanting meanwhile,
>
>> Bishop, bishop Barnabee
>> Tell me when my wedding be;
>> If it be tomorrow day,
>> Take your wings and fly away;
>> Fly to east, fly to west,
>> Fly to him that I love best [5].

Fungi came early to his notice. Boys in Norfolk used to 'puff in each other's faces the fine, brown, snuffy dust which fills the interior of ripe puff-balls or, as they are called by some, devil's snuff-boxes' [6], and wonder at the huge giant puff-balls, or bulfers as they were known in Norfolk, which were later to become one of his favourite dishes. A source of great mystery to Mordecai and William were the pieces of soft, rotten wood that they found fluorescing in the dark, and which they knew as 'touchwood'.

> In our schoolboy days we have still a vivid recollection of first becoming acquainted with 'touchwood'. . . . we had often heard of [its] marvellous properties and therefore recognised it at once from the description. It was the hollow of an old tree into which we had penetrated, as a hiding-place from our fellows, some dark evening, and were surprised upon discovering that the chips when disturbed exhibited a distinct pale fluorescent light. Naturally we took such pieces as exhibited the luminosity most strongly, and pocketed them for further experiment. When in bed the 'touchwood' was taken from our pockets and tested, under the bedclothes, after the lights had been taken away, and we, two boys, had been left alone. Truly the chips were phosphorescent, with a light strong enough to outline their form, and distinguish the letters upon any book or printed paper on which the fragments were laid [7].

Only later were they to discover that the source of the luminosity was the mycelium of the fungus which had rotted the wood.

When the time came for school the two boys were sent to the local dame school a few doors down the street from the shop. The house was very much like their own in appearance except that it had only two windows on the first floor (Fig. 2.4). The school was run by Miss Whaites and her niece, Sarah Obee, the latter being renowned as a penwoman, and both possessing 'attainments much above the average of the then village schoolteachers' (CH). It may be that it was from Miss Obee that Mordecai learned his bold, beautiful handwriting, but some sources suggest that he was indebted to his mother for this. The boys stayed at the village school until Mordecai was about

Fig. 2.4 Cottage where M. C. Cooke attended dame school. (About 1960).

10 and William 8, but then a great change came over Mordecai's life. His Uncle James Cubitt, Mary Cooke's younger brother, was a Baptist minister and therefore an educated man, and at this time he was the Pastor of the High Road Church at Ilford. Thither Mordecai was sent, travelling on the top of a stage-coach, to continue his education. It must have been hard for Mary and her husband to part with him, for he was very young to leave home, even to go to a relative, at a time when travel and communications were so slow. The Cooke's decision is a reminder of the great store which was set on education by the dissenting sects of the time.

His Uncle James was only the first of Mordecai's Cubitt relatives to have a profound influence on his life, unlike his Cooke relations who do not appear in his story at all. James Cubitt had received his call to the Church early; at 14 he was already teaching in the 'Sabbath school' at Norwich, and when he was 21 he was accepted as a student at Stepney College, a Baptist seminary. In 1834, when he was 26, he was invited by the church at Ilford to become their Pastor and it was here, probably after his marriage to a widow the following year, that his nephew joined him. The church ran a day school in its own schoolrooms and, as no certified teacher was appointed there until much later, it is probable that the Pastor's duties included that of schoolmaster. Whether Mordecai joined the school or was taught privately by his uncle, there is no doubt that it was James Cubitt to whom Mordecai owed his future career. Many years later he was to pay tribute to his uncle in an article in the *Morning Post* [8] in which he said that during the time he was with him:

was laid the foundation of every quality I was able to display in the way of work in after life. My uncle grounded me well in the rudiments of Latin, Greek, algebra etc. and he even made my daily walks contributory to the acquisition of useful knowledge, for he used to send me down the Barking-road with instructions to make collections of the plants and flowers that grew by the wayside, and to classify and name them afterwards with such assistance as I could get from his little botanical library.

Elsewhere Mordecai notes that the 'library' consisted of *Pinnock's Catechism* and McGillivray's edition of Withering's *British Bontany*. The *Catechism* is a small book, only 6″ × 3″ in size, containing 72 pages. In the form of leading questions and detailed answers it takes the student through the structure of plants and the logic of the Linnean classification, then considers from one to three examples from each of the 24 Linnean Classes including the lower plants. Consisting as it does merely of strings of facts, it is exceedingly dull, but was written, according to its 'Advertisement', or preface,

... with the sole view of rendering more easy the study of a science which ... is cultivated by all those who have any pretensions to a polite education and, if it be considered in a *moral* point of view, this study is well calculated to furnish us with instruction, and conduct us by gentle steps to the knowledge of that Great Being, who has condescended to form plants with so much delicacy, and grace them with such a variety of beauties.

Mordecai's very intimate and extensive knowledge of Greek and Roman legends and of the activities of the gods and goddesses is constantly in evidence in his writings and is doubtless attributable to his uncle's teaching. Later he loved to describe the social events of the clubs and societies to which he belonged in long, comic parodies of the legends, and is forever tracing the classical derivations of botanical names and terms. Occasionally he refers to his childhood training as, for instance, when he is discussing the origin of the name of the pond hydra, when he says: 'There is an old story of the Greeks in which we were interested in our schooldays . . .' [9].

Those were troubled times in the High Road Church, and in 1836 a section of the membership broke away to form its own, much stricter, congregation. The active part that James 'was compelled to take against erroneous doctrines and practices in the church made it desirable that he should seek another sphere of labour' [10]. So a year later he acceded to a request to move to Stratford-on-Avon, taking his nephew with him. Mordecai's only surviving recollection of his time there is of watching the proclamation of the accession of Queen Victoria. Soon afterwards, when he was 13, he returned to Horning.

James Cubitt had never been strong, and shortly after Mordecai's departure his health failed and he had to leave Stratford. After several more moves and much illness he took the post, at the invitation of the founder, of classics tutor at Spurgeon's Baptist College in London, but his health deteriorated further and he died in 1863 at the age of 55.

Meanwhile both Mordecai and his brother William were sent to a commercial academy at Neatishead, run by one William Moore, 'in order to obtain a good instruction in writing and arithmetic which had been neglected for the classics' (Aut.). It may be surprising to us to find such an establishment in a remote village in the country, but they were not uncommon at this time and were often run by teachers of mathematics. The instruction given was usually geared to practical applications, particularly surveying, for which at that date there was no formal training, and the teacher would inevitably practice that profession as a sideline to pedagogy. Such schools would be largely patronised by dissenters such as the Cookes, for they were independent of the established church to which the grammar schools of the day were closely linked. In Moore's school the instruction seems to have had a strongly practical bias, as Mordecai spent most of his time helping his teacher with his land surveying work and with calculating the rates, for Moore kept the Rates Book for the parish overseer.

His 18-months' stay at the Academy concluded Mordecai's education. He and William parted, William eventually to train as a teacher and 15 year-old Mordecai, after he had spent a short time helping in the shop, to be apprenticed to a business friend of his father's who had a wholesale drapery business in Norwich. Here Mordecai stayed for the next five years, the experience being best described in his own words:

> I was soon ensconced in a high desk in the entering room, to enter the goods in the Day book and make out the invoices, as the goods went out to the customers, and this was nearly all the instruction I received during my term.' (Aut.)

However, there were compensations. This was the heyday of the Temperance Societies, which were making strenuous efforts both to attract new supporters and to provide innocent but engrossing entertainment to keep potential topers far from the Demon Drink. To this end Temperance Bands and Choral Societies flourished all over the country and Mordecai, who as a Baptist had naturally been brought up as a strict teetotaller, was soon recruited to both. He was a competent if not an inspired musician, and mastered several wind instruments, including the flute, which he continued to play throughout his life. He also loved singing, and his resonant tenor voice must have been a considerable asset to the Norwich Choral Society. In later life both his family and his mycological colleagues were to remember his penchant for singing popular ballads, at home, at society dinners, and when out in the field. Clearly he took his temperance activities a lot more seriously than he did his apprenticeship, for he himself comments that his 'five years at Norwich were for the most part musical.'

There was another diversion from the tedium of the drapery business. Mordecai's employer was an amateur ornithologist and he encouraged his apprentice to take up the hobby. The Victorian mania for collecting anything and everything in the world of nature, and for showing off the trophies in elaborately mounted displays, was rapidly gaining ground at this period and, photography being in its infancy, the only way to collect birds was to shoot them. Mordecai joined enthusiastically in the carnage on the Norfolk marshes, where birds were slaughtered in their thousands to swell private collections and decorate drawing-rooms. Unlike some of his fellows, however, he was interested not only in the rarer birds, some of which were being shot to near extinction, but also in the more common species, for his interest in comparative morphology was already awakening. It must have been from this time that the stuffed birds that Leila Cooke describes in their home in Kentish Town dated.

> Under glass covers were birds he had shot himself in Horning. He had stuffed and mounted them himself; they were beautifully done. I remember a falcon, an owl, a grebe, a heron and a bustard. In Morris's book of birds he says that my father's was the last bustard seen on the Norfolk Broads. He gives the date when it was shot.'

Leila was wrong about the bustard. When listing the last of these birds seen in Norfolk, F. O. Morris, in his *History of British Birds* [11] mentions no-one by name. A bird shot near Norwich in 1831 is presumably the one to which she is referring, but as her father would only have been six at the time he could hardly have been responsible for its death. The last date on which bustards were seen elsewhere in the county was seven years later when Mordecai would have been 13, so it is highly unlikely that if he did possess a stuffed bustard it was one he had shot himself.

At some time during his stay in Norfolk, Mordecai also worked as an apothecary's assistant, though only two slender threads of evidence for this remain. In her biographical notes on her father, Leila Cooke writes of this time:

> Then he became assistant to an apothecary. He must have stayed there for some time and acquired a wide knowledge of chemistry and the mixing of drugs . . . . He made pills that were sold in little, flat round boxes with a blue lid on which was printed a Maltese Cross in red. These lasted until my time with two grooved wooden boards which were used for shaping the pills.

Also, in the manuscript of an unpublished book on economic botany for children (p. 295), during a discussion of opium, Mordecai remarks that in his youth he had known many old ladies who had become addicted to the drug through taking laudanum regularly: 'I know it for certain because I had to sell it to them.'

Mordecai's lively and enquiring mind must have been stifled during his five years in commerce, and it had not escaped his notice that even as a trained assistant in the drapery business his wages would only be £40 a year. So at the end of his apprenticeship, when he was 20, he joined the mass migration of population from the countryside to the cities, then at its height, and moved to London to start his life afresh. Possibly he travelled by train this time, for the new line between London and Norwich opened on June 28th, 1845, with three journeys a day each way. The trip took about five hours, the third class fare in a stopping train being 10/6, compared with 18/- for a second class seat on the express. Though Mordecai was never to return to Norfolk to live, his roots there were very deep; he visited his mother almost every year during her lifetime, and named several of his homes 'Norfolk Villa'.

Records of the next six years of his life are scanty and difficult to place in any chronological order, as only a few dates are known. On first moving to the capital in 1845 he stayed with another Cubitt uncle, his mother's elder brother, William IV (1801–1888), who was a manufacturer of plush, velvet and coach trimmings in Hoxton, near Shoreditch. Like the rest of his family he was a committed Baptist, and was one of the compilers of the *Baptist Yearbook*. Mordecai found a temporary job as a copying clerk in a solicitor's office. His swift and beautiful hand-writing would have made him much sought after in such employment, but the life could have been little more to his liking than that of a draper's assistant.

He did not remain very long with his uncle, for on Jan. 4th, 1846, he was living at Park Place, Greenwich, the particulars being given on the certificate of his marriage to Sophia Elizabeth Biggs, spinster, who lived at the same address. The wedding took place at nearby St. Alphege's Church (Fig. 2.5). As the two witnesses were certainly not Mordecai's relatives, and neither bore the name of Biggs, it seems likely that the wedding was not a family affair. The only unofficial acknowledgement of Mordecai's marriage now in existence is a bare mention of the fact, with Sophia's name and dates, in *The Cookes of Horning*; Mordecai himself never once refers to his wife in any of his voluminous writings. All information about her has therefore been gleaned from the few official documents available. She was 23 and Mordecai 21 when they married. Sophia had been born in the parish of St. Andrews, Holborn, and her father, who was dead by 1846, was a tobacconist. On May 18th, 1844, when she was living at Southwark Bridge Road, and nearly two years before her marriage, she had given birth to a daughter, Elizabeth Annie Thornton Biggs, known as Annie, whose father, David Thornton, was a clerk in the Excise Office. Nothing more is known of Thornton, though as the child was given his surname it may perhaps be assumed that Sophia was very attached to him. Why she moved to Greenwich, how she supported herself and her daughter, and how she met Mordecai, are unknown, though the fact that they both gave the same address in the marriage register suggests that they may have been living in the same lodging house. There may well have been moral disapproval of Sophia on the part of the families of both bride and groom, but particularly Mordecai's, with the strict morals of their sect. There is also a shadowy tale among Mordecai's living relatives that, before he moved to London, he was betrothed to a Norfolk

Fig. 2.5   St. Alphege's Church, Greenwich, scene of the marriage of M. C. Cooke and Sophia Biggs.

girl, Sophia Savage, whom he jilted. It was this girl who, at a ceremony at Neatishead Chapel in 1850, became the wife of his brother William.

Park Place, now Park Vista, is a terrace of plain but well-proportioned, three-storeyed, brick houses, with a pleasant outlook on to Greenwich Park. One end of the terrace has now been replaced by a row of small shops, but the rest remains. The centre three houses are rather more imposing than the rest, especially the middle one, which projects slightly and is topped by a fourth storey bearing a circular plaque with the legend 'Park Place'. Whether the name refers to the centre house alone or to the whole terrace is unclear, but even the smaller houses would have been far too expensive for Mordecai or Sophia to rent in their entirety; some part of the terrace must have been let out in rooms.

When his job as a copying clerk had finished Mordecai had taken a post as clerk to a solicitor in Tooley St. and this may well have been when he moved to Greenwich, a residential district which would not have been too far from his work. Probably it was while he was at Tooley St. that he turned his hand to a wholly new occupation, writing poetry, or more probably verse. Perhaps this was part of a romantic phase in his life which also inspired his marriage. He published his efforts in pamphlets with blue paper covers, selling at $4\frac{1}{2}$ d each, but these his earliest recorded publications, are lost for ever; the British Library has no copies, and Mordecai's own were destroyed with his other effects in the London blitz. It is on record that their titles included 'The Struggle for Freedom', 'The Flight of Thought', and 'The Course of Love', and we are told–and can believe, in the light of his later efforts–that they were of no great merit. Nevertheless, he is said to have lectured on poets and poetry at about this time, and it would be reasonable to suppose that it was now. Certainly in later life he could always produce an appropriate quotation from the poets for any occasion, whether he was speaking or writing. In 1849 he was consolidating his literary efforts by editing a periodical called *The Monthly Repository of General Literature*, of which there remains no trace in any archive. It is said (JR) that 'he wrote practically the whole of this short-lived periodical under various names or permutations and combinations of his own initials.' These excursions into writing, publishing, editing and lecturing were Mordecai's first, tentative entry into a world which, for the rest of his life, he was to make very much his own.

After he had been with the Tooley St. solicitor for about two years a disaster befell Mordecai. In his own words:

> ... occasioned by the close confinement at the desk till eight o'clock at night, I was prostrated by an attack of gastric fever and kept my bed for some weeks, to find myself on recovery without a situation, for my place had been filled.'

Exactly what he did after this we cannot be sure, but with a wife and stepdaughter dependent on him he must have been under severe financial strain. It seems likely that it was now that he moved to Birmingham to become an usher, or junior master, in a private school for boys. At that time anyone who cared to could set up a school, there was no system of inspection, and the qualifications, or lack of them, of the teachers were entirely at the discretion of the proprietor. Mordecai Cooke, with his versatility and wide interests, would have had little difficulty in obtaining such a post, though why he should have moved to Birmingham we do not know.

Chapter Three

Holy Trinity School, Lambeth. 1851-1860

State education did not come into being until 1870. Before that date schools for the children of the poor were the jealously guarded responsibility of the religious denominations, though by the 1850s they were being aided by ever increasing government grants given on condition that they must submit to regular inspection. Predictably, the majority of schools were owned by the Church of England, which ran them through the National Society for Promoting the Education of the Poor in the Principles of the Established Church, or the National Society, for short: its schools were known as 'National Schools'. The National Society was in active competition with the Non-conformists' British and Foreign Schools Society and its chain of 'British Schools'. By the 1840s both Societies had opened training colleges for teachers of their own denominations, though by no means all serving teachers had the opportunity to attend them.

Instruction in most of these elementary schools, where little more than religious knowledge and the Three Rs was taught, was by means of the monitorial system, under which only one teacher was needed for every 100 or so children, a system which served the double purpose of costing very little and making the most intensive use of the very limited supply of experienced teachers then available. To achieve this astonishing pupil-teacher ratio, all learning was by rote. The teacher would pick out promising senior pupils as monitors and instruct them in the next days' lessons, which they would learn by heart. The following day the monitor would recite the lesson, broken up into short sentences, in front of part of the class, and the children would repeat it after him in chorus until they too had memorised it. The system was already being condemned by the school inspectors in the 1840s, and monitors were later replaced by pupil teachers who might, after a few years, go to college and qualify for full teacher status. However, the master still had to instruct his pupils from day to day not only in how, but what, to teach, and the children in great measure still learned by rote under an authoritarian system which took little account of the special needs of youthful minds.

Some influential members of the dissenting churches, realising the missed opportunities of the current methods of instruction, began to look to the Continent for fresh ideas, for there experiments in new teaching techniques were yielding exciting results. In particular, the philosophy of J. H. Pestalozzi had gained wide acceptance. He considered that the education of the child begins with its mother and should include intellectual, physical and moral instruction. All the child's faculties should be cultivated simultaneously, the most important principle being spontaneity, or 'self-activity'. In particular, teaching should never be by rote–the child must be allowed and encouraged to perceive, to reason and to judge for himself. These were indeed revolutionary ideas in Britain, and especially in the Church of England. The National Society had laid down that:

> ... one of the most important duties impressed on them [the children] will be resignation to their lot ... By the very constitution of society the poor are destined to labour, and to this supreme and beneficial arrangement of Providence they must of necessity submit.'

Also, such methods could never be carried out under the monitorial, or even under the pupil-teacher, systems; many more fully qualified teachers would be needed, and for the Establishment to spend money on such luxury for the poor would be entirely against the ethos of the times. It had not yet been fully realised that an educated working class would be essential to an economically successful country in the modern world.

If largely rejected by the National Society, Pestalozzi's method rapidly gained ground among non-conformists and was increasingly used in their British Schools. Cooke's aunt, Naomi Cubitt (1811–1903), his mother's younger sister, had married a Pestalozzian teacher from Coventry, William Treen. In 1846 both were teaching at a newly opened British School in Tennant St., Stockton-on-Tees, which was run on Pestalozzian lines and had over 200 children on its roll. It was sometime in 1850 that Naomi Treen invited her nephew to visit Stockton to learn about the new system, and he accepted; perhaps he went straight there from Birmingham. He stayed for about three months 'in intimate contact with the system and its practical application'. Then he replied to:

an advertisement seeking a master and mistress to open new National Schools in Lambeth. In response I now received an invitation to an interview with the clergyman who was the chairman of the School Committee. Armed with all the credentials that I could procure I soon paid my visit, and after that two or three others, and was ultimately solicited to undertake the schools as an experiment to see how we could work together, and in 1851, [on September 1, to be precise] the year of the Great Exhibition, we were located in the School House, and waited in the empty school room, for the advent of pupils. We had not long to wait, and for some weeks the admission of pupils occupied much of our time.*

In having the Rev. James Gillman (Fig. 3.1) as its vicar the parish of Holy Trinity, Lambeth, was fortunate indeed. He was a Fellow of St. John's College, Oxford, and until 1847 had held the living of Barfreystone, a small village in Kent. But he had a driving urge to serve to the limits of his capability the less fortunate of his fellow men, finding no fulfilment in the slow pace of the countryside. So he moved to Lambeth, to a large, new, and desperately poor parish, without even a school, where there was greater scope for his tireless energy. It had a population of 6500, and Gillman visited every family in his flock at least once a year, finding his way through the appalling, rat-infested

Fig. 3.1 The Rev. James Gillman, Vicar of Holy Trinity, Lambeth (from a portrait by N. Macbeth).

*The wording of this paragraph of his autobiographical notes, with its reference to a mistress for the schools, and the subsequent use of 'we' rather than 'I', is the only hint Cooke ever gives of his marriage.

courts and lanes which crowded this bank of the Thames. All water, including drinking water, was of course obtained from the river, and inevitably the outbreak of cholera in 1849 devastated Lambeth. Gillman, ministering untiringly to his stricken parishioners, dared not go to his own home for three or four weeks for fear of infecting his family, so he slept on a sofa in the house of the local doctor.

Very often in a cholera epidemic it was the breadwinner of the family who was struck down, leaving his dependants destitute and a burden on the rates, and this outbreak brought home to Gillman the need for cheap life insurance accessible to the poor. So, with the help of a businessman, he started a scheme with a premium of only 1d a week, becoming its chairman in 1850; by 1877 its income was £2 million a year in weekly payments. So began the mighty Prudential Insurance Co. [1, 2]

Holy Trinity Church had only been built eight years when Gillman was inducted as Vicar in 1847, the building having been erected during the height of the evangelical fervour which had swept through all denominations in the 1830s. It stood at the far end of Lambeth Park from Lambeth Palace and the much older parish of St. Mary's, where Bligh of the Bounty and the Tradescants, father and son, the great Elizabethan plant collectors, are buried. Neither church now exists as a place of worship; Holy Trinity was badly damaged by bombing in the 1939–1945 War and has since been demolished, while St. Mary's has been deconsecrated and is now the headquarters of the Tradescant Trust.

Holy Trinity School was built close to its church with money from two local charities, Walcot's and Hayle's, and a grant from the Committee of Council on Education. It seems that Walcot's charity was administered by St. Mary's vestry, for in May 1853 there was an acrimonious exchange of letters in the *Lambeth Gazette* between the Rev. James Gillman and Mr. Churchwarden Taylor of St. Mary's, in which the vicar accused the churchwarden of the maladministration of the fund and of sheltering from the consequences of his misdeeds under the wing of his vestry.

The school building was simplicity itself (Figs 3.2 & 3.3). It consisted of one large schoolroom with rafters up to the roof, intended to seat 600 children. The children would not have had desks, of course, but were ranged on closely packed benches and wrote on slates (Fig. 3.4). The building is perhaps best described in the words of the Rev. W. H. Brookfield, in his Inspector's Report for 1853–4. [3]

> This is a very ample and substantial building, with the aid of a considerable grant of public money. From the circumstances of its closely adjoining the palace grounds at Lambeth, it has been necessary to leave one side of the school entirely without windows; nor is the style of architecture or the construction such as to admit of this privation on one side being compensated on any other; and the room is very inconveniently dark. It is

HOLY TRINITY SCHOOL, LAMBETH. 1851–1860

Fig. 3.2   Holy Trinity School, Lambeth, not later than 1930.

Fig. 3.3   Holy Trinity School, Lambeth, in 1975. The master's house was attached to the gable end. In the background is St. Thomas's Hospital.

Fig. 3.4   Holy Trinity School, Lambeth, in 1975. Score marks on the bricks where previous generations of scholars sharpened their slate pencils.

divided into boys' and girls' schools by a moveable partition 5 ft high. The mechanical arrangements are not generally what I should consider the most eligible.

However, the floors of the building were of wood rather than stone, a matter on which the Education Department was very particular, though it was prepared to accept any other discomfort which might be inflicted on teachers and pupils. Built out from the window side of the schoolroom, and so reducing the light even further, were the entrance porches for boys and for girls, and a small room to house the infants. Projecting from a gable end was a small, 2-storeyed house for the master and his family. This was to be the rent-free home of Mordecai and Sophia Cooke–and presumably of little Annie Biggs, now seven years old–for the next nine years. The school is still in use, though it has been extensively altered and added to; the schoolroom now has windows looking on to the Archbishop's Park, the playground has been greatly enlarged into the Park, and the master's house pulled down.

From Cooke's account of his appointment it is clear that Gillman was determined to find the right man as his schoolmaster. Two or three interviews, and only a provisional appointment in the first instance, suggest that he was not altogether sure about Cooke. We can probably take it that they got on at a personal level, or Gillman would not have pursued the matter after the first interview; so what was the difficulty? It is doubtful if Cooke's lack of a teaching qualification was critical, for this was by no means unusual at the time and could always be rectified while he was in post; in any case, this new school could not have afforded to pay a qualified teacher. Could the problem be his interest in Pestalozzian methods? Probably not, for Gillman was himself a man of liberal views who had already broken new ground in helping the poor to help themselves. Cooke's opportunity to use the new methods would anyhow be severely curtailed by having to work under the monitorial system. One important consideration must have been his religious convictions, or perhaps his lack of them. The fourth clause of the Terms of Union binding a school to the National Society reads straightforwardly: 'The masters and mistresses are to be members of the Church of England', and this Cooke clearly was not, though we know nothing of Sophia's religious convictions. It is very unlikely that Cooke was even a member of the Baptist Church, for though he had lived in Norfolk until he was 20 years old, by which time, as a convinced Baptist, he would certainly have been baptized, there is no record of the event in the register of Neatishead Chapel. If he was ever baptized into the Church of England the most probable occasions would have been at about the time of his wedding, or before or during his scholastic career. However, his name does not occur in the Baptismal Register of St. Alfege's Church in 1845 or 1846, or in the Register of Trinity Church, Lambeth, from 1850 to 1860 inclusive. The possibility of baptism elsewhere cannot be completely excluded, but it would seem very unlikely.

HOLY TRINITY SCHOOL, LAMBETH. 1851–1860

A final difficulty about Cooke's appointment as schoolmaster would have been the question of Annie. If her history were known, Gillman might well have had a problem in appointing the Cookes, whatever his own views. It would seem likely that, if she was indeed living with them, her identity was concealed and she was accepted as Mordecai's own child.

Holy Trinity was a small school–there were only about 70 boys and 60 girls paying 2d a week each, and 95 infants paying 1d a week. There were no pupil-teachers; the number of children was probably too small to allow of this, but in any case Cooke's lack of qualifications would have precluded him from having pupils. So boys, girls, and infants must each have been handled as a single class, probably with the help of monitors. Mordecai's salary was £45 per annum (the possession of a teacher's certificate would have doubled this) and Sophia's £25. There was also an infant teacher who, although she had a bigger class than either of the Cookes, earned only £8.10.0. The school's income and expenditure account for 1856 was as follows:

Income		*Expenditure*	
Endowment	27.10.0	Teachers' salaries	78.10.0
Voluntary contributions	8.17.0	Books and apparatus	8.10.0
School pence	65.15.0	Fuel and lights	14.10.0
Other	13.0.0	Repairs	3.10.0
		Other	12.0.0
	£114.12.6		£116.10.0

Gillman himself made up the deficit of £1.17.6, which he had reason to believe would be fully repaid him, and the total cost per child was 10s $3\frac{3}{4}$ d.*

We have only brief glimpses of the Cookes' day-to-day life in the school. As all learning was by rote, the noise must have been deafening, with 130 children being taught by two teachers in the same large room, albeit they were separated by a 5 ft. partition, though the conditions were nothing like so appalling as those which would have been produced by the much larger numbers for which the school had been built. The Rev. W. H. Brookfield, in the report which had been so critical of the school buildings, continued:

> The boys are taught by a master, who is in many respects an ingenious person, and the girls by his wife. There was a good deal of noise, partly perhaps, to be attributed to the construction of the room, and partly to the unexpectedness of my visit. I found the attainment of the boys pretty much in keeping with the general aspect and arrangement of the school.

*Information about the running of the school is taken from a grant application to the Committee of the Privy council on Education made in 1856. There is no reason to suppose that conditions differed materially in earlier years.

That discipline was a problem is clear from the report in the *Lambeth Gazette* of Sept 1st, 1853, on the School's anniversary celebrations. In his opening address Gillman noted that:

> The conductors of the school were most anxious to train the children in good habits by judicious discipline. Complaints had been made of the chastisement which the children received, but he could assure them that the children were chastised only when they deserved it, and then in a proper manner. He was made acquainted with every instance of punishment, so the parents of the children need be under no apprehension.

There is no reason to doubt the reassurance about corporal punishment, for everything we know about Cooke points to his being genuinely fond of children. The following year's celebrations were also noticed by the *Lambeth Gazette*, especially the fact that 'the children performed several evolutions with a precision that reflected considerable credit on the discipline kept by Mr. Cooke'. But a price had been paid for this success, for the Rev. Gillman 'published an amnesty to all boys who had been expelled the school and whose parents were desirous of having them replaced'.

The Committee of the National Society made quite clear the object of the instruction given in its schools:

> ... the sole object in view [is] is communicate to the Poor generally by means of a summary mode of education ... such knowledge and habits as are sufficient to guide them through life, in their proper stations, especially to teach the doctrines of Religion, according to the principles of the Established Church, and to train them in the performance of their religious duties by early discipline.

At Holy Trinity Gillman would have been responsible for morning and afternoon prayers, for Religious Instruction as such, and for the general religious tone of the school, but it would have fallen to the Cookes to ensure the proper religious content of the lessons. For instance, passages for reading and writing would be taken from the Bible or would point some improving moral; and history, geography and nature lessons for the older children would be used to instil a proper appreciation of God's graciousness and mercy. There is little doubt that, with his intensely religious upbringing, Cooke would have been fully capable of this. Neither was a National School teacher's job finished on Sundays–the Cookes would have been expected to parade the children, dressed in the best clothes they possessed, and march them off to at least one church service during the day.

Both boys and girls would have had lessons in the mornings, and these would have continued for the boys in the afternoons, but the girls would have spent the afternoons doing plain sewing. Some of the garments which they turned out would have been given to the most destitute of their own number, but many simple articles would have been ordered and paid for by the local gentry. Part of the £13 'other' income noted in the school's accounts may well have come from this source. As well as her afternoon sewing lessons, Sophia ran a 2-hour sewing class for girls on three evenings a week, but it is not know whether she was paid extra for this.

At this period the skill of the teacher and the progress of the children were tested annually by means of an examination conducted by the Inspector. The results at Holy Trinity in 1853 were most satisfactory, for the children answered 60 difficult questions out of 100 without knowing beforehand what they would be asked.

In his autobiographical notes Cooke writes:

> Very soon H.M. Inspector of Schools made his appearance, and, from the first it was evident that he was not favourably inclined towards us. We had not come from any College or Training School, or through the legitimate channels of communication, and he urged that I must at once present myself for examination, in the orthodox way. This I did as early as possible, with little preparation, and secured my Certificate, evidently a little to his Surprise and chagrin.

The implication of this passage is that the events occurred soon after Cooke's appointment to the school, but his memory so long afterwards must have been a little confused, for we know from the 1856 grant application, in which Gillman sought a higher salary for his teacher, that he had not qualified by then.

The examination which he would take would have been that set for pupil-teachers at the end of their 5-year apprenticeship and would have included:

1. An essay on a subject connected with the art of teaching.
2. The rudiments of algebra or the practice of land surveying and levelling [the latter a familiar subject for Cooke].
3. Syntax, etymology and prosody.
4. The use of the globe or the geography of the British Empire and Europe.
5. More completely, Liturgy and Chatechism, in Church schools the parochial clergyman assisting in the examination.
6. Ability to give a gallery lesson [i.e. to stand in front of a large class and give a lecture on a particular, often extracurricular, subject] and to conduct the instruction of the first [top] class in any subject selected by the Inspector.

There was very little organised science teaching in schools of any type before 1851. In the public schools it was considered that only a classical education could provide a suitable intellectual discipline for the country's future leaders, while in elementary and charity schools it was feared that children would be given ideas dangerously above their station should they be taught more than the basic essentials of religious knowledge, reading, writing and arithmetic. However, a few forward-looking educationalists took advantage of the excitement and interest generated by the Great Exhibition of that year to begin to introduce some simple science into school curricula, despite vigorous opposition from the religious bodies running them. Sir Lyon Playfair, Secretary of State for the Science and Art Department in the Board of Trade, and a chemist by profession, sanctioned the teaching of a subject called 'Knowledge of Common Things', which expounded the how? and why? of everyday objects, and books and wall diagrams were in due course made available to the schools.

A few years later other eminent scientists and educators, notably Thomas Henry Huxley, put forward the idea that the main objective of science teaching should be to enhance the children's powers of observation; that is, they should be taught Natural History (the subject we now know as Biology*) rather than Common Things. In a public lecture entitled 'On the Educational Value of Natural History', given in St. Martin's Hall in 1854, Huxley used all his skill and prestige as a lecturer to put forward the case for teaching the subject to children, summarising his main conclusions thus:

> Leave out the Physiological sciences from your curriculum, and you launch the student into the world undisciplined in that science whose subject-matter would best develop his powers of observation; ignorant of facts of the deepest importance for his own and others' welfare; blind to the richest sources of beauty in God's creation; and unprovided with that belief in a living law ... which might serve to check, and moderate that phase of despair through which ... he will assuredly sooner or later pass.

As to the age at which children should start natural history, Huxley said:

> ... it appears to me that ... the *common facts* [his italics] of biology–the uses of parts of the body–the names and habits of living creatures which surround us –may be taught with advantage to the youngest child. Indeed the avidity of children for this kind of knowledge ... is something quite marvellous ... On the other hand, systematic teaching of Biology cannot be attempted with success until the student has attained a certain knowledge of physics and chemistry.'

Cooke began his teaching career at precisely the time that Playfair was introducing lessons on Common Things, and the innovation must have had an enormous appeal for the new recruit to the profession with his strong

---

*The word 'biologie' had originally been coined by Lamarck in 1801, and was used by Huxley as a synonym for physiology (which included natural history) in 1854. It was not until 1874 that he specifically preferred it to natural history as the term for the science of living things, and from that time on it was generally adopted for this purpose. Natural history then took on its present meaning of the study of plants and animals in their natural surroundings, usually by interested amateurs.

scientific bent. Additionally, Gillman's progressive outlook would have ensured that Cooke met no opposition from the Church to the introduction of the new subject. However, he was not content with teaching science at his own school. As was to happen time and again later in his life, a missionary zeal overcame him and he was impelled to spread his ideas among his fellows. He was not long in finding a way. By the 1850s elementary schoolteachers, who were mostly of very lowly social origin, were becoming aware of themselves as professionals and had begun to band together in societies aimed at widening their outlook as teachers and at improving their standing with outsiders. All over the country they were forming lively Associations which arranged meetings, talks, discussions and outings for members, and published journals to which they themselves contributed most of the articles. Cooke was soon a member of the London society, the Metropolitan Church Schoolmasters' Association, and in 1854 a talk which he had given to fellow members in August was published in *The School and the Teacher*, a journal 'for the use of Masters, Mistresses and Pupil Teachers in Elementary Schools, Conducted by Church Schoolmasters'. The article is a remarkable one on several counts. It is four pages long, and the first $1\frac{1}{2}$ pages are entirely devoted to justifying the teaching of natural history on religious grounds, in language so fulsome that it is almost impossible to read it today without concluding that he was writing with his tongue in his cheek. Yet the article was the first of a long series, and one must presume that the editor of the magazine believed the author to be entirely sincere. After listing nearly every natural object mentioned in the Old Testament, from rainbows to deserts and from lions to locusts, Cooke continues:

> In the New Testament too, what a wreath of beauty does the 'Great Master' gather from the works of nature. How inimitably does he work them into parables, how admirably do they assist him in his work of tuition. The language of 'Him who spake as never man spake' glows with the warmth of the sunset, and warbles with the song of the birds.

Then, after many New Testament quotations, he proceeds:

> Not only in its relation to God, but also in its relation to man, would we claim for Natural History a place in the subjects of the teacher, and a more important, or indeed SOME place as a science in the Schedules of Her Majesty's inspectors, ...
>
> Enough has been said to suggest that the study of Natural History leads in one direction upwards to God, and in another downwards to the wants and necessities of man. The evident conclusion also suggests itself, that it is a suitable element in elementary instruction.

As an example of how correct religious bias can be applied in the teaching of natural history he takes the text, 'The righteous shall flourish as the palm tree', and shows how 'particulars' of the tree can be compared with those of a Christian:

its upright stem, its verdant crown, its leaves bowing as if in humility to its Creator; the water it requires for its nourishment and the immense quantities of delicious fruit it bears, the refreshing drink which may be prepared from it; its evergreen foliage, and the beautiful aspect it presents in the middle of so much desolation. These offer but a more minute picture of the same christian now looking upward for that victory of which the palm is a symbol, whilst humble before God, drawing all his nourishment from the well of living water, and bringing forth the beauty of a christian life; the salutary consolation that springs from him, with his leaves of hope ever green, presenting in this desert world an appearance as refreshing as that of a palm tree in the wilderness.'

When Cooke at last comes to the secular benefits of a knowledge of natural history the first he mentions is Huxley's point, that it induces in children the habit of observation; and to illustrate this he uses a lesson on birds, clearly based on his observations in Norfolk. The object of the lesson is to show how the forms of the beaks and feet of the birds reflect their way of life, and the article is accompanied by a beautiful, full-page engraving which must have been drawn from the birds he shot and stuffed himself (Fig. 3.5). It seems

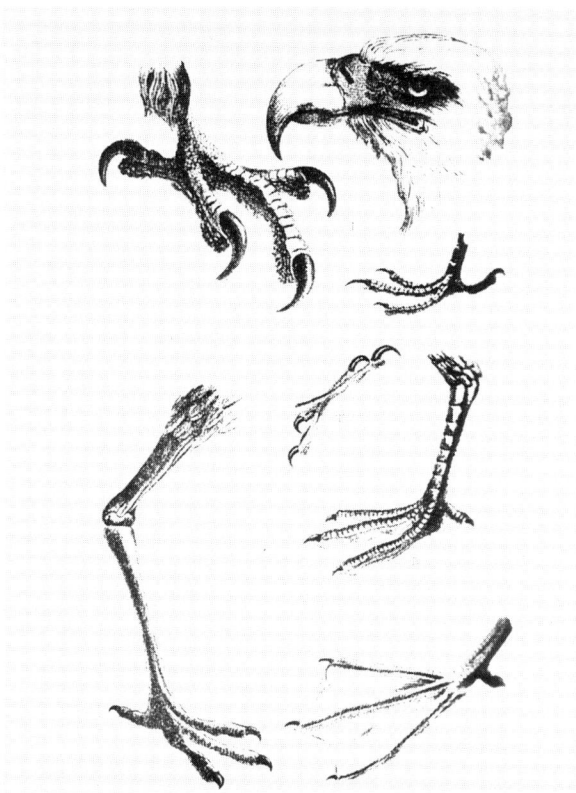

Fig. 3.5 Engraving by M. C. Cooke, illustrating his article in *The School and the Teacher*, Vol. 1, 1854.

probable that these engravings were studies for a fine series of watercolours of birds that he made at about this time, but which are now lost; for he had recently won prizes and certificates at classes for teachers on free-hand drawing, run at South Kensington by the Department of Science and Art. The list of 45 suggested lessons which follows this talk is based on animals and plants carefully chosen either for their familiarity to the children or their utility to man, and he makes a strong plea that, where at all possible, specimens should be brought into class for the children to handle for themselves, a revolutionary idea put forward by Huxley, but otherwise unheard of at the time.

Though given to very different audiences, the lectures by Huxley and by Cooke, who incidentally, were exactly the same age, were given on the same subject in the same year, and it is inevitable that they should be compared, for despite Cooke's extravagant style and sycophantic attitude towards the Church (which institution Huxley virtually ignored in his lecture) the message put over by the two men is strikingly similar. It is recognised that it was Huxley's towering influence, and his alone, that revolutionised science teaching at that time, but did Cooke arrive at similar ideas independently? We cannot be sure; they may already have been under discussion in magazines and periodicals, and Cooke might, depending on the timing, have attended Huxley's lecture before he gave his own, though he makes no mention of this. But even if his ideas were not original he, an elementary schoolteacher, led most professional educationalists in putting them forward at this date.

There remains the question of Cooke's attitutde to religion. Does the extravagant obsequiousness of this talk express what he really felt, or was it carefully contrived to win over the Church authorities and achieve the end he so much desired? He was 29 at the time, and in view of his known lack of religious belief some years later, and of the fact that he had never been accepted into the Baptist Church, the sincerity of his religious outpourings must be in some doubt. In later life he certainly enjoyed delivering declamatory and histrionic after-dinner speeches and presidential addresses, and this lecture was quite possibly a forerunner of those.

Cooke's next articles in *The School and the Teacher* are, in fact, unsigned, but there can be little doubt that he was the author. They consist of a series of 'Lessons' each dealing with a major division of the animal kingdom. The influence of Pestalozzian thinking on his teaching methods is clear–there is to be no learning by rote in the natural history class–the first words of the first lesson are: 'Write the words "Natural History" on the blackboard and elicit by questions the meaning of the term'. Cooke uses Cuvier's classification, he was familiar with the writings of Lamarck, and the detail into which he goes in an accompanying article shows that his reading must have been extensive.

The series of lessons, which were, by our standards, exceedingly formal and dull, continues until 1857, the author still taking account of the susceptibilities of the Church. For instance, after pointing out the marked similarities between men and monkeys (*The Origin of Species* was not published until three years later), he is careful to note that the former are distinguished 'above all by having an eternal soul capable of *bliss* or *woe*' (his italics). These articles are among his last writings on the animal kingdom–for the rest of his life he was to confine himself almost exclusively to the world of plants.

Apart from the fact that he had visited the 'Museum of Surgeons' we do not know for certain how or where he obtained the information on which he based his teaching, but he must have made full use of the many facilities which were beginning to be available in London to help the lower classes to educate themselves. For instance, it is very probable that he attended the evening classes for artisans at the Government School of Mines and of Science Applied to the Arts, which was located in Jermyn St., backing on Piccadilly. Huxley became Professor of Biology there in 1854 and played a prominent part in the lectures. Later, too, Cooke almost certainly made use of the Natural History Museum at South Kensington, which first opened its doors to the public in 1857. At this early date it was housed in some of the corrugated iron buildings 'of a plain and economical character' which covered most of the site. From its inception it had been intended to be a teaching institution for artisans and their instructors, the teachers' manual on botany which it made available having been prepared by the eminent Cambridge botanist, Professor J. S. Henslow. Public libraries were only in their infancy at this date, but Mechanics' Institutes and similar organisations had their own technical libraries which Cooke might well have used; and if suitably sponsored he could have obtained a reader's ticket for the great Library of the British Museum, where such rich sources as the *Philosophical Transactions of the Royal Society*, with its discussions of all the most recent scientific ideas, would have been at his command.

At the end of 1857 he received what was no doubt a very welcome addition to his finances. Earlier in the year a benefactor had offered the sum of £10 to the Metropolitan Schoolmasters' Association to be divided between the winners of a competition for the best essays on kindness to animals. There were only two entrants, Cooke and a teacher from a neighbouring school at Kennington Oval: Cooke won the First Prizes for lessons to both higher and lower classes, a sum of £5 in all.

Cooke made his botanical debut, again in *The School and the Teacher*, in 1857, with an article on Economic Botany, and followed this up in 1859 with a series of 'Notes and Lessons' on the same subject. This new interest marks an important turning point in his career and will be examined more fully in the next chapter. The articles were the last he would write for *The School and the Teacher*.

The Department of Science and Art was set up by the Government in 1853 with the object of raising the level of scientific, technical and art education throughout the nation, and it was responsible for schools of art and 'industrial design' all over the country including the School of Mines in London. Huxley worked closely with the Department, and his lectures in St. Martin's Hall

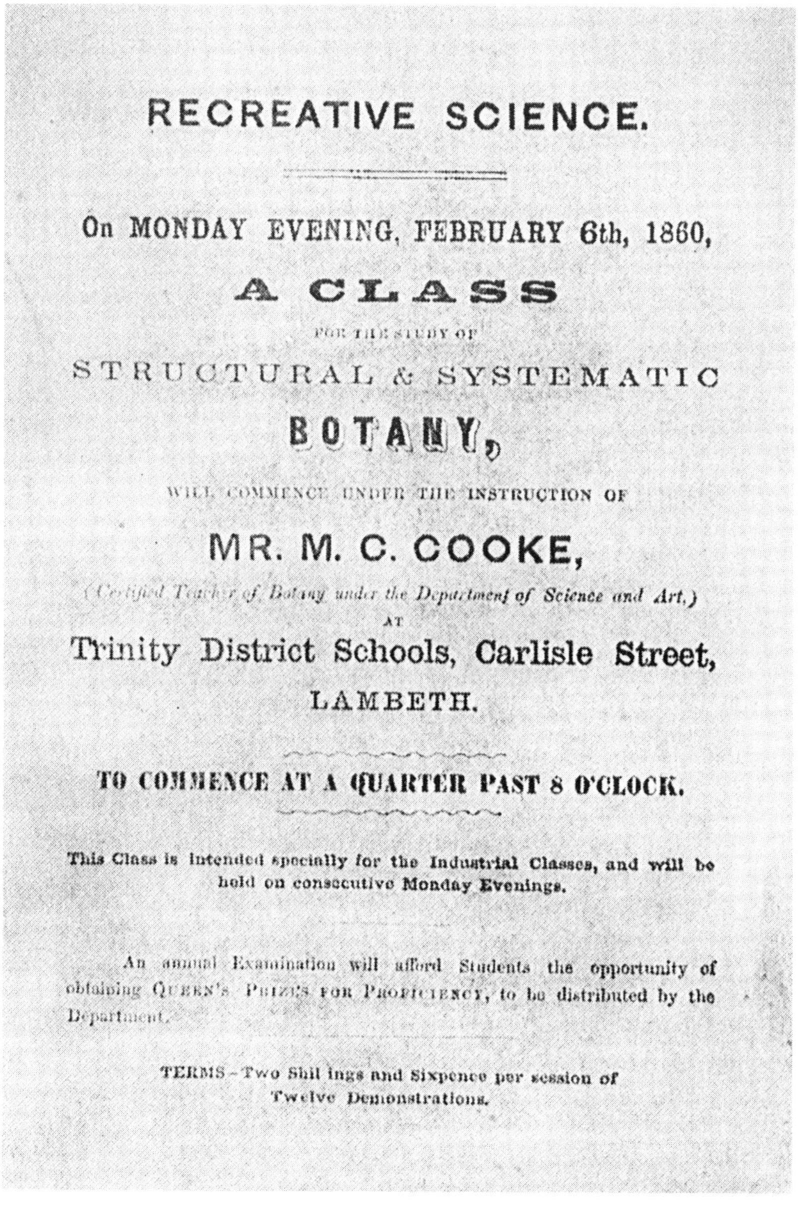

RECREATIVE SCIENCE.

On MONDAY EVENING, FEBRUARY 6th, 1860,

A CLASS

FOR THE STUDY OF

STRUCTURAL & SYSTEMATIC

BOTANY,

WILL COMMENCE UNDER THE INSTRUCTION OF

MR. M. C. COOKE,

(Certified Teacher of Botany under the Department of Science and Art,)

AT

Trinity District Schools, Carlisle Street,

LAMBETH.

TO COMMENCE AT A QUARTER PAST 8 O'CLOCK.

This Class is intended specially for the Industrial Classes, and will be held on consecutive Monday Evenings.

An annual Examination will afford Students the opportunity of obtaining QUEEN's PRIZES FOR PROFICIENCY, to be distributed by the Department.

TERMS—Two Shillings and Sixpence per session of Twelve Demonstrations.

Fig. 3.6   Handbill announcing Cooke's evening classes.

marked the beginning of a great drive by himself and others to encourage the teaching of science in schools, and to ensure that there were teachers suitably equipped to do the job. In June 1859 the Department announced that examinations in science for teachers would be held for the first time in November and December of that year, in the hope that they would 'stimulate the public to do the work of scientific instruction for itself as much as possible, the state being simply an auxiliary' [4]

Cooke entered for the Botany papers, which included both systematic and structural aspects: it is strange that he did not sit for Zoology too, for he would surely have passed with flying colours. Because of the short interval, only five months, between the announcement of the examinations and the date they were to take place, no candidate had time to prepare himself–all were examined on their existing knowledge. In any case, there were no suitable botanical text-books at the time, a situation which Cooke himself was to remedy later. In the circumstances it was very commendable that of 57 candidates taking one or more of the whole range of subjects, 42 passed, obtaining 65 certificates between them. Cooke obtained a First Class pass, but the only other candidate taking Botany failed. One of the examiners in the subject was Professor Huxley himself; the other was Dr. Edwin Lankester, a professor and lecturer in biological subjects, especially botany, at a number of London medical schools, and Superintendent of the Food Collection at the South Kensington Museum. Soon after this Lankester was to have a considerable influence on Cooke's career.

As to any material reward that might accrue from success in the examination, the Science and Art Department had intended that teachers' salaries should be augmented by up to £20, depending on the grade of the pass. But Huxley's untiring efforts to promote science in schools had only met with partial success–the Education Department consistently refused to allow these special payments to be made to teachers in elementary schools on the grounds that science was an inappropriate subject for their pupils. So, although Cooke obtained enormous personal satisfaction from his triumph–it was the only paper qualification in science that he ever obtained–he never benefited from it financially. However, like other teachers who felt equally cheated, he set about trying to capitalise on his new status as best he might. He had a circular printed (Fig. 3.6) advertising a series of 12 evening classes in structural and economic botany, to be given by himself at Trinity Schools. They would be for the Industrial (working) Classes, and would be held on Mondays at 8.15 pm, starting on Feb. 6th, 1860, the charge for the course being 2s 6d. He also tried to attract his fellow teachers by placing a similar advertisement in *The School and the Teacher*. But on the day insufficient pupils presented themselves and the classes were never held, though the few who did turn up were, as we shall see, to form the nucleus of the field excursions he would soon be inaugurating.

Chapter Four

Out of School-Hours. 1851–1860

Mordecai Cooke's brother Ebenezer was his junior by twelve years, but despite this difference in age the two had much in common. Both had red hair and brown eyes; both had inherited their mother's gift for drawing and her love of nature, though in Mordecai's case the latter was to dominate his life while Ebenezer's bent was overwhelmingly towards art; and both, though at different times in their lives, had been strongly influenced by Pestalozzian principles of education. At some time between 1845 and 1850 their parents, Mordecai and Mary Cooke, had moved to the Cubitt stronghold of Neatishead where, as in the Cookes' native Horning, they ran the village Post Office and store. So as a boy Ebenezer had attended Neatishead school where he was fortunate in encountering a teacher who was to transform his whole life. His experience is best described in his own words:

> Between 1845–1850 the transition from the old school of our forefathers ... to the new school of trained teachers took place in our retired parish. The old dame and the severe schoolmaster passed away, and among the new teachers one came ... with a new method, Pestalozzi's. ... The school was transformed, its tediousness vanished and unknown powers awoke. He joined us at play ... taught us to think and to learn from our own thought. It was a revelation, and the impression of it lives still, an example of what a teacher may do who enters into the thought of a child [1].

Throughout most of Ebenezer's life he and Mordecai were very close to one another, the older brother having a considerable influence on the younger. Ebenezer's gifts as an artist were such that on leaving school at 14 it was decided that he should become a lithographic draughtsman, to which end Mordecai found him an apprenticeship with a London lithographer, a Mr.

John Lane. This was how, in 1852, after Mordecai had been at Trinity School for a year, the 15-year old Ebenezer came to settle in with his brother's family at the school-house in Lambeth and begin to learn his trade.

Before the application of photographic reproduction methods to printing, lithographs were, with engravings, the means by which art work was reproduced commercially. Posters, advertisements, labels, prints, whole-page book illustrations, any design which was not combined on the same sheet with type-face, was reproduced by the new process, which had only been discovered at the end of the eighteenth century and was introduced into Britain in 1823. Lithography differs from printing and engraving in that reproduction is from a flat surface, not from raised marks, and is based on the fact that oil and water do not mix. The design is either drawn directly on to the stone in an oil-based medium or transferred to it from a key drawing on paper. The stone is then coated with a gum etch which reacts with the oil medium but is not held by the stone surface. When the whole stone is then wiped over with a damp cloth and inked by means of a roller the ink only adheres to the greasy lines of the drawing, so that when the printing paper is laid on the stone and pulled through the press, only the drawing is transferred to it. For such a process it is obvious that the quality of a commercial lithographer's work depends heavily on the skill and artistry of his draughtsmen, and for Lane to have taken on Ebenezer as his apprentice the boy must have shown considerable promise.

On his arrival in London, Ebenezer tells us, his brother was 'book-keeper or accountant to the Working Tailors' Association, Westminster Bridge Rd., a branch of the Castle St. Association' [2]. These and other self-help organisations were part of a larger movement for working people which had been started by the Christian Socialists in 1850. We have no idea how Mordecai was introduced to them, though we might guess that the Rev. James Gillman, with his reforming zeal, had brought them to his notice; but his involvement, though brief, was to have a profound influence on his younger brother's career (Fig. 4.1).

That band of dedicated men, the Christian Socialists, were mostly Cambridge graduates who, in their various ways, had devoted their lives to an attempt to ameliorate the lot of the working classes, mainly through offering them improved educational opportunities. The leaders of the movement included Tom Hughes, F. J. Furnivall, Charles Kingsley and, above all, F. D. Maurice. Maurice was Professor of English Literature at King's College, London until 1853, when he was forced to resign because of what were considered to be his near heretical religious beliefs, not to be tolerated in a godly establishment such as King's. It was he, together with John Ludlow, a lawyer, who first organised working men of the same trade into co-operative Associations where they could pool their resources for their mutual benefit, the first, in February 1930, being the Working Tailors' Association at 34, Castle St., Oxford St. But most of the Associations were very short-lived, surviving for a

Fig. 4.1 Ebenezer Cooke, 1837–1913.

year or two and then collapsing through internal stresses or financial incompetence. This was the fate of the Working Tailors which, by November 1851, was being referred to as 'the *late* Association'[3]. In fact it seems to have been replaced as early as May of that year by the Castle St. Tailors of the same address, but by August the financial affairs of the new group too were in chaos. It staggered on until May 1852, when the Working Tailors' Association of Westminster Bridge Rd. is first mentioned, but whether this was a branch of the Castle St. Tailors, as stated by Ebenezer, or its successor, as seems more likely, is uncertain. No record of the fate of the Westminster Bridge Rd. group, or of what part Mordecai's book-keeping played in its demise, has been traced.

One of the functions of the Associations was the provision of educational lectures and classes for their members, a field in which the Working Tailors were particularly active–Ebenezer recalls that their premises both at Castle St. and at Westminster Bridge Rd. included halls in which these classes were held. It was on the educational activities of the Associations that Maurice and his colleagues continued to build after the organisations themselves failed, their dreams culminating, in 1854, in the opening of the Working Men's College at 31, Red Lion Sq. (the premises of the Needlewomen's Association, which had failed the previous year) with Maurice as its Principal. The College offered three categories of classes to its evening and weekend students–Theology, the Humanities, and the Natural Division, the latter including Physiology, Drawing, Machinery, Music and Arithmetic. The teachers were all to be drawn from among the founders and their friends until such time as the College itself should produce enough well qualified students to take over: teaching was voluntary and unpaid. Emphasis was to be placed on the Humanities and a liberal education rather than on technology (for the latter was already catered for by the Mechanics' Institutes) and an active social life was to be encouraged.

On October 30th, 1854, Ebenezer, according to his diary, went to St. Martin's Hall, Long Acre, to hear Maurice give the Inaugural Address for his new College; presumably he was taken by Mordecai, who would have attended as an officer of one of the founding Associations. Ebenezer was deeply impressed by Maurice, but decided that the College did not quite fit his own requirements at the time. However, he cherished all his life a pamphlet by John Ruskin entitled *The Workman and his Art*, which was given to him at the door of the Hall that night. It may have been the drawing classes given by Ruskin, assisted by Rosetti, which finally prompted Ebenezer to join the College in July, 1855, enrolling for Latin, English Grammar and Geometry as well as Art. From that time his association with the College, first as student and later as teacher, was continuous for nearly 40 years. He took an active part in its social life, including indulging his interest in natural history by joining the Field Club, though in his first year he could not afford to go on its outing. Like his brother, he had a good voice and was a member of the Singing Class which, on one occasion, visited Down House to sing for Charles Darwin.

From 1854 to 1858 Ruskin took regular classes at the College in Elementary Art and Landscape, then continued intermittently until 1860, his aim being to teach his students to 'see', for he believed that 'seeing' was the beginning of all art and thought. He tried to make them happier people rather than more efficient tradesmen, and Ebenezer Cooke is mentioned as being one of his most outstanding pupils whose 'native bent was too strong to be denied' [4]. Ebenezer was very strongly influenced by Ruskin, whose philosophy reinforced the Pestalozzian ideas which had affected him so profoundly in his schooldays. Both the Cookes recall with pride their contact with that almost legendary group of men, the Christian Socialists.

Up to now Mordecai Cooke's interest in natural history had been intense but very general, for he delighted in all aspects of Nature but had studied none in depth. At some time during his very early years in Lambeth an incident occurred which was to change all this, and set his feet unswervingly on the mycological road for the rest of his life, though still with occasional excursions into the wider world of natural history. The conditions of his job were such that he would not even have had Sundays free in term-time because of the compulsory church parade for the children, but somehow, presumably during the school holidays, he still managed to visit Norfolk. Many years later he wrote an account of one of these visits:

> It was my good fortune to be introduced to an East Anglian gentleman, who resided in a small agricultural village not ten miles from Norwich. I had been invited to give a gossiping lecture to the rustics in the schoolroom, and was asked to take a preliminary tea

with the squire. It soon became manifest that the hobby of my host was 'edible fungi', a subject of which I was then profoundly ignorant, but I became greatly interested in the discovery that there were other fungi beside the mushroom which might be eaten, and I had the pleasure of looking over his portfolio of coloured drawings, and hearing his explanation and encomiums. This was my first inspiration to turn my attention to 'toad-stools'. I had never seen them before, or at least with an appreciative eye, and the subject came upon me as a revelation. At first I did as so many others have done, restricted my interest to their edible qualities, and had no ambition beyond being able to recognise, collect, and devour some half-dozen different kinds of 'toad-stools' which, in all my surroundings, I had been taught to regard as 'rank p'isen'. Since that eventful evening I have never abandoned the pursuit and it has been my solace. (Aut)

The East Anglian gentleman who had such a profound influence on Cooke, and therefore on British mycology, was Richard Ward, one of the local landed gentry. His estate was at Salhouse, and the Hall stood 'in a well-wooded lawn' and was 'richly embellished with paintings of art and virtue'.* Ward must have been of good standing in the small mycological circle of his day, for many of his fungi were identified by the Rev. M. J. Berkeley, the Father of British Mycology (p. 150), and none other than Sir Joseph Hooker, the Director of Kew Gardens, sought specimens from him for the Kew Museum. Ward's drawings of toadstools, some of them copied from early botanical illustrations and others drawn from life, passed to his grandson, Ivor Fanshaw Ward, who has lent them to the Castle Museum, Norwich. Cooke recalls that at that time the only books in print for the identification of toadstools were one volume of Withering's *Arrangement of British Plants* and the Rev. Berkeley's account of them in Hooker's *English Flora*.

Some years after this encounter, Cooke visited another well-known East Anglian naturalist, an old man, Mr. Daniel Stock, bookseller of Bungay (a small town on the Norfolk-Suffolk border) who had for many years been writing articles on the fungal flora of the district for the local press, and who was the founder of the Bungay Botanical Society. The only time in his life that Cooke ever found the highly poisonous white Amanita, *A. virosa*, was in the Bungay woods in the company of Stock, and the painting that he made then eventually became Plate I of his monumental *Illustrations of British Fungi* (Fig. 4.2).

In January 1854 the first of a series of 173 articles on 'Vegetable Pathology' appeared in the *Gardeners' Chronicle*, written by Berkeley, and covering all forms of plant disease including that due to pathogenic fungi; after the completion of the series Berkeley continued to publish individual articles on a wide range of fungal diseases of plants, and it was presumably to these two sets of articles that Cooke referred when he wrote:

*White's Directory of Norfolk, 1854.

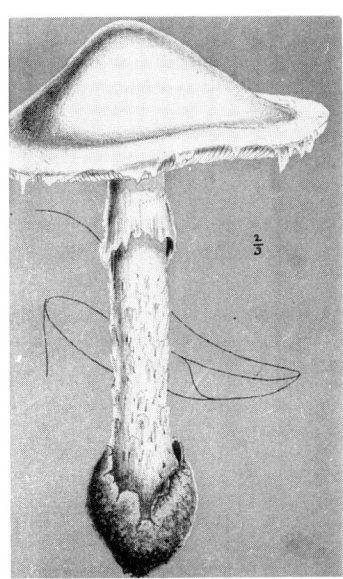

Fig. 4.2 M. C. Cooke's illustration of *Amanita virosa*, later used as Plate 1 of *Illustrations of British Fungi*. Original in colour.

> I cannot determine in what year it was that I discovered a series of chapters on Mycology contributed by the Rev. M. J. Berkeley to the pages of the *Gardeners' Chronicle*, illustrated by woodcuts. I set myself to get all these papers, until I got a complete set, which I extracted, and ultimately bound into a volume, and used as a guidebook to the knowledge of British Fungi, and found to be invaluable. Many years after I made a present of this volume to one of my most promising pupils, as a treasure to be preserved. (Aut.)

Though we hear no more about fungi during the rest of Cooke's time at Trinity School, his acknowledgement of his debt to Berkeley's articles and his sudden blossoming as a mycologist immediately he left, make it certain that he was devoting a great deal of time to his chosen subject at least from 1854.

In the 1840s, stimulated by pride in Empire and our dominant position in world trade, there was much interest in Britain in the nature and provenance of those natural products of the animal, plant and mineral kingdoms that were of use to man. They became known collectively as 'economic products', and lectures were given and books written about them from scientific, commercial and popular points of view. Sir Joseph Hooker, before he became Director of Kew Gardens in 1841, had lectured on Economic Botany at Glasgow University, and had amassed a large collection of material for demonstration to his students, which he brought with him when he moved south. In 1848 this collection was set out at Kew for display to the public and became the first Museum of Economic Botany in the country. Enthusiasm for economic products was further stimulated by the Great Exhibition of 1851, and soon afterwards small educational museums were started in many places, a flurry of popular lectures was given, and numerous books and articles published.

One of the most energetic of the publicisers of the subject was Peter Lund Simmonds (1814–1897), and he found an enthusiastic disciple in Cooke. The latter gives us no hint as to how he met Simmonds, but a paragraph of his autobiographical notes indicates the significance which he attached to the meeting:

> Another important feature of my scholastic career was the study and tuition of the economic products of the vegetable kingdom, to which I was urged by my friend Peter Lund Simmonds, author of a well-known volume entitled 'The Economic Products of the Vegetable Kingdom'.

There is some mystery about Simmonds' origins although all sources agree on one thing, that he was born in Denmark in 1814. One obituary [4] has it that he was the eldest son of Lt. George Simmonds, R.N.; another [5] that he was born Peter Lund, but was adopted and brought up by George Simmonds whose name he added to his own. Like Cooke, he received little formal education, but on all other counts their youthful careers could hardly have been more different. Simmonds joined the Royal Navy as a midshipman at the age of 12 and then, at 17, was sent by an uncle to Jamaica as a sugar planter. There he remained for three years, witnessing the end of slavery in the colony, an experience which later led him to compare the lot of the English poor unfavourably with that of Jamaican slaves. On returning to England he seems to have plunged immediately into the world of journalism, writing articles on all sorts of subjects, from slavery to the newspaper industry, for a wide selection of periodicals. The first we hear of his association with economic products is in 1841, when he joined the staff of the *Agricultural Journal* and the *Mark Lane Express* and began to contribute extensively to their columns. Mark Lane and Mincing Lane, in the City and close to the Docks and East India House, were the hub of Britain's commerce with her Indian dominion, and the trading centre of London. Brokers in all the main commodities–tea, coffee, sugar, cocoa, rubber, tobacco, spices, wine, ivory, silk–had their headquarters there, and samples of their wares were always to hand. The *Mark Lane Express* was the brokers' journal. Simmonds' interest in the subject must have been aroused during his sojourn in Jamaica, but his intimate knowledge of it could only have been self-acquired, for there is no evidence that he received any training, or was bound apprentice, at any time. The book mentioned by Cooke, *The Commercial Products of the Vegetable Kingdom*, was a volume of 658 pages which came out in 1853, ran to several large editions and became a standard text. Altogether Simmonds wrote about 50 books and innumerable articles on every aspect of trade and commerce, and it may well have been Cooke's association with this prolific writer that set his feet on the same path.

On the subject of economic products Cooke continues:

> In furtherance of this study ... was the collection and construction of a ... Museum of such objects, for which purpose Mr Simmonds introduced me to brokers in Mark Lane, and Mincing Lane, and explained my object, soliciting them to give me small samples of any new or interesting product which came into the market, to which he added a great number from his own collection or Technological Museum.
> This collection was arranged in glass bottles in glazed wall cases in the school room ....

His classroom must have been a very unusual one for those days, when the occasional wall chart or globe were the only teaching aids normally used. He could enliven his lessons on Common Things by passing round the contents of his glass jars for the children to see, handle and smell for themselves, while he explained where the objects came from and what they were used for.

He wasted no time in putting his ideas over to a larger audience. Early in 1857 he was elected to the committee of the Metropolitan Church Schoolmaster's Association, and on May 2nd of that year he:

> ... read an interesting and instructive paper on 'The Study of plants in relation to commerce and manufactures'. The room was hung with specimens of fibres and drawings of plants; while in various places there were arranged about 500 bottles containing vegetable productions such as gums, nuts, spices, waxes etc. ... At the close of the paper Mr. Cooke intimated his intention of handing over to the Association a large number of the specimens to be added to the Scholastic Museum originated last year by Mr. Myers, ... he also advocated the desirableness of at once forming a committee to collect and arrange the specimens, receive collections etc. This suggestion was acted upon. ... It is intended to have boxes, containing an illustration of the manufacture of particular articles, and to lend these boxes to members for use in their schools.
> ... A room has been set apart ... in the Westminster Rd. for the present reception of the Museum; and donations in vegetable, animal or mineral productions may be sent there. ... Donations in money for the purchase of bottles, cases etc. will also be thankfully received.

The idea of museums as aids to popular education was a new and exciting one in the 1850s, and its immense possibilities were only just being explored. Up to then they had been crammed and dusty treasure-houses overflowing with extensive collections made by specialists for specialists, and hence with no attempt to display or explain them. But if ordinary people were to be taught about the wonders of science, by which every aspect of their lives would increasingly be affected, one way to do it would be by means of attractively arranged museum displays using objects carefully selected to illustrate specific points and labelled with accurate descriptions and explanations. Such museums were burgeoning on all sides, and the idea must have had an enormous appeal for Cooke, for he was early in the fray and was involved with them in one way or another for over 20 years. He even gave a talk on planning a local museum for Epping Forest in 1896, some years after he had retired.

He threw all his boundless energy into making a success of the Scholastic Museum. In June he was writing to the authorities at Kew Gardens on the Museum's headed notepaper, giving one of the City brokers as reference, to ask if, without damaging their own collection, they could spare a sample of 'the Hog Gum of Jamaica' for the Museum. Though he must undoubtedly already have visited Kew's Economic Museum as a member of the public, this letter is, as far as is known, the first time he was in personal contact with that organisation. He ends his letter:

> ... we hope, by creating a taste for such studies in both masters and pupils, to enable them to appreciate more highly such collections as that with which you are so intimately connected at Kew.

In October, doubtless at the request of Cooke,

> ... P. L. Simmonds Esq. gave a lecture [to the Metropolitan Association] on the economic uses of shells. Specimens of various shells and shell work were exhibited and created much interest. A vast amount of information on the structure of shells, as well as their uses was given by the lecturer.

By November there was good news:

> It was announced that the Royal Commissioners of 1851 had made a grant of animal, vegetable and mineral substances, for the purposes of the Scholastic Museum now in the course of formation ... and that nearly 600 specimens had been collected, chiefly through the exertions of Mr Cooke, who has so ably seconded Mr Myers in the establishment of the Museum.

Little progress was made with the Museum in 1858 as much of the teachers' time was taken up with preparations for a great choral festival in which all the London schools were to take part, but at the end of that year:

> On Friday evening ... the members and friends met at Trinity Schools, Lambeth, to witness the diorama painted by Mr. Cooke entitled 'Loiterings in the Levant'. The object of the exhibition was to pay off the debt incurred in establishing a museum in connection with the Association. Mr. Cooke deserves great praise for the manner in which he has painted the scenery and for the clear and lively manner in which each picture is described.

Once again, as had happened all through his teaching career, Cooke's skill as a draughtsman was put to effective use, but this time on a grand scale. The paintings for the 'diorama' must have been considerably larger than anything he had tackled before, and one wonders whether Ebenezer had a hand in their planning and execution.

For some months in 1859 numbers at meetings of the Association dropped off badly, but in September it seemed to come to life again when a very well attended conversazione was held at which Cooke regaled the company with a humorous talk on 'Five Cups of Tea'. The year ended with another talk by him, reprinted in full in *The School and the Teacher*, on 'The history of a cotton gown, from the seed-pod to the rag-bag'. The subject is tackled from every conceivable aspect and in astounding detail, in a way which his audience must now be expecting of Cooke and which would only have been possible if he had access to specialist sources of information. First he discusses the derivation of the word cotton, and of the Latin name of the plant, *Gossypium*; then its cultivation and use, from ancient times to the present day, in the Old World and the New. He then goes into its botany and present day cultivation, the quantities produced in various parts of the world, and the price paid for it once it reaches these shores; then the whole process and economics of spinning, weaving, bleaching and printing, including the chemistry of the last process, are covered in detail. The making up of the finished material into the gown of the title is treated at length with the punning humour that is now characteristic of Cooke. But the talk did not end there. The gown having reached the rag-bag, the speaker proceeded to a discourse on paper-making, and only after a final mention of the economics of the cotton trade does he finally draw his marathon discourse to a close. Nearly 50 years later, when reminiscing to a friend, he wrote:

> I remember an early lecture of mine given several times and call[ed] A cotton gown from the seed pod to the rag-bag–which I am sure was once in print but I know not where. It was so long ago [7].

Soon after this talk, reports of meetings of the Metropolitan Association become very irregular and, after its few years of vigorous life, it seems eventually to have faded away, though not in all probability entirely due to loss of enthusiasm on the part of its members. Ever since 1857 there had been much discussion in the columns of the *Journal of the National Society* on the pros and cons of forming a National Church Schoolmasters' Association of which all the presently independant local Associations would become branches. The Metropolitan Association voted stoutly for independence but the scheme was nevertheless carried through, and it was at about this time that the Metropolitan began to fail.

In the archives of Yale University there is a thick, leather-bound, manuscript volume of about 350 pages which was presented to that Institution by Cooke's daughter Leila after her father's death. It is entitled *Illustrations of*

Oeconomic Botany. Vol. 2. Diclinous Exogens.* *M. C. Cooke, London. April 1858.* The whereabouts of Volume I are unknown. Into this book Cooke copied out, in his beautiful, flowing hand, descriptions of economically important plants belonging to families with unisexual flowers, together with their history, provenance and uses, leaving a space for an illustration of each though only a few of these have been completed. The origin of each description is acknowledged and Cooke must have read widely to find them, for they are taken not only from the *English Cyclopaedia* and the *Yearbook of Facts, 1841*, but from such specialised sources as the Catalogue of the Madras Exhibition, books on medicinal and economic botany by the great botanist John Lindley, and Jamaican, American and British technical journals. As few of these sources would have been readily available to an elementary schoolteacher, Cooke probably gained access to them through the good offices of Peter Simmonds. The purpose of the compilation is obscure. Though obviously related to Cooke's enthusiastic support for the Scholastic Museum, its exclusively botanical contents arranged according to the then current ideas of botanical classification show that it could not have been intended as a Museum catalogue. Perhaps it simply served to satisfy his own interests as a collector and classifier which were now becoming an increasingly important driving force in his career.

Of the numerous societies, British and foreign, to which Simmonds belonged, one of the most august was the misleadingly named Society of Arts (now the Royal Society of Arts), founded in 1754. The name is, in fact, an abbreviation for the Society for the Encouragement of Arts, Manufactures and Commerce, a much more accurate description of its functions. Up to 1850 these had consisted chiefly of offering prizes at all levels for the furtherance of the activities in which it was interested; then the rapidity of technical advance rendered the prize system obsolete and the function of the Society changed. Paper-reading meetings became an important activity, and enquiries were initiated into problems and matters of moment raised at them. In 1852 a *Journal* was started with the object of recording the papers and discussions. The Society did much to encourage money-saving inventions, such as cheap heating systems, which might improve the living conditions of the poor and, inspired by the Great Exhibition of 1851, it took a lead in establishing local museums for the education of the working classes.

Simmonds joined the Society in 1855 and applied, unsuccessfully, for the post of Assistant Secretary the following year. He was a frequent contributor to its paper-reading meetings and to the *Journal* on every sort of subject, he introduced Karl Marx to membership, and eventually, in recognition of his being 'eminent in the application of abstract science to the Arts, Manufactures and Commerce', he was made a life member. It must have been on Sim-

---

*Exogen = Dicotyledon. The word refers to the method of thickening of the stem in these plants.

monds' recommendation that Cooke began to write articles for the Society's *Journal* in 1858, for he was never himself a member. His first contribution, 'Notes from the Gold Coast', remarks culled from 'a series of letters received during the last years from a correspondent at Cape Coast', is mysterious, for there is no indication at any period of his life that he had contacts in that part of the world. Probably the letters were lent to him by Simmonds as the basis of an article, or he may have read about the Gold Coast and himself recast the information in the form of letters. His three other contributions to the *Journal*, the last in 1860, were on 'New paper Materials' (there was a shortage of rags for paper-making at the time), 'Naphtha' and 'Pulu, or Vegetable Silk' (a Hawaiian fern), the contents of all being useful digests of a scattered literature.

The year 1860 was a momentous one for Cooke for several reasons, and not least because it marked the start of what was to be his highly prolific career as an author. The book with which he made his debut, giving as his credentials, 'Director of the Scholastic Museum', was *The Seven Sisters of Sleep*, 'a popular history of the seven prevailing narcotics of the world'. Like his articles on economic products, it was a compilation from the literature, with the addition of statistics supplied by Simmonds, all of which are acknowledged in a flowery introduction headed 'Prefatory Admonition'. We have no idea why Cooke should have chosen such a subject for a popular book except that the first and longest section is on tobacco, of which he was a lifelong addict–the final chapter, too, consists of a wordy apologia for smoking. About one third of the book's 350 pages are devoted to The Weed, folowed by sections of decreasing length on Opium, Hashish, Betel, Cocaine, Belladonna, and the Siberian cult of the agaric *Amanita muscaria*. It seems that Cooke was anxious to include a narcotic from every climatic region of the world, and the use of *A. muscaria* was the only example he knew from a cold climate.

The book starts with a long and elaborate dedication to lovers of each of the narcotics he is about to discuss. Then he shrugs off possible criticism of the format with the words: 'Why I should have chosen such a title for my volume, and wherefore I should have invested it with a legend is a matter of little importance. It was a fancy of my own, and if any think fit to quarrel with it, they may do so without disturbing my peace of mind'. Chapter 1, 'Somewhat Fabulous', is devoted to his newly invented legend:- The Queen of Sleep has seven sisters, all envious of her throne. She instructs her Minister of Sleep to endow them with power over man's waking hours, giving him splendid dreams, illusions and ecstasies. The sisters take over different parts of the world–Morphina is allocated Tartary and Mongolia; Virginia (tobacco) has four-fifths of the earth; Gunja takes the Nile, Ganges, Indus and Niger; Siraboa has Malaya and the neighbouring countries; and Erythroxalina, Bolivia and Peru.

> Two less favoured, less beautiful and less successful sisters, pouting and repining at the good fortune that had attended the others, secluded themselves from the rest of the

world and rushed into voluntary exile. Datura ... fled to the Northern Andes ... The pale and dwarfish Amanita, turning her back on sunny lands, and glowing skies, sought and found a home and a refuge, a kingdom and a court, in the frozen wastes of Siberia.'

Tobacco is treated in quite a different way from the other narcotics, discursively and lightheartedly, with numerous literary allusions, lengthy stories and much punning. Accounts of the others, though never technical, are much more straight-forward and serious, the long section on opium–very topical then, for the Opium Wars with China were only just over–making interesting reading today. Cooke quotes De Quincey's published experiences with the drug approvingly and at length (when his library was sold after his death it contained five volumes of De Quincey's works), and while pointing out the tragic results of addiction, he is convinced that the pleasures of opium are such that it is no more wrong to indulge in it than in tobacco smoking. He sees no justification in the opposition then being mounted to the trade in the drug that Britain was forcing on an unwilling China, regarding it as innate in the Chinese character that 'coolies' will somehow or other manage to procure drugs and to succumb miserably to them.

Two of the characteristics of the *Seven Sisters* were to recur again and again in Cooke's later works. A great many of his books are compilations from the existing literature–'scissors and paste' jobs–often skilfully done and serving the important purpose of opening up for the lay public the wonders of the natural world which would otherwise have remained hidden in inaccessible scientific books and journals. Throughout his life he also loved either to invent a legend to fit some current event, or to adapt a classical myth for the purpose.

Sometime in 1859 the Rev. James Gillman, to whose liberal and broadminded outlook Cooke owed so much, moved to another parish, and the living at Holy Trinity, Lambeth, passed to the Rev. W. E. Green, 31 yrs old and a graduate of Oxford, but otherwise quite unremarkable. The immediate result of the change of incumbent was, as Cooke records:

> ...that as usual, all the officials at the church had been changed, from the pew opener upwards, until none of the old staff were left except at the Schools'.

He continues:

> The Irish curate was commissioned to visit the Schools nearly every day, which he did persistently, and interfered so much that I recommended him to attend to his duties at the Church and leave me to my duties at the Schools. This, of course, led to my being advised to resign which I did forthwith.

There seems to be some confusion in Cooke's mind as to whether he was allowed to resign or was, in fact, sacked, for in the next paragraph of his autobiographical notes he says: 'The reason alleged by the clergy for my leaving the Schools was that the education was too secular'. Another source (JR) suggests that Green considered Cooke's botanical activities to be a waste of time, while Cooke himself gives yet another reason for his sudden departure. In a very brief curriculum vitae written in 1902 for the Philadelphia Academy of Sciences he states that 'family troubles' led to the change. This is the first of only two references we have in the entire corpus of his surviving writings which suggests that all may not have been well with his family affairs. Probably there was no one reason for the abrupt end of his career as a schoolmaster; all of those given could have played a part in bringing an unhappy situation to a head.

Now, after nine years of steady employment, Cooke found himself without a job. He must by then, through his activities with the Metropolitan Schoolmaster's Association, have been well-known in the teaching profession in London, and it is perhaps strange that he did not move on to another school, possibly one run by the Non-conformist Churches with a more liberal outlook than that of the Church of England. But with the Department of Education still actively discouraging the teaching of any but the most basic school subjects Cooke, strongly biased towards science as he now was, may well have felt that the scope was too limited for the profession to hold any further attraction for him. He would probably have endorsed the sentiments of his brother William, who threw over teaching as a career at about the same time saying that he was 'disgusted with much work, scanty pay and clerical domination'. William obtained a highly unsatisfactory job as a clerk with a firm of ironmongers in Norwich, but Mordecai, because of his experience in setting up the Scholastic Museum, his qualification in Botany and, most importantly, through the patronage of Simmonds, could hope for a more congenial future. He even managed to sell, for the large sum of £100, the museum he had arranged in his schoolroom, thus starting his new life with modest financial backing.

Mordecai's departure from Lambeth meant a move for Ebenezer too. Ebenezer had, by this time, finished his apprenticeship as a lithographer and set up as a small master on his own. We do not know whether he continued to live with his brother's family when they first moved, but soon afterwards he married Ellen Lane, the daughter of his former employer, and the couple eventually shared the Mordecai Cookes' home in Kentish Town.

Chapter Five

Interregnum. 1860–1862

When Mordecai Cooke lost his job as a teacher he lost his home as well, so it must have been with some relief that he accepted Simmond's kindly suggestion that he should move into his own, overlarge home. Simmonds had moved four years earlier from Bucklersbury, near the Docks and his work, to an elegant terraced house in newly developed Pimlico. No. 8, Winchester Street (Fig. 5.1), like the rest of the long terrace, still stands. As recently as the 1840s Pimlico had been an area of marshes and market gardens along the bank of the Thames, but the land had been gradually bought up by Thomas Cubitt (1788–1856)*, the great speculative builder who had been transforming London by covering great estates and waste places alike with high quality houses for the upper middle classes and the wealthy. Simmonds moved to Pimlico in the year that Cubitt died, when Winchester Street was only partly built. It was completed to Cubitt's designs, but No. 8 may not actually have been built by him.

Aided by Simmonds, Cooke, with great drive and energy, set about making full use of his enforced idleness. Simmonds had recently taken on the editorship of a new but short-lived journal, *The Technologist*, to which he invited Cooke to contribute; between 1860 and 1862, Cooke wrote some 16 articles from one to twelve pages long on a wide selection of economic products, mostly of the vegetable kingdom. One 'On some edible fungi' is of interest in that it tells us that dried *Morchella esculenta*, *Helvella crispa*, *H. lacunosa*, and

---

*Thomas Cubitt's family, like that of Cooke's mother, came from Norfolk, but from the village of Buxton, due north of Norwich. There are several lines of Cubitts in the county, each originating in a different village, and there seems to be no close relationship between Thomas Cubitt and the Neatishead family.

53

Fig. 5.1 No. 8, Winchester Street, Pimlico, the home of P. L. Simmonds, seen in 1974.

the Beef-steak Fungus, *Fistulina hepatica*, were all proffered for sale in this country at the time. It is doubtful whether Cooke would have been paid for these articles, but they would have been a means of making his name known in the right circles.

From the date-lines of two of Cooke's books, both published in 1862, we know that while he was working on them he was in some way connected with the Twickenham Economic Museum, but as this establishment is mentioned nowhere else in his writings, and the Museum's records no longer exist, we can only guess how he became associated with it and in what capacity. Its founder was Thomas Twining (1806–1895) of Perryn House, Twickenham, a wealthy philanthropist belonging to the family of well-known tea-merchants. In his youth he had travelled a great deal, then for ten years had suffered severe ill-health, largely as the result of a serious accident, and it was not until he was forty that he began to put all his energies into an endeavour to better the condition of the working classes by opening up for them opportunities for further education. To quote *The Times*, he was 'devoted to economic questions' and was 'one of the earliest and most strenuous advocates of technical education'.

As we have already seen, one of the new functions undertaken by the Society of Arts in the 1850s was the promotion of educational museums, and Thomas Twining, then a vice-president of the Society, was in the forefront of the movement. In 1856 he:

... began to collect at the House of the Society of Arts his repertory of objects and information relating to domestic and sanitary economy, which, from the union of these attributes, took the name of 'Economic Museum'. The food department was exhibited in embryo at South Kensington in 1857, and gave rise to Dr. Playfair's admirable food collection, subsequently developed by Dr. Lankester. However, due to Twining's continuing ill-health, the Museum was moved in 1860 to a building specially erected at Twickenham in the grounds of Mr. Twining's residence' [1].

At the time of the move, a long article about the Museum, unsigned but probably written by Twining himself, appeared in *The School and the Teacher*, giving a fascinating insight into the attitude of the wealthy and educated towards the lowly beings beneath them on the social scale.

For several years the Council of the Society of Arts has been engaged in a comprehensive scheme for raising the physical, intellectual and social condition of those industrious classes whose welfare is so essential to our national prosperity. One of the features of this scheme is the establishing of Economic Museums, or collections of useful things popularly classified, attractively illustrated, familiarly explained by printed labels, and calculated to teach the working classes ... what sort of dwellings they should live in; ... what fabrics they should wear ... what food they should eat and how it ought to be cooked ...[2].

The writer then suggests that every educational establishment, from Mechanics' Institutes to the humblest village school, should have its own 'Economium or Educational Cabinet', however small and incomplete (Trinity Schools under Cooke had, of course, been far better equipped than this four or five years before Twining's article was published). The Twickenham Museum was to be the nerve centre of the new movement, supplying ideas, and offering instruction on collecting, classifying, lay-out and labelling. It was to have a Curator, a Mr. W. Freeman, a workshop and a library, and it would lend material for the illustration of lectures. It was being set up '...*strictly for the purposes of social improvement and Christian benevolence*' and;

to afford practical guidance to Clergymen, Medical Men, Schoolmasters, and others entrusted with the bodily care or intellectual development of the people, and especially to benevolent persons or societies engaged in the improvement of the dwellings and household comforts of the poor.

Twining's activities at the Society of Arts were taking place just at the time that Simmonds introduced Cooke to that organisation and the latter was writing on economic products for its journal, as well as setting up museums himself for his pupils and fellow teachers. He could not have failed to be aware of Twining's interests and would quite likely have been introduced to him by Simmonds and have asked his help. It would, therefore, have been an obvious move for Cooke to have sought paid or unpaid employment at Twickenham when the Museum opened, since this almost coincided with the loss of his job as a teacher.

Sadly, the Museum's life was a short one; it was burned to the ground in 1871, and Twining, feeling himself too old to 'begin afresh the main labour of his life', devoted himself indefatigably to writing books, lectures and tracts about and for the technical education of the poor.

Sometime during the autumn of 1860 Cooke at last found paid employment as 'lecturer on Botany on fixed evenings in the week in the gardens of Holly Lodge, to the gardeners and other servants in the employ of the Baroness Burdett Coutts' (1814–1906) (at that time, in fact, she was still only Miss Angela Burdett Coutts). This remarkable woman, the richest in England and one of the great Victorian philanthropists, had inherited her money from her grandfather, the banker Sir Thomas Coutts, and had resolved from the first to use it for philanthropic purposes, which eventually reached such vast proportions that they touched almost every aspect of Victorian life. Two of her greatest interests were the welfare of destitute women and girls, and the education of the lower classes of all ages, among which she did not neglect her own employees.

Her estate at Holly Lodge, Hampstead, covered about 60 acres; as well as the Lodge itself, her country home, there were 18 houses for estate employees, stables, a farm, a nursery, a kitchen garden and greenhouses.* The number of outdoor staff employed on such an estate must have been fairly large. Miss Burdett Coutts had always been interested in science–she had joined the Society of Arts as early as 1843, becoming a life member–and in 1860 she began to take a special interest in botany and natural history. It is not surprising, therefore, that this should be the moment at which she began to arrange for the botanical education of her gardeners. Whether Cooke replied to an advertisement or whether, as is perhaps more likely, Miss Burdett Coutts heard of him through the Society of Arts, or some other educational activity in which she was engaged, we do not know. The only record that remains of his employment at Holly Lodge are some remarks he made at the Working Men's College the following winter: 'I once had a winter class of gardeners, men as ignorant, by the way, of botany as any in the world, for they know the Latin names of plants and no more.' It would appear from this that the classes only lasted for one season; all the same, they altered his way of life sufficiently to induce him to move from Simmonds' house in Pimlico to a home of his own in Kentish Town, in order to be nearer his work.

Unemployment gave Cooke a splendid opportunity for embarking in earnest on his career as an author, and characteristically he wasted no time in

---

*The estate was broken up and sold for building in 1922, but is still a very exclusive unit with strict regulations governing its management and development.

doing so, producing three small books by the end of 1862. We do not know his reasons for publishing his first with Blackwood's, but his meeting with his next publisher, Robert Hardwicke of Piccadilly, was no accident, and the close association which developed between the two men–at one time Cooke was visiting Hardwicke's shop almost every day–was to be extremely important to Cooke's future career. He was introduced to Hardwicke by Dr. Edwin Lankester (1814–1874), his examiner, with Huxley, in the Department of Science and Art Examination. Lankester, as we have seen, took over responsibility for Twining's Food Museum when it was transferred from the Society of Arts to the South Kensington complex, taking the job very seriously, including writing a popular guide to the collection. Whether Cooke had made his acquaintance through the Society of Arts or through attending his lectures at the School of Mines we cannot be sure, but in either case, Lankester's well-known kind-heartedness and geniality would have ensured that he would willingly use his influence to help a deserving younger man.

Lankester (Fig. 5.2) had known struggle and hard times in his own youth in East Suffolk, for though he was the son of a prosperous builder, his father had died when Edwin was very young, the family's affairs had been mismanaged, and the boy had to leave Woodbridge Grammar School when he was only twelve years old to be apprenticed to a surgeon. When he was 20 his friends lent him the money to go up to University College, London, which had recently started a new course in Medicine, and there he obtained his preliminary medical qualifications of M.R.C.S., L.S.A. Finally, raising the finance

Fig. 5.2   Dr. Edwin Lankester, 1814–1874.

himself, he finished his training in Germany, obtaining his M.D. at the age of 25.

On returning to England he set up his consulting rooms in Saville Row, off Piccadilly and very close to Hardwicke's publishing house, but despite the effort he had put into qualifying, he actually practiced medicine very little. His income for some years was obtained chiefly by teaching the biological sciences, which then formed a large part of the medical curriculum, at various London medical schools. Afterwards, the greater part of his life was to be devoted to a sustained effort to help in various ways London's poor and underpriviledged. First, he too joined the ever-growing movement for the improvement of the technical education of the working classes, his special contribution being in human physiology and the causes and prevention of disease. His teaching in this latter field could hardly have been more immediate for, as an experienced microscopist, he worked with Dr. Snow who, during the London cholera outbreak of 1854, had been the first to make the vital discovery that the disease is water-borne. Thanks to the pioneering efforts of Dr. Chadwick and others, the matter of public health now began to be taken very seriously and, as a direct result of his work on cholera, Lankester was appointed the first Medical Officer of Health for Westminster, the position he held when he introduced Cooke to Hardwicke. Then in 1862, after a prolonged struggle against entrenched opposition, he was elected Coroner for Middlesex, and set about exploiting his verdicts to publicise the appalling living conditions to which the poor were condemned, and because of which so many died tragically unnecessary and often violent deaths.

By the nature of his interests this tall, portly man, with brown hair and eyes, a luxuriant beard and a beautiful voice, met and became friends with large numbers of like-minded people from all walks of life, among whom was the Marx family; indeed he was one of the few mourners at Karl Marx's funeral [3]. Lankester had two passions besides philanthropy, namely microscopy and botany, which latter 'ruled his life.' His wife, Phebe, was a botanist in her own right and, like Edwin, she published her books with Hardwicke. Such was Lankester's devotion to the subject that he became a Fellow of the Linnean Society, and the great botanist John Lindley, whose student he had been at University College, named a new genus of tropical plants after him, *Lankesteria* (Acanthaceae). As to his eminence in microscopy, by the time that Cooke knew him he had been joint editor of the *Quarterly Journal of Microscopial Science* for seven years and president of the Microscopical Society for a year.

When Lankester introduced Cooke to Hardwicke in 1860 or 1861 the latter's publishing house, to which he had only moved from neighbouring Duke Street in 1856, was at No. 192, Piccadilly (Fig. 5.3), a street which was not then the centre of the retail trade that it is to-day. In 1857 the Royal Society moved into its new premises at Burlington House, followed by the Linnean Society in 1858, and close by also was the School of Mines

INTERREGNUM. 1860–1862

Fig. 5.3   No. 192, Piccadilly in 1840. (From Tallis' *London Street Views*).

in Jermyn Street; many doctors besides Lankester had their consulting rooms in the vicinity. In fact, it was a most convenient situation for a publisher specialising in medical and technical books as Hardwicke did. His establishment was a plain, 4-storeyed, terraced house with a large shop window of small, leaded panes at ground level. It stood between St. James' churchyard and Fortnum and Mason's shop.*

Robert Hardwicke (Fig. 5.4) and his doctor brother, William, came from a very old Lincolnshire family, but had joined the mass migration from the countryside to the cities which was taking place at the time and had settled in London. Though the records of the publishing business are lost, Hardwicke seems to have run it, on the whole, very successfully. An advertisement in a book he issued in 1864 lists works by nearly 80 authors published under his imprint, and these do not include Sir William Hooker of Kew, Charles Darwin or T. H. Huxley, all of whom used Hardwicke at one time or another. The great majority of his authors were medical men, quite possibly guided towards the Piccadilly premises by Dr. William Hardwicke, and almost all his books were on medical or biological topics. He printed at least one catalogue for the Royal College of Surgeons (1874), and in an advertisement in a publication ten years earlier he claims to be 'Publisher by Appointment' to that

---

*The terrace was demolished in 1881 to make way for a new headquarters for the Royal Society of Painters in Watercolours. The building still occupies the site today but is now used as a showroom for Pan-American Airways.

Fig. 5.4 Robert Hardwicke, 1822–1875

body. The College, however, has never issued its own Warrants and considers that Hardwicke's claim 'has no real meaning and can only have been intended to inspire confidence and boost trade'. The College Librarian has further pointed out that Hardwicke's was:

> not the only firm to practice this mild deception, for we have a copy of the 1851 catalogue of Samuel Highley [a friend of Hardwicke's] and Son, who not only stated that they were 'Booksellers, by Appointment, to the Royal College of Surgeons of England' but incorporated a very tasteful wood-engraving of the College into the design of the cover of the catalogue as well.'

In addition to books, Hardwicke also published periodicals such as the *Popular Science Review* and the *Journal of Botany*, and was always willing to try out new ventures and special commissions for his authors.

In appearance he was shorter than Lankester, and somewhat corpulent, and though his expression was kindly and his hair and beard luxuriant, he was not a handsome man, for he had somewhat protruding eyes and thick lips. However, he was exceptionally genial and warm-hearted with a very wide circle of friends, among whom he counted many of his authors as well as his fellow businessmen. The key to his friendships seems to have been his own intense interest in natural history, especially botany. He gathered around himself a large band of enthusiasts and was prepared to assist their projects materially in every way he could. He and Cooke must have taken to each other immediately for Cooke was soon absorbed into Hardwicke's circle and the two men worked closely together for what were to be perhaps the most crowded and exciting ten years of Cooke's life.

The first book that Hardwicke undertook for Cooke was the compact, soft-backed *Manual of Structural Botany* (Fig. 5.5), only 123 pages long. In it each item is illustrated with a small, clear wood-engraving, over 200 in all, drawn on to the block by the author himself but prepared for engraving by the professional, W. M. Ruffle, who was also a keen amateur naturalist and beer tippler, and who had probably met Cooke when he attended the latter's botanical evening classes. Explaining his reasons for writing the *Manual* Cooke says:

> When I passed the examination in Botany held by the Science and Art Department I was greatly astonished to find that there was no handy book available which could serve as a text-book for such an examination, and I at once prepared one, the syllabus of the examination before me. (Aut.).

Its purpose was to:

> . . . refresh the memory of the student concerning the past, rather than, of itself, instruct him in the future of his studies. No attempt has been made to popularise the subject; this

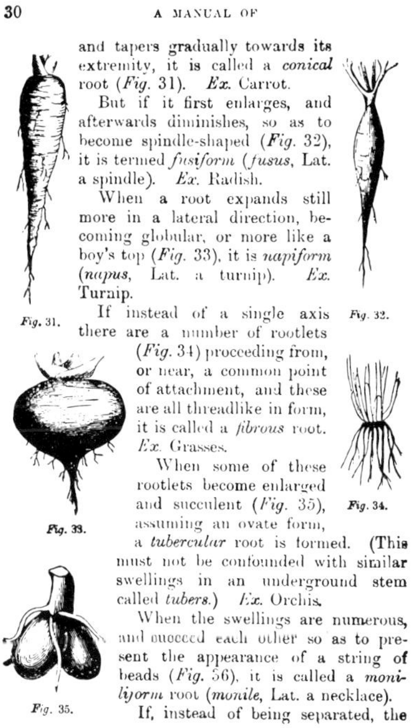

30 A MANUAL OF

and tapers gradually towards its extremity, it is called a *conical* root (*Fig.* 31). *Ex.* Carrot.

But if it first enlarges, and afterwards diminishes, so as to become spindle-shaped (*Fig.* 32), it is termed *fusiform* (*fusus*, Lat. a spindle). *Ex.* Radish.

When a root expands still more in a lateral direction, becoming globular, or more like a boy's top (*Fig.* 33), it is *napiform* (*napus*, Lat. a turnip). *Ex.* Turnip.

If instead of a single axis there are a number of rootlets (*Fig.* 34) proceeding from, or near, a common point of attachment, and these are all threadlike in form, it is called a *fibrous* root. *Ex.* Grasses.

When some of these rootlets become enlarged and succulent (*Fig.* 35), assuming an ovate form, a *tubercular* root is formed. (This must not be confounded with similar swellings in an underground stem called *tubers.*) *Ex.* Orchis.

When the swellings are numerous, and succeed each other so as to present the appearance of a string of beads (*Fig.* 36), it is called a *moniliform* root (*monile*, Lat. a necklace).

If, instead of being separated, the

Fig. 5.5 Page from *A Manual of Structural Botany*.

must be considered rather as a skeleton of dry details, to be filled up and attired according to the taste of the demonstrator.

It was admirably suited to do this, not least in that it was priced at just one shilling. It proved to be exceedingly popular, running into 16 editions, the last in 1877, but Cooke profited little financially for he had sold the copyright for £10.

Structural Botany was quickly followed by the alphabetically arranged *Manual of Botanic Terms*, of which the author says in his preface:

> The production of another work of this character ... may at first appear to be presumptuous ... Without attempting to detract from the labours of my predecessors, I think that the present will be found to possess advantages for those who have not had the benefit of a classical education. The Terms are written in their anglicised forms, under which they are commonly employed; and the derivations are recovered from the mysteries of the Greek alphabet. Both these innovations, it is confidently hoped, will commend the work to such of the operative classes as are cultivating the study of BOTANY.

In an age when all educated persons were familiar with Latin and Greek, and would appreciate without difficulty the meaning of botanical terms derived from them, the translation of these terms in textbooks would not normally be considered necessary, so that working class students would immediately find themselves at a disadvantage. Cooke, whose own classical education was so fortuitous, was in a unique position to appreciate this problem and to attempt to remedy it. The first edition of his book was concerned only with flowering plants, but in the second edition, which did not appear until 1875, he added terms used in cryptogamic botany (the study of seedless plants such as ferns and fungi).

It was Hardwicke who suggested to Cooke that he should write his first book on fungi, a group to which he must have devoted much attention since his introduction to it by Ward some years earlier. The publisher had recently launched a series of 'Plain and Easy' books on various aspects of natural history–Mrs. Lankester had contributed one on ferns–and he 'proposed [to Cooke] to publish a popular introduction to the study of Fungi, which I was to write and submit to him, and this I immediately prepared.' The *Plain and Easy Account of British Fungi* was the first popular book on the group, which at that time was 'a singular and despised family'. In his Introduction Cooke describes the sorry situation:

> Many of those who would merit the title of 'good botanists' know little or nothing about them [fungi]. That part of our scientific literature which is devoted to them is remarkably scanty; and the young student, or the operative [working class] botanist, whose means are limited, enquires in vain for assistance in gaining even a slight knowledge of a very interesting section of our Flora.'

Hardwicke's enterprise in tapping this virgin territory paid off; the *Plain and Easy Account* ran to 16 editions, the last as late as 1898. An anonymous reviewer of the first edition writes:

> ... we have no hesitation in saying that it is the very best and cheapest manual that can be placed in the hands of the beginner. Mr. Cooke writes for those whose education and means are limited, and with pre-eminent success. It is really a pleasure to read the manuals he has published, for they are up to the mark, and so complete as to leave hardly anything to be desired. This new work on the fungi appears to be equally valuable with those which he has already printed [4]

The book is written in a chatty but very readable style, with plenty of anecdotes, but the author does not shrink from concise descriptions of the structure of the different species of toadstool in sufficient technical detail to make identification possible. Only the more common of the larger fungi are included, but mycophagy (the eating of fungi) is encouraged and recipes for cooking the edible species are given. Here, too, we learn how he himself pronounced the vexed word 'fungi', when he tells his readers in a footnote that it 'has the *g* soft, as *fun-ji*', and it is interesting that, even at this early date when the plural 'fungusses' was in very common use among mycologists, his uncle's teaching leads him to protest against 'the barbarism'. The beautiful, delicately coloured illustrations were a family effort: they were drawn by Mordecai himself, the proofs were hand-coloured by Ebenezer's wife, Ellen, and the lithography was by Ebenezer himself, now set up in his own business (Fig. 5.6). In

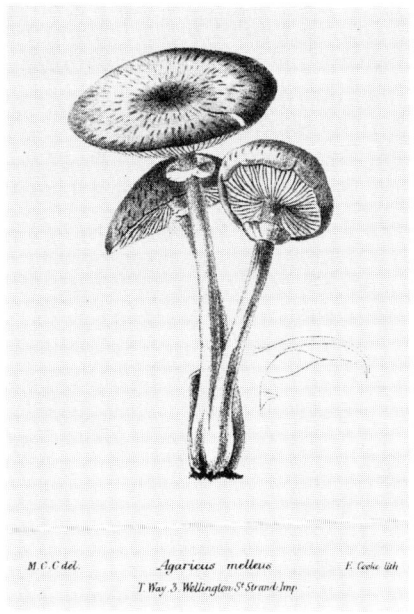

Fig. 5.6 The honey fungus (Latin name now *Armillaria mellea*), an illustration from *British Fungi* drawn by M. C. Cooke and engraved by Ebenezer Cooke. Original in colour.

later editions published by W. H. Allen, the plates are a travesty of the originals, having been redrawn to a larger size in a coarse, mechanical style, with crude colouring.

Cooke's mycological interests must by now have moved beyond the larger fungi to the microscopic species, for very shortly he was to be publishing on these too. His active work on them, at any rate on those causing plant disease, seems to date from his field excursions with the students who enrolled for his stillborn evening classes early in 1860, and if this is so he must have procured his first microscope at about this time. Microscopes then were comparatively unsophisticated instruments, and were proportionately less expensive than their modern counterparts; in a few years' time most amateur naturalists would be able to aspire to one of their own. However, for someone in Cooke's financial position even a comparatively small sum would have been a major setback and would hardly have been possible without some lucky monetary windfall, especially as he describes his first purchase:

> When we were still a novice ... our pride and ambition being fostered by friends and acquaintances, we came to the conclusion that we should require a large-sized binocular microscope, with all the most modern appliances. Being then young and foolish, we accordingly expended fifty pounds on a beautiful instrument ... [5].

He could hardly have afforded £50 from his schoolmaster's salary of £45 a year, and certainly not when he was unemployed; perhaps he used the money from the sale of his school museum for the purchase.

Meanwhile, other opportunities were opening up for him. In 1857 the Working Men's College had moved from Red Lion Square to new premises in Great Ormond St., but Mordecai Cooke had kept up his contacts with it, if not directly, then through Ebenezer. In 1861 he accepted an invitation to lecture there, taking a class in Physiological Botany from 8–9 p.m. on Tuesday evenings, starting on October 29th. At a General Meeting of the College the preceding Thursday the teachers for the session gave 'explanations as to their respective classes, the subjects of study and methods of work'.

> Mr. Cooke, a new teacher, said, in reference to his Botany class ... The winter is, in fact, the best time to begin, though not to continue the subject. Before examining numbers of plants, you must know a little of vegetable physiology and the morphology of plants. I have devoted as many as 40 evenings to this introduction. I disagree so far with the German botanists, that I think physiological botany more important than what we call 'hay-making', or collecting dried specimens. We shall use the microscope, and next spring go out together into the fields [6].

This is the first of a number of occasions on which Cooke, whose mycological activities were in practice almost exclusively devoted to 'hay-making', stresses his belief that 'physiological botany', a term which in those days seems to have included all aspects of the subject except systematics, was in fact more important.

Natural history walks had been an established and popular part of the social life of the College since its inception, and took place regularly on alternate Saturdays. They were often strenuous, but they were the best known and most enjoyed of the activities of the summer months. So most of Cooke's students would be well prepared for the botanical excursions of the spring term which their teacher, from his own experience as a beginner, and from the other evening classes in which he had been engaged 'in distant localities from each other' [7], now regarded as an essential part of a course in Botany.

However, only a few weeks of the spring term had passed when the classes came to an abrupt halt. According to Cooke himself:

> The Exhibition [the 1862 International Exhibition. See below.] year had kept me fully employed much, I fear, to the loss of some of my botanical pupils, and one of the consequences of this pressure, combined with other private reasons, was the resignation of the position as teacher of the Botanical Classes at the Working Mens' College [6].

Further comment is supplied by Ebenezer in a letter written in 1875,* years after the event, to R. B. Litchfield, the Principal of the College. Ebenezer wrote that he had just heard that:

> the Executive of the College had asked my brother [Mordecai] to become the Curator of the Museum. Did you know this? Do you approve? I am puzzled about it. If it was right that he should withdraw from the College some years ago, nothing has happened since to reverse that. The request to withdraw he will not have forgotten. I don't think he will come back and I fancy you would not wish it–I will meet you about it if you wish.

That the dispute was bitter and deeply felt on both sides is clear from the unyielding attitudes of the protagonists 13 years later, but what it was about neither Ebenezer nor the College archives give any hint. In another letter to the Principal written four days after the first, Ebenezer returns to the topic, asks again for an interview, then, remarking, 'I never see him myself and therefore can say nothing to him', continues 'My impression is you will not care to have him back. Should like to know if this is correct, but you must hear my statement first.' Three days after this another curator was found for the Museum. Why the brothers were not seeing each other at this time is another mystery, for they were living less than a mile apart.

No teacher could be found to complete the Botany course, so rather than see it lapse, one of the students, Alfred Grugeon, that remarkable figure in the history of the College, was persuaded to carry on until the end of term. He

---

*In the archives of the Working Men's College.

wrote later: 'M. C. Cooke was an excellent teacher and most helpful to members of his class, and was not above learning from them. I owe him much' [8]*. It would seem that Cooke's dismissal was in no way a reflection on the quality of his teaching. He in his turn wrote a few years later that at the College 'I must in candour state that I met with the most earnest students, all 'good men and true', that I have ever met together under similar circumstances. The teacher left his pupils with regret' [7].

Alfred Grugeon (1826–1913) was a year younger than Cooke and, like him, had been passionately interested in natural history from childhood, but in Grugeon's case there was no doubt from the beginning that it was the plant kingdom to which he was drawn. This is the more remarkable in that he was born in Spitalfields, the son of a silk weaver, so could have had little opportunity to botanize as a child. As he grew older his travels and observations extended to suburban fields and lanes, but it was not until he was 21 and a wood-turner by trade, that he knew of the existence of any botanical literature or system of plant classification. Then he began to collect plants in earnest and to teach himself botany from books. Hearing of Cooke's class at the Working Men's College, he had joined it and, in his own words:

> ... sat for the South Kensington Examination [not the Teacher's Examination taken by Cooke] and, wonderful to relate, I passed and took a prize. After a few weeks in the following session, the teacher, M. C. Cooke, left us, and as the Executive could not get another teacher, the class was likely to disperse, and as they did not wish to do that, they induced me to preside at the class meetings and treated me as a teacher. Still, though I acted in that capacity, I felt in a false position, so in 1862, at the November examination for teachers, I sat for two days, and got two certificates [the same qualification as Cooke] and so became a recognised teacher, although I never ceased to be a student [8].

Despite the credit Grugeon later gave to Cooke's teaching, the classes lasted for such a short while that he must have reached a very high standard by his own unaided efforts before attending them to have been successful in passing the examinations at such short notice. His association with the College was to be life-long, as lecturer, conductor of botanical excursions, curator of the museum and frequenter of the common-room, of which he became known as 'the Father'. He was much loved by both staff and students for his kindness, good-humour and independent views.

Nearly a year after leaving Trinity School, Cooke still had no full-time job, though evening classes and other casual employment would have brought in a small income, and Sophia and Annie could have been contributing by taking in dress-making or doing domestic work. But now, once again, Simmonds came to the rescue. Following the outstandingly successful International

---

*However, according to Grugeon's obituarist, Dr. Ramsbottom, he was not entirely uncritical of Cooke and 'freely admitted that the mycologist was not everything that could be desired'.

Exhibition of 1851 another was to be held in London in 1862, and through the good offices of Simmonds, Cooke was appointed by the Commissioners for New Zealand to compile the catalogue of their exhibits, a job for which his experience with the Scholastic and Twickenham Museums would have fitted him well. It was not, however, a very long or arduous task, for only 115 exhibits are listed and there are no descriptions or illustrations.

The catalogue must have been completed to the satisfaction of the Commissioners, for no sooner was it finished than the Commissioners for India asked Cooke to perform a similar service for them. The Indian catalogue was on a much larger scale and must have been a daunting undertaking in the short time at his disposal. It has 286 pages and occupies the whole of the third volume of the Exhibition catalogue. Almost half the exhibits consisted of economic products from the plant and animal kingdoms, and the food, pharmaceuticals etc. produced from them, a subject which Cooke had already made his own. The illustrations include two engravings borrowed from Hardwicke and one by Cooke himself, and many of the entries are annotated, some, especially fungal products, in considerable detail. For instance, 'Exhibit No. 1680. Mushrooms collected from the stumps of trees', must have been sent in unidentified. Cooke decides that it is a new species of the genus *Pleurotus* (the Oyster mushroom group), states his reasons for this, and names the fungus *P. subocreatus*. Only then does he describe the uses to which the fungus is put in the Straits Settlements (a region included with India). Quite possibly this is the first new fungus species he ever described. In his notes to *Hirneola auricula-Judae* (the Jew's ear fungus) from Singapore (in fact, probably *Auricularia polytricha: H. auricula-Judae* is a north temperate species) he remarks that 'it is sent as a food product, but, we should imagine, of very little merit': clearly he had not yet read of the eating habits of the Chinese who formed, even at that time, a large proportion of the population of the island.

The 1862 Exhibition was even larger than its predecessor of 1851, but it never captured the imagination of the public in the same way. It came at a time when Victorian self-assurance was at its height, as can be seen from the introduction to the Catalogue:

> We may not be more moral, more imaginative or better educated than our ancestors, but we have steam, gas, railways and power-looms, while there are more of us and we have more money to spend.

The Exhibition was housed in buildings specially erected for it on the site of the present Natural History Museum in Cromwell Road (Figs 5.7, 5.8) and extending northwards to Kensington Gore. They were intended to be permanent, but all the moveable parts were dismantled soon after the Exhibition closed and used to build the Alexandra Palace, while the remainder was blown up in 1877 to make way for the new Natural History Museum.

Fig. 5.7   The South Kensington building for the Great International Exhibition of 1862, with the gardens of the Royal Horticultural Society in the foreground.

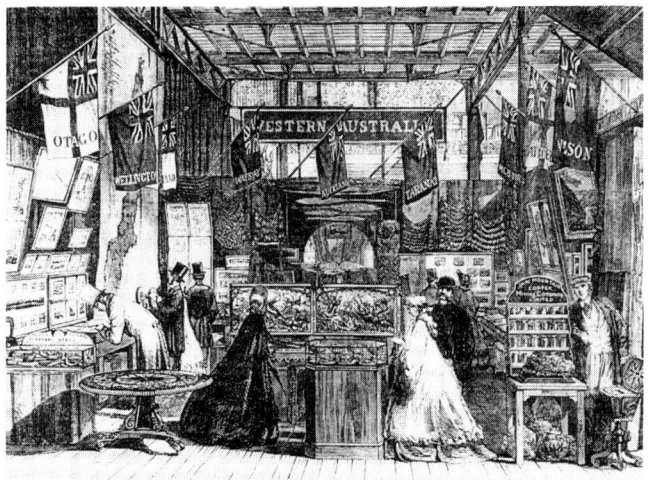

Fig. 5.8   The New Zealand Pavilion at the 1862 International Exhibition.

Great importance was laid on the Indian section, which was given 10,000 sq. ft. of space on the ground floor. According to the Catalogue:

> The management of this large and important part of the Exhibition has been given to Dr. J. Forbes Watson, the Reporter on Indian Products, who has devoted so much attention to a due development of the staples of India, and has lately been so closely occupied in re-arranging the valuable East India Museum at Fife House, Whitehall Yard ... The articles for which space cannot be given have been moved to the India Museum ... there to constitute a supplementary collection. ... The India Museum itself contributed from its varied resources a collection of very considerable interest.

Thus did Cooke come into contact with Dr. Forbes Watson, and secure employment for himself for the next 18 years.

It must have been late in 1860, when Mordecai was employed by Miss Burdett Coutts, that the Cookes first moved to Kentish Town, but no record of their first address there survives. We do know, however, that early in 1862, probably in January, but certainly before April 1st, they moved again to 6, Montague Place in the same suburb, and that Ebenezer and Ellen, the latter heavily pregnant with her first child, moved in with them. Mordecai's financial situation still being very precarious, the Ebenezers' contribution to the rent must have been most welcome.

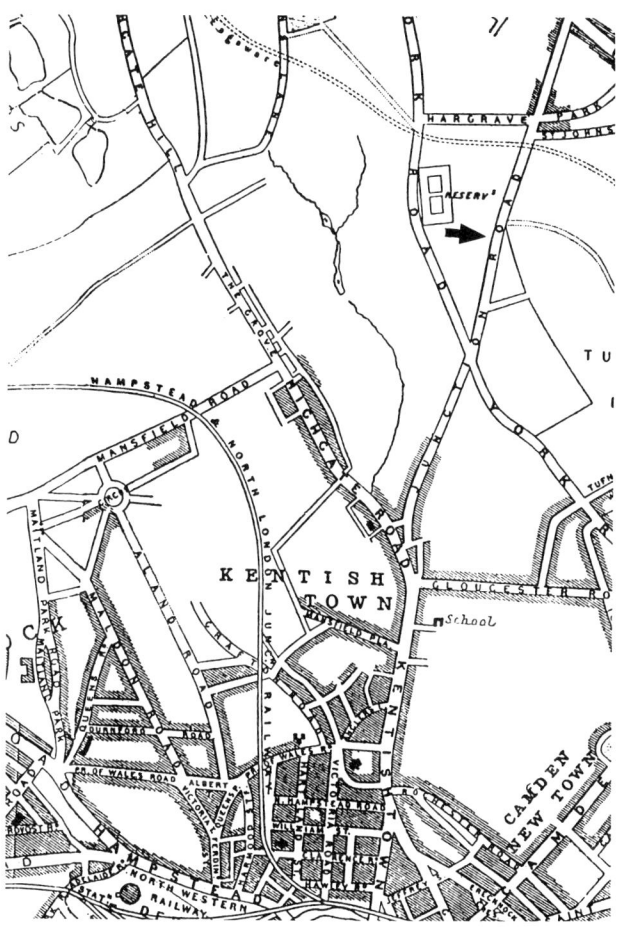

Fig. 5.9 Kentish Town in 1862. Cross-hatching indicates the extent of the built-up area. The Cookes moved to 146, Junction Road in 1870 (approximate position indicated by arrow), by which time the house had already had one occupier.

Kentish Town in the 1860s was changing rapidly from a suburban village surrounded by fields into the North London inner suburb we know today (Fig. 5.9). It was beset by ribbon development along its historic, older roads, and new residential streets were rapidly covering the areas between them with lower middle class urban housing. Building was entirely speculative, often by entrepreneurs of very moderate means, with the result that there was no attempt at over-all planning and standards in the different developments varied enormously: the contrast between Kentish Town and Thomas Cubitt's Pimlico was complete. For the next few years, however, Kentish Town would still be quite a respectable address; only in the following decade would demolition of slum housing for the building of Euston Station and the great new Inner London railway complex, decant thousands of homeless and destitute families, evicted without compensation, into Kentish Town and the adjacent suburbs. Then the existing residents, in an effort to dissociate themselves from a district with so bad a name, began to change their address to 'Upper Holloway', a custom which, in due course, the Cookes were also to adopt.

The small builders who were responsible for the development of Kentish Town could only raise enough capital to put up a few houses at a time–hence Montague Place (the Cookes' new home), Springfield Terrace and Junction Villas (to which they moved later), and many others, a different builder being responsible for each group. The confusion arising from such a system of addresses must have been appalling, and eventually all were changed and the houses numbered according to the road in which they stood, No. 6, Montague Place becoming No. 286, Kentish Town Road.

Fig. 5.10 Annie Elizabeth Thornton Biggs, 1844–1920.

At 6, Montague Place, on April 6th, 1862, Mordecai's step-daughter Annie Biggs, (Fig. 5.10), now aged almost 18, gave birth to Harry Linnaeus Cooke, Mordecai's first child. The only evidence for this event is Harry Linnaeus' birth certificate; no whisper of his true parentage, or of that of any of Mordecai's other children, is to be found in any existing letter or document, or has reached me from any descendant. It is therefore only possible to speculate as to the events leading to this extraordinary situation, and as to the resulting personal relationships between members of the household.

Mordecai and Sophia, now aged 37 and 39 respectively, had been married for 15 years when Harry Linnaeus was conceived, and were still childless despite the fact that Sophia had successfully given birth to Annie so many years before. We know that Mordecai was fond of children, and he must have longed to have some of his own. In view of later events it may well have been that Annie was not unwilling to submit to him this first time. But what of Sophia? Did she condone the deed? Unlikely as this may appear, it seems to me that it is at least possible that she raised no objection if she believed strongly enough that Mordecai had a right to children of his own. In any case, there is little doubt that she continued to live in the house after the event; she was certainly there at the time of the 1871 and 1881 censuses, she died there, and her presence is formally documented on other occasions. But no informal record, written or verbal, of a resident 'Grandma', 'Aunt Sophia', or other such euphemism remains. I believe that a *ménage à trois*, with Annie accepted as Mordecai's wife and Sophia playing the part of an older relative is a strong possibility. Annie must surely have taken the style of 'Mrs. Cooke'–she signed herself thus on the birth certificates of her younger children. If Sophia reverted to her maiden name of Biggs she would have been accepted by all, perfectly correctly in fact, as the children's grandmother. Though there was no blood relationship between Mordecai and Annie there could still be no question of the legalisation of their union, for divorce was difficult then, and was the prerogative of the rich, and in any case, neither Church nor State would countenance a marriage between step-father and step-daughter.

As to the reaction of Mordecai's family, and of friends and neighbours, to the unconventional situation, this again can only be surmised. The children of the union need never have been told of the relationship between their parents, even if Sophia did continue to live with them. Ebenezer and Ellen, however, must have been fully aware of the whole affair, as must William, and the family in Norfolk, for Mordecai remained in close contact with them. Because they had only just moved to Montague Place when Annie was pregnant, they were unlikely to have known their neighbours well enough for local gossip to be a problem, and it is also probable that few of Mordecai's friends visited him at home or knew much of his private life. Nevertheless, contemporary morality, anyhow as publicly expressed, being what it was, Mordecai must, from then on, have been in constant dread of discovery and exposure, and have tended to become turned in on himself. It is my belief that this is part of the explanation of his future loneliness and inability to meet professional colleagues as a social equal.

The news of their son's unorthodox behaviour would have been a terrible blow to his devoutly religious, not to say bigoted, parents, and their reaction may have been the origin of the (untrue) story told in the family that Mary, at about this time, broke with Mordecai and never saw him again. In fact, Mordecai was at Neatishead for Christmas that year (though we do not know whether Sophia, Annie and the baby were with him). One reason for his presence was that he was negotiating to buy his parents' shop, home and garden in Neatishead for them from its present owner and next-door-neighbour, John Banham, shoe manufacturer. Negotiations were completed on December 31st, the younger Mordecai obtaining the property for £370 and mortgaging the money on the same day to Henry Andrews of Ipswich, gentleman, for £200. We must assume that his parents paid rent to him thereafter. It is also recorded that during Christmas week he found a group of 20 puff-balls (*Lycoperdon pusillus*) on a bank at Neatishead [9].

A fortnight after the birth of Harry Linnaeus, Ellen also gave birth to a son, Arthur Ebenezer, and the Cooke brothers, despite the 12-year difference in their ages, became fathers for the first time almost simultaneously and while living under the same roof.

From this time onwards Mordecai Cooke would move into new worlds; he would learn, though with painful and unconquerable feelings of inferiority, to mix with colleagues higher in the social scale than himself; and he would throw himself with demonic energy into a variety of missionary adventures aimed first at propagating the gospel of natural history among the public at large, and later at advancing his chosen field of mycology in every way he knew. His activity in these years was so intense that to follow events in strictly chronological order would lead to an account too chaotic to comprehend. The following chapters therefore describe the various aspects of his life in parallel streams, but to appreciate the intensity of his living, it must always be remembered that several of these streams were flowing at one and the same time. (Appendix A)

Chapter Six

The India Museum. Family Matters. 1862-1880

> On the completion and publication of the Indian Catalogue, I was offered a position for a period uncertain, in the India Museum then located at Fife House, Whitehall. I was supposed to have charge of the Vegetable Products, and to superintend the disposal of the surplus raw products which had been sent over from India for the Exhibition. Thus undoubtedly the knowledge I had acquired in the collection and organisation of my Museum of common objects, and virtually of the Economic Products of the vegetable Kingdom, brought me into the Civil Service. (Aut)

So wrote Cooke of his next job, again a temporary one in the first instance, but this time to last. The India Office records for the period are missing, so we do not know his salary or status, but they were probably humble, as his obituarist describes his position as that of 'a sort of handy man at the Museum'.

The India Museum had originally been set up by the East India Company at its headquarters, East India House, in Leadenhall Street in the City, but in 1858 the building had been sold and the Museum taken over by the India Office and removed to Fife House, Whitehall, (Fig. 6.1), where Dr. Forbes Watson had been appointed as its Director and Reporter on the Products of India. Like most of the scientists of his time Watson (1827–1892), a Scot, born and educated in Aberdeen, had graduated in Medicine. No University in Britain offered a degree course in Science until 1860, when London became the first; the older Universities did not follow this lead for many years, though they did allow specialisation in a science subject in the final year of a B.A. course.

However, medical curricula included various branches of science as compulsory subjects, with the result that Chairs in these subjects existed long before Degrees could be taken in them. A medical graduate with a strong leaning towards science, such as T. H. Huxley, might hardly practice medicine at all, while others such as Edwin Lankester might move freely between the worlds of Science and Medicine. Watson became an Army surgeon serving in India, but after a while he left active service and, remaining in India, became lecturer in Physiology at a medical college there until he fell ill and was sent home on sick leave. Back in London, he was first given the job of investigating the sanitary uses of charcoal at the School of Mines in Jermyn St.; then he was set to discover the nutrient value of Indian food grains. When the India Office took over the India Museum, Watson, with his first hand knowledge of the country was an ideal choice for the Directorship.

The space allocated to the Museum was on the top floor of Fife House, hardly an ideal situation for a display open to the public. At the time of its opening the magazine *Leisure Hour*, describing it as badly catalogued and cramped but serving a useful purpose, continued:

> Ascending the stairs to the top floor we are among the natural products of India . . . They are classified and arranged in tolerable order . . . cereals . . . tea . . . tobacco . . . fruits dried or preserved . . . coloured models in plaster of such that cannot be preserved . . . dyes . . . drugs, chemicals and medicaments without number [1].

Fig. 6.1 Fife House, Whitehall Yard, in 1805. In Cooke's time the India Office and India Museum were housed here.

Three years later, just before Cooke took up his post, the *Illustrated London News* [2] published a large picture of the interior of the Museum (Fig. 6.2) and commented 'There is no branch of history or of manufacture, and scarcely any description of raw produce, which is not illustrated in this most interesting Museum.' Cooke's own handiwork remained in evidence for many years, for when his daughter Leila visited the Imperial Institute (to which parts of the collection were later transferred) in 1925 she noticed that 'on some of the exhibits of grain from India, shown in small glass frames, the labels were still in his hand-writing'.

Cooke's appointment probably remained a temporary one for five years, and he must have continued his part-time teaching activities during this period, for when his second child, Ada, was born in 1864, his occupation is given on her birth certificate as 'Teacher of Botany'. At some time during his first two or three years he wrote to Dr. Hooker, the Director of Kew Gardens, asking to be considered for the appointment of examiner (he does not mention to whom), humbly describing his qualifications and experience of adult teaching, and offering to 'undergo a personal examination as a test of ability'. As there is no further mention of this post is would seem that his application was unsuccessful.

The Museum was open to the public from 10 a.m. to 4 p.m. on three days a week, and by ticket on a fourth day, but Cooke himself worked a surprisingly short 5-hour day, probably including Saturdays, which was presumably the standard Civil Service practice at the time. During the whole of the 18 years he spent there we hear nothing at all of his day-to-day life in the job, though some idea of the size and scope of the collections of which he eventually had charge can be gained from the description of their transfer to Kew in 1880:

Fig. 6.2   The India Museum, Fife House, in 1861.

of rice alone there were about 2000 samples from the most widely distributed districts of India and weighing in the aggregate about 3 tons ... a series was separated showing every type of variation to which Indian rice is subject. The amount of this variation in form, colour and texture is almost inconceivable, and the trouble and expense which must have been involved in the accumulation of the specimens is amply justified by the clearness with which the fact is now brought out [3].

Cooke's job would have been no sinecure. We do not know whether or not he enjoyed his work, and we know nothing of his relationships with his colleagues or with Watson. We can only assume, from his long tenure of the post and his steady promotion, that he carried out his duties to Watson's satisfaction despite the fact that his real interest lay in other, quite different fields.

A good idea of the wide range of vegetable products with which he had to deal, and of the type of information which he was expected to provide, can be obtained from two official Reports and a series of articles which he wrote for the Museum. The Reports were entitled *On the Gums, Resins, Oleo-Resins, and Resinous Products in the India Museum, or Produced in India* and *On the Oil Seeds and Oils in the India Museum or Produced in India*, dated 1874 and 1876 respectively, each with a forward by Watson. They are foolscap-sized, and each is about 150 pages long. The sub-divisions into which the products are grouped are based on the 'function of the substances under report', an innovation of Cooke's which he hoped would be an improvement on previous publications in which the products had been listed in 'mechanical', or alphabetical, order. The materials included had been sent in either for the 1851 or 1862 Exhibitions, or by merchants trading from India, and the author complains that the information sent with them was not always adequate. In length, the reports vary from half a page to four pages. A full report consists of the English and native names of both plant and product and the Latin name of the plant; a description of the raw product, including such characteristics as its solubility and chemical analysis, odour, and the method of collection; a discussion of any matters of special interest, including lengthy quotations from other writers; and the import/export figures. If the product is well-known, e.g. myrrh, its etymology and history from the most ancient times is given in detail, information which, to judge by some of his previous publications, it would have given Cooke enormous pleasure to provide. In addition to these two volumes he was responsible for much of the compilation of a massive work by Watson entitled *Index to the Native and Scientific Names of Indian and Other Eastern Economic Plants and Products*.

The articles written during this period, 16 of them, appeared in the *Pharmaceutical Journal* between 1870 and 1873 and were devoted to Indian herbal medicines and the plants producing them, a subject in which he was particularly interested (p. 208). He dealt with it in much the same way as he had the gums and oils. In one article he showed that a fine powder, Hasaṅ-y-yusaf, which until then, had been thought to be of diatomaceous origin, consisted in fact of the spores of the quillwort, *Isoetes*. In another an edible fungus growing on beech trees in Kashmir, and known as the Beech Morel, is identified as a species of *Cyttaria*, a genus newly discovered by Darwin in Tierre del Fuego in the 1830s, during his voyage in the 'Beagle'. Cooke also returns to the subject of the Jew's ear fungus, some dried specimens of which had been sent to India from China. By now he had realised that the fungus is highly esteemed by the Chinese, and his comments on national eating habits are perhaps worth quoting:

> So peculiar are the tastes of the Chinese that we are ready to believe it possible for them to relish anything whatever, provided it can be manipulated in any way so as to be capable of mastication.

But:

> It never seems to have entered into the heads of Englishmen, not even Professed mycophagists, that the leathery-looking Jew's Ear could be converted into food [4].

The reports and articles described are the only publications resulting directly from Cooke's 18-year stay at the India Museum, but at Kew there is an unpublished, undated volume containing 70 full-page engravings of *Indian Plants* with a manuscript note on the fly-leaf explaining that it was 'prepared under the supervision of Dr. M. C. Cooke for the India Museum'. In it is an engraved card recording its presentation to the Royal Botanic Gardens by the Secretary of State in Council of India. The majority of the plates are by W. M. Ruffle, Cooke's friend and the engraver of the illustrations for his Manuals. Probably the book is one of a series described in the autobiographical notes as having been prepared for free distribution to Chambers of Commerce in Britain and India, along with 'portable collections of raw products, duly named' for which Cooke was also responsible.

In the archives of the India Museum are three hard-backed exercise books entitled Volumes 15, 16 and 17 of the *Catalogue of Economic Products in M. C. Cooke's Collection*, filled with several thousand hand-written entries of the common and Latin names of economically valuable plants from all over the world, grouped into sections and classes. They clearly have nothing to do with the collections of which Cooke had charge, but are probably explained by a reminiscing letter he wrote to a friend in his extreme old age:

> Years and years ago I employed myself in making a collection of Economic Botany for which I had an affection as it first led me into the India Museum and secured me with a satisfactory income. Later years I have been almost as assiduous in distributing it as I was in collecting–Except for one collection of about 200 specimens which was sold to the India Office of Products not the produce of India–[5].

It seems incredible that he would ever have had the space to store a collection of the magnitude implied by the Catalogue, even if he had been able to obtain such exotic specimens, so perhaps the entries represented the specimens he would have liked to have collected, while the 200 mentioned in the letter were the reality. He sold the remainder of the collection 'by advertisement' and gave 'lots away to school teachers and others'.

The two Cooke families were still sharing No. 286, Kentish Town Road in January, 1864 when Annie gave birth to her eldest daughter, Ada Gordon. When registering her baby she used the same style as previously, making no pretence of marriage to Mordecai, but simply calling herself Annie Elizabeth Biggs; frequently from now on she would omit her third name, Thornton, and always she entered her first two names in the reverse order from that on her birth certificate. It was soon after Ada's birth that the Midland Railway Co. bought up the whole group of houses of which No. 286 was one, preparatory to pulling them down to make way for Kentish Town Station, so forcing the Cookes to move. Both families had grown, for Ellen too had a daughter, born at about the same time as Ada, and this may have been one of the reasons why the two households parted company. In any event, the Ebenezers went to 4, Sussex Place, Camden New Town, while the Mordecais moved only about a mile into Junction Road, a continuation northwards of Kentish Town Road. Though they were still in the Borough of Islington, they were now just outside run-down Kentish Town and could give as their address the rather more select Upper Holloway. No. 6, Springfield Terrace was to be the first of the family's three addresses in Junction Road, and the birthplace, in May 1886, of their third child, Willie Cubitt. In calling his son after his brother, albeit in abbreviated form, Mordecai was continuing the Cubitt tradition of naming one son William, as well as ensuring the continuing recognition of the Cubitt connection. With the arrival of Willie, Annie's position must have been tacitly accepted by all, for her name appears on this child's birth certificate as 'Annie Elizabeth Thornton Cooke, formerly Biggs', the style of a married woman.

In Upper Holloway at this time development was proceeding rapidly, as London's northern suburbs pushed ever further outwards, following the new local railway lines which would carry increasing numbers of middle class residents quickly and cheaply to their work in the city. Railway building in the district had only begun about ten years earlier, and the local people must have suffered considerable dislocation of their daily lives while Gospel Oak and Upper Holloway stations were being built and the line between them accommodated in a deep cutting across the south end of Junction Road. No sooner was this over than a start was made on linking the local lines with the new Paddington to Farringdon Street Underground, causing further difficulties for residents. However, trains were not the only form of public transport available. Horse-drawn omnibuses plied up and down Junction Road carrying the more highly paid manual workers (the lower paid walked), clerks, business men, shoppers and theatre goers to the metropolis. In 1870 a new form of transport, the horse-tram, made its appearance and was an immediate success, one result being that a tram depot wth car-sheds and stabling came to occupy a large area on the east side of Junction Road* To the west of the road, behind the houses, there was already a cattle market which, combined with the depot, would have ensured that Junction Road remained the preserve of the lower middle classes–those with higher social aspirations moved out. Nevertheless, there were still green fields nearby.

On July 12th, 1867, Cooke was promoted to be Reporter on the Economic Products of India at the excellent salary of £327 a year. His duties at the Museum included not only preparing its contributions to international exhibitions abroad, but accompanying them and remaining in charge of them while they were on show. Thus he had been at the Dublin Exhibition in 1865, and in 1867 he made his first trip overseas to the Great Exhibition in Paris. His British chauvinism is very evident in an article written during his stay there for a magazine of popular science:

> Let the stranger cross the Pont de l'Alma, and obtain his first glance at the extraordinary mausoleum in which lie entombed so many of the fondest hopes and rarest achievements of Science and Art which the Great Fair of 1867 has gathered together from all parts of the civilised globe, and he may well ask himself if that is indeed the temple which the most artistic of nations has dedicated to universal genius. It might have been designed for a big gasometer, or an elliptical railway station, but, spoilt for both these purposes, it has become the Great Exposition. What is it possible to say, even yet, of the contents of such chaos as the interior exhibits? . . .
>
> There is certainly much to amuse, and there are many sources of instruction . . . and it may be added withall that it deserves the honour of being the ugliest and best abused of all the great collective fairs of the present century [6]

The purpose of the article is to compare the state of the science of microscopy in the exhibiting countries. The host country he disposes of briefly: 'Microscopy is *not* one of the social institutions of France'. He considers that the best instruments she has on show would compare with the cheapest British ones. However, in a second article written a month later he concedes that, after further study, the Continentals deserve praise for one or two items. In particular:

> I have also seen the working of high powers [microscope objective lenses] constructed on the 'immersion' principle, at a comparatively cheap rate, and certainly excellent results. I am disposed to think that we know too little of these objectives in England, and have hitherto regarded them with a trifle too much of prejudice.[6]

He is also scathing about photomicrographs, then in their early days; those from Belgium especially are 'by no means the equal of those we are accustomed to see'.

*There is a London Transport depot there to this day.

The Cookes lived at Springfield Terrace for four years, then in the autumn of 1868, when Annie was again in an advanced state of pregnancy, they moved along Junction Road to 2, Junction Villas, next door to the Boston Arms tavern, well beyond the limits of what had been the built-up area only 8 years previously (Fig. 5.9). The two 3-storeyed, grey brick terraces, one on each side of the Poynings Road turning, which are now called Nos. 118 to 128, Junction Road, are probably the original Junction Villas:* at the southernmost end is a two-storeyed, red brick house which could once have been the tavern. Soon after the family had settled in to No. 2 their fourth child, Ernest Frederick, was born, and this time Annie's name on the birth certificate was given simply as 'Annie Cooke, formerly Biggs'.

Back in Neatishead the death of Mordecai's father took place in February, 1869, and he was buried in the graveyard of Neatishead Chapel where he had worshipped for so long. Despite his chronic ill-health he reached the ripe old age of 70. By his will he instructed his executors, his wife and his eldest son, Mordecai, to sell up his estate and invest the proceeds to provide Mary with an income for life; on her death the capital was to be divided between all their children. The shop, of course, was not the elder Mordecai's property, but his son's, for the latter had bought it for his father seven years earlier. Now that his mother no longer needed the premises he could sell, thus releasing a considerable sum of money for his own use. It is safe to assume that since leaving Peter Simmonds' house, his various homes in Kentish Town had been rented, but now he was in a position to buy for the first time. His initial step, in 1870, was to redeem the £200 mortgage on the Neatishead shop; then, in the autmun of that year, he bought No. 2, Grosvenor Villas, only a few hundred yards down Junction Road from Junction Villas, and later renamed No. 146, Junction Road, where the family was to remain until it finally broke up some 20 years later.

Meanwhile arrangements were going ahead for the sale of the shop to one of Mordecai's younger brothers, Josiah Henry (1832–1889), a talented young man who as a youth had been apprenticed to a decorator and glass stainer, but had left the trade to go into the grocery and drapery business at Burnham Market, a small town near the Norfolk coast. William Cooke tells us that he was 'intensely fond of horticulture and built large glass-houses', but does not say where, and one of his grand-daughters told me that he was also an expert wood-carver. The sale of the shop was completed in May, 1871, when the premises changed hands for £420, a profit for Mordecai, over seven years, of £50. Josiah continued to run the shop and Post Office until his death at the early age of 57, and he is the only one of the Cooke brothers whose descendants still live in the district.

It may not have been entirely coincidental that the man from whom Mordecai Cooke bought 146, Junction Rd. was Henry Negretti, the founder 20

---

*All the houses now have single storeyed shop fronts built out over their front gardens.

years earlier of the well-known firm of Negretti and Zambra, precision optical and scientific instrument makers, who supplied, among others, the Palace, the Royal Observatory at Greenwich, Kew Observatory and the British Meteorological Society.* The two men could well already have been acquainted through their common interest in optical equipment.

No. 146 (Fig. 6.3) was one of a row of pleasant, 2-storeyed, detached houses, square and solid, built of yellow-grey London brick, with a low-pitched slate roof and a tall chimney at each end. It had a small front garden divided from the road by a low brick wall with a hedge behind it, and a long back garden sloping upwards to, according to Leila Cooke, some old lime trees. Only a few feet divided it from the houses on either side. Its nicely proportioned front door was surrounded by an elaborate pillared moulding with the words 'Norfolk Villa' carved across the top, certainly by Cooke. This house must have been at least the second of his homes to which he had given this name, for he also uses it on a letter dated January, 1867, at which time he was living at Springfield Terrace. On each side of the front door of No. 146 was a large window, windows and door being surmounted by matching cornices. On the first floor were three smaller windows with curved heads and a continuous cornice running over all. The corners of the house were decorated with long-and-short stonework, and the tradesmen's entrance was reached through a porch discreetly hidden at the side of the house. The back of the house had obviously been extensively altered in fairly recent times when I visited it, but the lean-to conservatory opening from the back sitting-room was undoubtedly the 'greenhouse' described by Leila Cooke as the one her father used as an extension to his study. In 1977 it was in a terrible state of repair, much of the glass having been replaced by plywood. I was unable to enter the house,

Fig. 6.3 No. 146, Junction Road, in 1974. The Cookes lived here from 1870 to 1898.

*The firm has had a long and successful life. In 1946 it became a public limited company, moving from Holborn to new headquarters at Aylesbury, Bucks., where it now specialises in aviation and monitoring equipment.

but according to Miss Cooke there were four large rooms on the ground floor, as well as the kitchen premises, five bedrooms on the first floor, and two maid's rooms in the attic, which must have been very low and cramped.* The district rate book for 1871 shows that the gross estimated rental of the property was £52, giving a rateable value of £42, and a poor rate, the only rate levied, of £4.14.6.

Some idea of the social standing of the district at the time can be obtained from the census returns for 1871, where the Cooke's neighbours are listed as a 'manufacturer employing labour', a singing teacher, a chief clerk, an attorney and a solicitor, each with a single, living-in, general servant. The Cooke's girl, Elizabeth Horne, a 23-year old from Westminster, must have been kept very busy in a family with four young children in a comparatively large house. There would probably have been gas lighting, for oil lamps were becoming things of the past in towns, but there would have been coal fires in all the rooms, a kitchen range, and a coke-fired copper for the large Monday wash. There was no bathroom–that had been added later over the kitchen extension–each bedroom would have had its flowered china basin and jug of cold water standing on the wash-stand, with a matching chamber-pot under the bed. Hot water would have been carried upstairs from the kitchen by the maid. The kitchen and a downstairs lavatory would have been the only rooms in which cold running water was laid on, hot water probably being obtained from a boiler built into the kitchen range. Baths would have been taken in a hip-bath, probably in the kitchen, for it was warm there, and close to the supply of running water. Primitive though these arrangements were by today's standards, they were a huge advance on the back-yard pump at Horning that Mordecai remembered from his childhood.

The garden, according to Leila Cooke, was largely the preserve of the children, no attempt being made to keep it up except that 'now and again a man with a scythe used to come in and cut the grass'. Leila also remembered that there were two pear trees in the middle of the garden and that a lot of *Agaricus arvensis*, the wood mushroom, used to come up under them, and 'we had them for breakfast every day'. When she was small she, and doubtless her older brothers and sister too, 'had a little garden like a child has'.

The story must now return briefly to Norfolk and the Neatishead Cubitts (Tree 1). It will be remembered that William Cubitt II (1724–1794) was a prosperous farmer who had a total of 14 children by his two wives. Mordecai's mother, Mary, was William II's grand-daughter by his first wife, Hannah

*In 1977 the Peabody Trust had recently bought No. 146 as a 'half-way house' for homeless families awaiting rehousing, and with three or four such families in occupation it was inevitably in a very derelict condition. Previously, the Trust told me, it had belonged to one family for 50 years. On my next visit in 1981, the house and most of its neighbours had been demolished, their front hedges were 10ft high, and the land behind lay derelict. Junction Road now carries heavy lorries roaring in and out of London in place of the horse-drawn omnibuses, trams and carts of over 100 years ago; the stench of diesel fumes has replaced that of horse dung.

(1724–1759). After Hannah's death William married Mary Bayes (1744–1823), one of whose children was James (1781–1821). James became a farmer, probably at Ivy Farm, near Neatishead, but he:

> ... departed this life insolvent ... after only a few days indisposition ... The deceased was a most respectable and industrious Farmer and his pecuniary difficulties were brought upon him, not from want of attention and economy, but from the distressed state of Agriculture [7].

By the time of his early death James and his wife, Clarissa, had already had five children, Mary Cooke's step first cousins, one of whom was William Quincey (Clarissa had brought the name Quincey into the family). William Quincey eventually became the owner of Ivy Farm and raised six children, the eldest of whom was John Quincey, born at the farm on Nov. 17th, 1846, and a step second cousin to Mordecai Cooke through his mother, Mary. Like his father, William Quincey ran into financial problems which were particularly severe in 1870, and it must have been at least partly as a result of this that at about that date 24-year old John followed the example of so many of his fellow East Anglians and migrated to London to seek fame and fortune.

Naturally on his arrival he would be in touch with his Cooke relatives who were living at the time at 2, Junction Villas; indeed, he may well have seen something of Mordecai at Neatishead while the latter was attending to his dead father's affairs and negotiating the sale of the shop. It is quite likely that he stayed with the Cookes for a while, while he looked for a job and lodgings. His coming could have brought a breath of fresh air into Annie's life which, with a 'husband' much older than herself and four children to care for, could have been rather dull for a lively and good-looking 26-year old. In any event, she and John were attracted to one another, and in August, 1870, Annie conceived John's child and as on every other occasion when the Cookes moved house, Annie was pregnant when they moved to 2, Grosvenor Villas. Most unfortunately we have no way of discovering what were the emotional relationships between the four protagonists in the this drama, Annie, John, Mordecai and Sophia, therefore we cannot tell what their reactions would have been to this shattering turn of events, but it is obvious at least that Mordecai was in no position to protest.

We do know that on February 21st, 1871, John Quincey Cubitt, ticket collector on the railways, then living at 23, Anne St., York Road, Lambeth, married Annie Elizabeth Thornton Biggs, spinster, of 2, Grosvenor Villas, Junction Road, at St. John's Church, Holloway, after the calling of banns (Tree 2). One of the witnesses was Isaac Biggs, clearly a relative of Annie's, but nothing more is known about him. Whether Mordecai or Sophia were at the wedding we cannot tell, but the fact that it took place locally suggests that they had a hand in the arrangements.

Now that Mordecai at last had children of his own it would obviously be

TREE 2

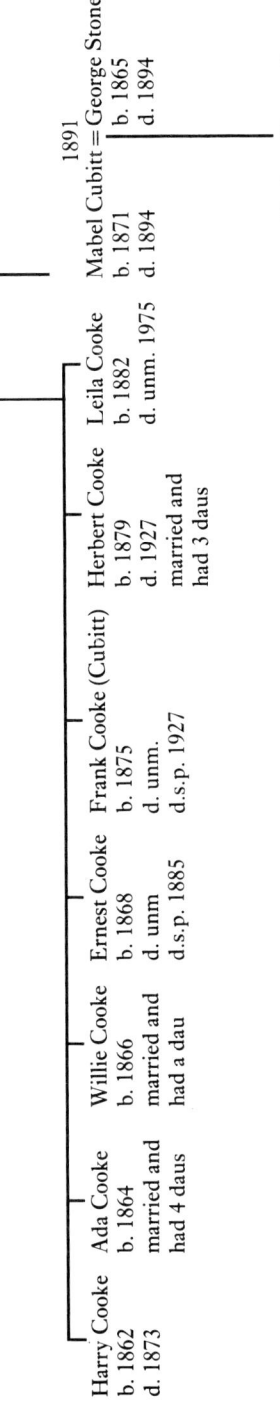

to Sophia's advantage if new arrangements could be made for Annie which would allow her mother to slip back into her rightful place as Mordecai's wife, and what better way of achieving this, to the greater happiness of all concerned, than by regularising the *de facto* relationship between Annie and John and setting them up in a home of their own? If this was indeed the plan it was to fail in the end because those concerned did not understand, or would not take into account, their true feelings for one another. For in the light of subsequent events it seems that Mordecai was, in fact, very deeply attached to Annie and she to him, though her devotion was not to be as long lived as his; the attraction between Annie and John seems to have been far more superficial. However, given the extremely limited nature of the surviving evidence concerning this extraordinary marital tangle, my interpretation of the known events is only one of many possibilities.

On May 21st, 1871, three months after the wedding, Annie Elizabeth Thornton Cubitt, formly Biggs, gave birth to a daughter, Mabel, at 4, Emerson Street, Southwark, a house in multiple occupation where the couple had settled after the wedding. Annie was the only Londoner among the six occupants, the others being a millwright from Gloucestershire and his wife, together with their lodgers, a tailor from Preston and herbalist. Clearly by marrying John Cubitt, who had had to make his own way in London with no support from his impoverished father, Annie had descended several rungs of the social ladder, and she may not have found it easy to adapt to the crowded conditions and reduced circumstances in which she found herself.

Meanwhile the household at 2, Grosvenor Villas was settling down to its new situation. Sophia now stepped fully into her role of wife and mother–this much is revealed in the census for 1871. Fortunately for the Cookes, it was only necessary to enter on the census form the relationships of the members of the family to its head, Mordecai, so the children's relationship to Sophia could remain hidden. The three eldest, Harry (8), Ada (7), and Willie (4), were at school. It would be interesting to know how the sudden disappearance of their mother was explained to them, and whether she paid her young family occasional visits. Whatever Mordecai himself felt about the momentous events in his household, and it is unbelievable that some part of him was not deeply shaken by them, there is not the slightest hint that he let them interfere with his work, his mycology, or his numerous other activities outside the home, as will be obvious from the chapters which follow.

Much of his work, and at this time all his mycology, was carried on in his home, and the power house for all the activity was his study, or library, at the back of 2, Grosvenor Villas. Leila Cooke has vivid childhood recollections of that study and of the red-haired man who dominated it; and as, amid the frequent and bewildering crises which overwhelmed this family, the one stable element was the absolute priority her father gave his work, her memories of the 1890s are likely to hold true for the whole period from the 1870s, before she was born. She writes:

One of the ground floor rooms had a large greenhouse attached with a fountain in the centre. This was my father's study. In the study, two sides of the room had bookshelves from floor to ceiling. All the poets from Chaucer to his day, all the essayists from Francis Bacon to his day. All the known books on Fungi and Botany, including rare ones like Gerard's Herbal, Couch's Fishes, Morris's Book of Birds. Layard's Nineveh, Champollion's Egypt. Fungi by M. J. Berkeley, with whom he collaborated [Chap. 9], de Bary, Tulasne, Quelet, Lamarck and De Candolle that I can remember. He was an omnivorous reader.

In one corner of the room stood a nest of drawers filled with foreign shells, corals and sea urchins, the tiny ones in old red and blue pill boxes. In another corner were a dozen or so cabinets for storing microscopic slides, and more shelves for holding portfolios of his own drawings. There was a very large book filled with his own drawings of birds [probably those mentioned in Chap. 3].

Standing on the Mantlepiece itself were a number of small specimen glasses in which were diseased ears of grain. He was very neat and methodical in all his work. Never a book or a paper out of place.

'A place for everything and everything in its place'.

However, things were slightly different when, as frequently happened in the autumn months when the toadstool season was at its peak, large collections of these plants arrived for him to identify. His friend and fellow mycologist, Worthington Smith (p. 98) has left a vivid record of one of these occasions:

> One day I called on Cooke at Kentish Town. He had a very large collection of fresh larger fungi on his table and floor. Someone had sent him some hampers full. He was hastily looking over them and as he discarded them, he threw them into the fire place, there were scores of rotten agarics sticking all over his fire-place, grate and fender. A nasty job for some one to clear up [8].

History does not relate to whose lot it fell to clean the grate, but clearly it must have been a more urgent matter than that of the general upkeep of the room as described to me by Leila Cooke many years later:

> Nobody was allowed in and the room was never dusted. As he went out he always locked the door in case the maidservant went in. He had books all over the place and all round the floor were magazines like *Grevillea* [p. 178] in piles. The whole library was lined with books but there was never enough space and they were always on the floor. Nobody was allowed in ever. *Never.* Dust stayed there for five or six months or more. *Nobody* was allowed in. He used to have it cleaned up about once a year, I suppose. He always took the key of the library with him in case the servant should go in and want to sweep a little. Now and again he would clean a corner patch out and Jane, or whoever she was, would just come and sweep that corner, but she mustn't go further away.

He would have needed to be neat and methodical to work in conditions such as these.

The Vienna International Exhibition of 1873 was the last which Cooke attended on behalf of the India Museum, being mainly responsible for the preparation of the sections referring to vegetable products [9]. The Exhibition opened on May 1st and remained open for six months, but Cooke was away from home from March 20th, presumably occupied in setting up the Museum's exhibit. The only records that remain of his experiences in Austria are scattered references in his popular books to the eating habits of the natives and the wild foods on sale in Viennese markets. For instance, mycophagist though he is, he cannot find a good word to say for the sauce made by his hosts from *Agaricus (Armillaria) melleus* (Fig. 5.6), the honey fungus [10], but he delights in the wild strawberries he found in the street markets [11].

The year after the Exhibition, the India Museum was closed to the public in preparation for its transfer from Whitehall to more suitable premises at South Kensington, where it reopened in 1875. Cooke benefited very materially from this change, for he was put in charge of the entire collection of vegetable products with the duties both of arranging them and of providing information about them, a job which carried the status of Assistant Curator at a salary of £400 a year. The appointment was for three years, and was renewed for a similar term in 1878.

While he was in Vienna, and only two years after Annie's marriage, a terrible tragedy befell the Cooke family. The eldest boy, Harry (Fig. 6.4), now 11 years old, attended Stranraer House College, a private day school 'for young

Fig. 6.4 Harry Linnaeus Cooke, 1862 1873.

Fig. 6.5 M. C. Cooke aged 48. Photograph taken in Vienna at the 1873 International Exhibition.

gentleman' in Abbey Road, St. John's Wood, run by a Mr. Barker and his wife. No description of the school remains, but as there is no mention of it in the local directory for 1863 it cannot have been established for very long. Probably it differed little from the many other similarly short-lived schools of the period catering for the less affluent members of the middle classes, and for which no legal standards were set.

In the afternoon of Thursday 29th May, 1873, Sophia received an urgent message to say that her 'son' had fallen off a swing and was seriously hurt. She rushed to the school, found him unconscious, and was with him when he died the next morning. By great good fortune the record of the inquest survives and tells the story. The boys were in the playground at lunchtime and some of them were clambering about on the swing, which was 6'7" high. Albert Beale, Harry Cooke and another boy were sitting on the cross-piece, but were not fooling about or pushing one another–five other boys were adamant about that. Suddenly Harry lost his balance and fell, crying out as he did so. He landed on his head and never spoke again. The Headmaster, Mr. Barker, also saw the accident and was equally certain that Harry was not pushed, but he did recollect that the lad had complained of 'a pain in the head' about three weeks earlier. He called a doctor immediately, and the latter found that Harry was still conscious and 'capable of being roused', but the next morning the boy died from 'convulsions following concussion of the brain'.

For Sophia, in sole charge of her four grandchildren and with the inquest and all its formalities to face unsupported, this must have been an agonising time; but at least the waiting was short for the inquest was held the day after Harry's death at St. Marylebone Workhouse. Sophia must have been in a cruel dilemma as to what relationship she should claim with her husband and her grandson when making her deposition, especially as the coroner for the case was none other than Dr. Edwin Lankester, now Coroner for Middlesex, to whom, of course, Mordecai was well known. However, it is unlikely that Lankester would ever have met Annie, so that his suspicions as to Sophia's relationship to Harry were probably not roused. It is worth while quoting Sophia's deposition in full:

> Sophia Lucy Cooke, having been sworn upon the day and year and at the place above mentioned, deposed as follows:— I reside at 2, Grosvenor Villas, Junction Rd. I am the mother of the deceased, 11 years old. His father is Mordecai Cubitt Cooke, Assistant at the India Museum. My boy went to school at the Barker's Stranraer College. I heard an accident (sic.) on Thursday at 4 pm. I saw him before he died. He was quite insensible. He died on Friday. I have no-one to blame. *The body seen is my deceased child.*

All the depositions were taken down in the same atrocious hand-writing, and errors in grammar can certainly be attributed to the clerk. So too, most probably, can the alteration of Sophia's second name from Elizabeth to Lucy, although Lucy may have been a pet name which she gave in error.

There can be no mistake, however, over her claim to be Harry's mother, and her decision to swear to this lie in a coroner's court must have caused her much heart-searching.

Harry was buried in Finchley Cemetery. The original gravestone is no longer in existence having been replaced with a new one by Leila Cooke when Mordecai himself was buried in the same grave: but if she copied the inscription on the original stone it read simply:

'Harry Linneaus Cooke, died 30th May 1873, aged 11 years' giving no indication of the child's parentage. On the present stone is also inscribed a (deliberately?) ambiguous verse which could well have been composed by Cooke himself.

> He is not dead, the child of our affection,
>   But gone unto that school
> Where he no longer needs our poor protection
>   For Christ himself doth rule

Nothing is known of Annie's reaction to her son's death, but Mordecai was deeply distressed by it and treasured a photograph of the boy to the end of his life. Two nature-study books given to Harry as presents have also survived and are now in the possession of Mordecai's grand-daughters.

It must have been some time after this tragedy that Annie's marriage to John Cubitt began to founder, though whether Annie precipitated the break by her unfaithfulness to John, or whether John had first deserted Annie, we can never know. All we are certain of is that on March 3rd, 1875, Annie, calling herself Annie Cubitt, gave birth to a son, Frank Harry, at 159, Vauxhall Bridge Road, Westminster, a son whose father's name she refused to divulge. The legal result of this would be that the child would take his mother's surname, Cubitt. Now William Cooke, in 'The Cookes of Horning', gives Mordecai's fifth child as 'Frank Harry, born 1875', and in his will Mordecai appoints 'my son Frank Harry Cooke' as one of his executors; but no birth certificate for Frank Harry Cooke' exists. It seems to me indisputable that Frank Harry was Mordecai's son, though Annie, for once, was not prepared to admit it, perhaps because she was still living with John Cubitt.

Four years pass before there is any further record of Annie, then in 1879 another son, Herbert Stephen (known as Bertie) was born to Annie Elizabeth Cooke, formerly Biggs, and Mordecai at 2, Grosvenor Villas. The clock had been turned back and Annie, who from now on was to use whichever surname she found most convenient in her immediate circumstances, had returned to her former status in the Cooke household, bringing with her not only Frank, but John's daughter Mabel too. We know this because Leila Cooke, in 1968, wrote to one of her nieces giving, without comment, a list of her (Leila's) brothers and sisters which includes Mabel–she obviously had no inkling that Mabel was not her full sister. William Cooke, however, was well aware of the facts, for Mabel is not listed among Mordecai's children in 'The Cookes of horning'. On Mordecai's side at least, the 'affection' of the verse on Harry's gravestone must have been deep indeed.

Despite Annie's return Sophia remained with the family–this much is shown by the census of 1881. The enumerator must have been one of the most bewildered men in London by the time he had completed his entry for No. 146, Junction Road, and the entry must have been one of the most extraordinary.

Name	Relation to Head of family	Condition	Age	Rank, profession or occupation
Mordecai Cooke	Head	Married	55	Civil Servi Secretary of [illegible] for India [illegible]
Sophia Do	Wife	Married	58	,,
Ada G. Do	Dau	Un	17	Scholar
Willie C. Do	Son	Un	14	,,
Ernest F. Do	Son	Un	12	,,
Anne E. Cubit	Wife's Dau	Mar	36	University Cambridge. U. States America.
Mabel Do	Wife's Dau	Un	9	Scholar
Frank Do	Wife's Son	Un	6	,,
Herbert Do	Wife's Son	Un	2	
Louisa White	Serv't	Wid	56	Servant

Obviously Mordecai had decided to explain his complex household by acknowledging two families, his own and that of his married step-daughter, Annie Cubitt. As at the previous census he was not required to name the mother of the three children he claimed as his own, but he was deceiving the authorities in changing Bertie's surname and in omitting to state that he was the father of Bertie and Frank. The extraordinary claim that the three 'Cubitt' children were his wife's (Sophia's) must I feel be a clerical error on the part of the enumerator, for Mordecai would surely have explained them truthfully as his wife's *grand*children. We do not know whether Annie and the 'Cubitt' children were using his surname in their daily lives, or whether it was merely a strategem for the purposes of the census, but it is certain that by the time they were adult the boys had adopted the name Cooke, for Bertie's children only know their father and their Uncle Frank by that name.

When the Census was taken Cooke, though still employed by the India Office, had been seconded to Kew as a botanist (Chap. 12) but this was altogether too complicated for the enumerator, whose entry, in as far as it is possible to read it, implies that Cooke was himself Secretary of State for India. A further virtually incomprehensible entry is that for Annie's 'Occupation'. My own opinion is that this is merely further evidence of the state of confusion to which this household had reduced the enumerator, and that it represents his attempt to cope with Mordecai's insistence on the inclusion of his American degrees (p. 182) so as to emphasis his professional standing. The other possible explanation, that it was an effort to invent an occupation for Annie's absent husband, seems even more implausible.

Mordecai now had a very large household dependent upon him, and it is perhaps not surprising that in his correspondence he frequently laments his shortage of money. Considering too that he was now 55, Leila's complaint, that he spent much of his time shut away alone in his study, working and writing, is understandable, for he must have found a lively young family a strain, but the necessity of earning the money to keep them all was urgent.

Some idea of his appearance at this time can be pieced together from surviving photographs (Fig. 6.5) and from Leila Cooke's notes. He was a slight man, about 5'8" in height, with the short, stubby fingers that run in the Cooke family. His eyes were grey, and he had thick, red hair and a luxuriant red moustache and beard which last he 'kept squared, not an imperial, but squared'. It was this red hair which, in conjunction with his Christian name, resulted in his being taken for a Jew, something which always caused him great distress. Usually he was a quiet man, and in his photographs he looks solemn, but he had a pronounced sense of humour and, if talking about fungi, he was eloquent. By the time Leila can remember him he had lost his Norfolk accent and spoke 'pretty good London', by which I think she meant Standard English, but:

> He was full of Norfolkisms, for instance, if a window rattled he would say, 'There's old Kett's bones', a reference to Robert Kett, leader of Kett's rebellion who was hanged in chains outside Norwich Castle in 1549. If you made an excuse for dirty hands, he would say, 'You are like old Betsy Green, who would say 'They'll come clean on Friday when I make the dough'.

Annie (Fig. 6.6) was a good-looking young woman, about 5'5" tall, of medium build, with dark hair and brown eyes, and she clearly had a mind of her own. Her photograph confirms her daughter's description of her as being a bit of an extrovert.

Of Sophia no record whatsoever remains except in a few official documents and the single entry in William Cooke's history of the family. Even her grave is unmarked. It seems that wherever possible every trace of that sad figure has been expunged from history.

It probably mattered little to Mordecai whether it was Annie or Sophia who was running his household at any given moment, for it would have been

Fig. 6.6   Annie (Cooke) Cubitt, née Biggs, aged about 40.

the duty of either of them to see that his routine remained undisturbed. All the evidence suggests that he was no exception to the universal belief of the time that a woman's place is in the home and that the children should be seen and not heard. Now that at last he had some of his own he had very little time to spare for them, and it was apparently left to Annie (or Sophia) even to choose their schools. Leila Cooke told me:

> He was always busy with his own thoughts and I don't think he bothered about us [the children] . . . He wasn't really interested in his children as children. I can never remember him getting anybody to talk about anything at meals. He had his meal, swallowed it down and went away to his library and let the children have their tea.

On the other hand:

> He always used to buy me strawberries. I had said on one occasion that I liked strawberries, and he always brought me strawberries.

It was the vivacous Annie who organised the family entertainments; songs round the piano and sets of Lancers. Mordecai

... seldom relaxed except at Christmas, when, unaccompanied, he would sing in his resonant voice a wide range of such old favourites as 'Sally in our Alley', 'Simon the Cellarer', 'The Vicar of Bray', 'Little Brown Jug' and 'Green grow the Rushes, O'.

He knew too 'all sorts of funny old Norfolk songs'.

Chapter Seven

Lovers of Nature. 1862–1866

We have followed Cooke's gradual advancement through his 18 years at the India Museum and seen something of his tumultuous private life during that time, but these events only form a background for the leisure activities which meant so much more to him than his paid employment, and through which he has left an indelible imprint not only on the biology of his own time, but on that of today. His years at the India Museum were marked, as we shall see, by a crescendo of activity on behalf of the two causes nearest to his heart, mycology, and a mission to provide for others opportunities for enjoying Nature as fully as he did himself. During the 1860s he was only finding his feet as a mycologist, but his efforts as a publicist for Nature reached their peak then; by the 1870s his contribution to mycology was becoming internationally recognised and was consuming most of his free time, resulting perforce in the winding down, though never the abandonment, of his activities as a field naturalist. These two aspects of his career will be traced separately, beginning with his contribution to the wider world of natural history.

When conducting evening classes in botany at the Working Men's College (p. 64) and elsewhere, Cooke set great store by field studies, encouraging students to find living plants on which to work out for themselves the application of the often tedious classifications and lists of facts which had been presented to them in the classroom. These excursions had been popular, but as he reported in 1863 [1].

> after the year's tuition and the annual examination the classes appeared to close and the members to disperse, as I often feared, almost forgetting Botany and their Botanical Studies. A desideratum seemed to be some association among the members of these classes which should bring together kindred hearts, and by mingling together warm

them up to the pursuit of the study they had commenced, and carry them still further by the mutual interchange of ideas.

When circumstances forced his resignation from the College:

the teacher left his pupils with regret, and they in their turn wooed till they won him back again in a new capacity. A social meeting over a cup of tea and 'fried puffballs' had two very important results, one, conviction by force, that fried puffballs are not poison but 'a dainty dish to set before a king',

and secondly, the idea of starting a field club for his old pupils. The social occasion referred to took place in 1862 in the rooms of a Mr. W. Martin of Great Turnstile, Holborn (of whom nothing more is known). Alfred Grugeon and his friend Mr. C. J. Savage, also of Dalston, were present, as were two others whose names are not recorded. At a further meeting of these six men, according to Cooke's Presidential Address the following year, they agreed that the:

six Botanists would join ourselves together into a Society, club or association, and meet together at stated periods, to talk of what we had done, arrange to go out on short trips together, call ourselves by a name, and invite others to join us who were really anxious to improve themselves or us, and finally, by means of a small subscription, just sufficient for incidental expenses, to prevent its being in any way a burden to the humblest, or an obstacle to the poorest who might join us.

On the third of December, 1862, we commenced in earnest . . .

The first prospectus of the Society of Amateur Botanists set out its objectives:

Established for rendering mutual assistance in the study of British Plants, by organised Excursions into the Country–the interchange of specimens–the communication of papers–the establishment of a Library, herbarium and Museum, or such other means as may from time to time be deemed expedient.

Candidates for membership had to be proposed in writing and to pay an admission fee of 2s 6d; thereafter there would be an annual subscription of the same amount. Meetings were to be held on the first and third Monday of each month at 8 pm and at them, among other things, papers would be read and excursions arranged.

The history of natural history societies has been ably told by David E. Allen. Local societies had been a well-known and flourishing feature of Victorian social life ever since the first were founded in the 1820s and 1830s, but the rather lofty objects of these were academic discussion accompanied by the publication of 'Transactions', and the setting up of museums and libraries; field excursions hardly featured among their activities. The expenses of, and therefore the subscriptions to, such societies were necessarily high, so that their membership was automatically confined to the moneyed classes. In certain industrial cities, however, there existed groups of working men with an intense interest in matters botanical, and they were forming themselves into clubs, often based on a local public house, for the purpose of collecting plants in the field. Members would spend Sundays, their only free day up to this time, on botanical excursions into the surrounding countryside, followed in the evening by discussion and the exchange of specimens over a convivial pint.

In the 1830s the first genuine field clubs, a cross between the aristocratic natural history societies and the pub-based botanical groups, began to appear. These were the creation primarily of the professional classes, but they were based firmly on field excursions which might last all day, the excursionists repairing to a local hostelry in the evening for a companionable meal, discussion of the day's finds, and perhaps the reading of a paper or two. Such an arrangement meant that the clubs had no need of their own rooms or of any regular means of publication, so subscriptions could be kept quite low. Field clubs were flourishing all over the country by the 1860s, and it is with these that the Society of Amateur Botanists must be compared.

Although it was a very humble body with a low subscription, and its avowed object was work in the field, it called itself a Society of Botanists, not merely a field club. It had, from the first, intended to hold fortnightly meetings, and to do this it was essential to have the regular use of a room, in the Society's case, the former St. Pancras Library. At an annual rental of two guineas this was a struggle, and had not Robert Hardwicke come to the rescue and offered the Society an assured meeting place on his own premises, free of charge, it could well either have folded or become too expensive for its working class members. Few other societies could have been so fortunate, and Hardwicke's generosity probably accounts for the survival of what was, at this period, the unusual social mix of its membership. For soon after its inauguration it began to draw members from the professional classes in increasing numbers, and although by the end of its brief life the latter seem to have taken the Society over, the fact remains that in 1863 it was one of the earlier groups devoted to field work in which the great Victorian divide between the professional and the operative classes was bridged, eligibility for membership depending solely on a commitment to the study of botany. It was to be 1870 before the Rev. Charles Kingsley, in Chester, would repeat this feat, though when he did so it would be on a considerably larger scale. He, like Cooke, had been conducting evening classes in Botany, and as he was a well-known and brilliant lecturer, had attracted audiences from all walks of life. So when he too converted his class into a society for holding field excursions and indoor meetings it was already socially mixed and, because of his own positive belief in the equality of all men under God, he saw to it that on outings, at least, there were no class distinctions, all members travelling second class on the railway.

It is very unlikely that Cooke had any such high-minded principles for his Society. It was simply that he felt a certain responsibility for his old students in a subject to which he himself was passionately devoted, that he enjoyed the conviviality of field excursions, and that the more recruits he could get, the more successful his Society would be; if they came from the professional classes, so much the better as far as he was concerned, for to advance his mycological interests he needed to break into that world himself. Nevertheless, he had succeeded in doing unintentionally what Kingsley did not do deliberately until 8 years later.

How did Cooke's new recruits come to hear of so obscure a Society? Partly, no doubt, through a very favourable notice that had appeared in one of Hardwicke's publications, the *Popular Science Review* (p. 159) in 1863; there can, indeed, be little doubt from an inspection of the membership list, that the publicity came mainly through Hardwicke and his circle into which Cooke had by now been completely absorbed, especially after the Society began to hold its meetings on the publisher's premises. Hardwicke himself was made an honorary member; his manager, Thomas Ketteringham, joined, as did Ketteringham's friend, Witham Bywater, the proprietor of a nearby saddlery business, both of whom were to play a considerable part in Cooke's next venture. Two engravers who already had connections with Hardwicke's firm became members, G. W. Ruffle and W. G. Smith, who was also a mycologist and would soon become a close associate of Cooke's. There was a considerably older man, the well-known botanist, the Rev. W. W. Newbould, and also Edward Jaques, F.R.M.S., a Civil Servant in the Woods and Forests Office. And there were four very young men, all of whom were to rise to prominence in the world of biology in years to come–Edwin Lankester's 16 year-old son Ray, who would become Britain's first marine biologist and a Director of the Natural History Museum; his close and life-long friend, William Dyer, then aged 20 and destined to become the Director of Kew Gardens; and Dyer's two friends, Henry Trimen, the same age as himself, and 18 year-old James Britten, both to become well-known botanists. Apart from his own students there were two other recruits for whom Cooke himself must have been responsible–his brother Ebenezer, who was sharing his house at the time, and Daniel Stock, lately of Bungay in Suffolk, with whom Cooke had been collecting fungi while still a schoolmaster. Stock had fallen on hard times and had moved to Stoke Newington, where he was now a solicitor's clerk.

The eagerness with which these middle class botanists joined an initially working class society at this early date is probably explained by Cooke himself when, many years later, he was talking about the Society. 'As far as I am aware', he said, 'this was the only instance, at that period of time, in which any periodical excursions were made by any society in or around London' [2]. No wonder London botanists were keen to join. In one further way the Amateur Botanists were unusual for their time–they welcomed women as well as men to membership, at least in theory. In practice, however, the active membership of 35 in 1864 was entirely male; only among the 13 corresponding members were there two women.

Cooke took his duties as President very seriously. From the first he was determined that the Society, 'instead of being merely a company for the collection of rare plants', as so many were at the time, would devote itself 'to a close examination of common ones' [3], and would preferably concentrate on one group each year. To this end he had, at the inaugural meeting, offered a prize, M. J. Berkeley's (p.150) book on *British Mosses*, for the best collection of those plants sent in by the end of the year. But mosses did not prove popular with the members, and at their suggestion it was decided rather to concentrate on Composite plants. Cooke changed his offered prize to Bentham's *Handbook of the British Flora* with his own *Microscopic Fungi* (p. 160), then in press, for the runner-up.

Organised excursions followed by 'a cosy tea' took place on alternate Saturday afternoons (the $5\frac{1}{2}$-day week was now coming in) from 3 o'clock to 7, but in addition Grugeon and Savage at least, and probably others as well, went out on their own initiative and reported their finds at the Wednesday meetings. In his Presidential Address for 1863 Cooke gives an account of the year's excursions:

> though not always so numerously attended as could have been wished, [they] have nevertheless been assiduously persevered in during the summer. Although on one occasion the Excursion consisted of your President alone, who might have been seen on a certain breezy Saturday afternoon hunting for members and plants, both unsuccessfully, on North Woolwich Marshes, yet the number generally exceeded unity, though seldom equalling that we had anticipated. The Annual Excursion to Epping Forest was the most successful of the year . . .
>
> What is to be done in these excursions? I think manifestly, it should be, first to collect composite plants, second to collect for the herbarium, third to secure such fruits as may be in season, fourth to take a record of all cryptogamic plants met with as are known to the excursionist.

Though his moss project had been voted down, Cooke was still determined somehow to include cryptograms in the Society's activities.

It was on an excursion to Highgate Wood that the life-long but uneasy friendship between Mordecai Cooke and Worthington George Smith (1835–1917) (Fig. 7.1) began. Smith came from a long line of Hertfordshire farmers but his father, a minor Civil Servant, had settled in the then residential suburb of Shoreditch where Worthington was born. His educational background differed only slightly from Cooke's, for it was confined to what his local Church school could teach him. If he received any higher education it was very scanty, most of what he knew being self-taught, something he was to boast of in later life. But unlike Cooke, Smith knew from an early age

Fig. 7.1 Worthington George Smith, 1835–1917.

exactly what he wanted to be–he intended to become an architect. To this end he was apprenticed to a firm of architects in the Strand, eventually becoming a skilled draughtsman and a member of the Architectural Association. However, he loved design, and his formal qualifications, though hard won, did not satisfy him; he began drawing flowers and leaves from life, and this led on to an interest in botany. Then a meeting with a professional engraver opened up yet another field for him and he quickly learned this art too. He was now fully equipped as an illustrator and, to quote his biographer:

> Although architecture in the abstract appealed to Worthington Smith's artistic senses, the practice involved him in things which he found frustrating and repugnant, for his firm specialised in dilapidations and drainage. Worthington dreamed of designing cathedrals: his employers demanded that he design drains, and in 1871, at the age of 26, he revolted, and devoted himself to freelance illustration and engraving. At first he drew buildings, then plants, and later archeological material as well.

Two years after his revolt he met Cooke through the Society of Amateur Botanists. According to his own account, Smith's objective on the Society's excursions:

> was almost solely the larger fungi, which I collected for drawing and dissection, I had been on this job practically all my life, as I began noticing fungi when I was quite a young child. The first fungus that attracted my attention was Coprinus atramentarius [an Ink-cap], in my father's garden. It was not present overnight but it was there the following morning. It filled me with wonder [4].

In the same article Smith gives a graphic description of his friend at these early field meetings:

> I was then 26 years of age and he 10 years older ... Dr. Cooke was an inveterate smoker, and when he was not pulling at his pipe he was singing. I was neither a smoker nor a singer. Dr. Cooke at that time had noticed the larger fungi but little [this was not entirely true]; he collected and studied the microscopic fungi, and was very keen on all animate objects. He captured toads, frogs, newts and snakes.

This last activity was to result in two years' time in his next book, *Our Reptiles* (see below).

Of the Society of Amateur botanists Smith says:

> The members met in the first floor room of Mr. Robert Hardwicke's bookshop in Piccadilly. ... There was an engraved brass plate on the side of Hardwicke's door inscribed 'The Society of Amateur Botanists'. I did this and presented it to the Society.

While Cooke continued to concentrate for the time being on the microfungi, Smith was rapidly becoming an authority on the larger species. But Cooke, when he had achieved recognition as a mycologist, would return to his first love, the agarics, while Smith would try his hand at the smaller species. It was probably at this stage that Smith became acutely jealous of Cooke, and the two men began to make vitriolic remarks about one another to third parties and, in Smith's case, in print. It is fair to add, however, that Cooke was not the only victim of Smith's pen, for the latter was to become well-known for what were euphemistically termed his 'incisive' book reviews. Cooke seems to have attacked Smith in kind only if first provoked by him. Apart from such occasions his attitude to his friend is shown by a letter written to the Department of Botany at the British Museum (National History) in 1908:

> Permit me to thank you for your interest in sending me a copy of Smith's synopsis of Basidiomycetes. The latter appears to me to reflect great credit on all concerned. It is admirably got up and I have faith in Worthington Smith, that it is irreproachable in the text, although we may not always agree in matters of detail.

After Cooke's death Smith made some very unpleasant and largely untrue remarks about him in a letter to their mutual friend (through the Amateur Botanists) James Britten, then attempted to make amends by adding:

> You must not feel I have any bad or jealous feeling against C. The opposite is the case and I know that he was often quite right when his opponents were quite wrong. I always valued his opinion which was never halting, but was always strongly outspoken [5].

The talents of the two men as artists also caused friction, though again the real ill-feeling seems to have been on Smith's side. Smith was a professional, and a very competent one, but he had been trained in the precise, mechanical style necessary for architectural drawing, so that throughout his life his illustrations remained somewhat stiff and formal. Cooke described his style thus, when writing to a friend towards the end of his own life:

> Most of Worthington Smith's work amongst fungi is decidedly diagrammatic, especially his earlier work. . . . they must be perfectly symmetrical, all true curves and all constructed upon the same model–but differing in size and colour. Take any one of his figures and there is nothing in it but could be drawn by dictation–an agaric, such and such a size, pileus plain or conical, stem long or short, bulbous or regular etc. etc. etc. The style is a failure because it does not try to represent the peculiarities of the species, does not try to make the drawing like that and none other, and, in my mind, it is that which makes Smith's figures so unsatisfactory. They are only 'woodcuts' [6].

More recently E. C. Large has made the same point in his own inimitable way:

> When he had occasion to draw a tomato in a hurry to illustrate some advertisement in the *Gardeners' Chronicle*, it looked exactly like an ornament off a wrought iron gate, and so realistic you could almost lift it off the page with a pair of tongs [7].

Cooke claims that he himself draws in a 'natural' or 'artistic' style, and advises others to do the same, but Smith, in the vicious letter to Britten quoted above, says: 'he was no draughtsman, his hand was far too heavy and shaky', and: 'his volumes on the Basidios (sic.) (p. 215) are of very little good. They have provoked a good deal of criticism, this because the drawings ill represent the subjects'. Both these statements are blatantly untrue. Though not all Cooke's *Illustrations* are of the same high standard, they are still used for reference purposes today; and his hand was steady enough, when he was well into his seventies, to produce excellent wall diagrams for schools at the request of a publisher. Smith's tirade probably expresses the anger of a professional artist at competition from an amateur. No wonder that Leila Cooke remembers that the two men were great friends in 'a *quarrelsome* sort of way'.

Smith married in 1856, and some time thereafter he and his family were living, like the Cookes, in Kentish Town. They seem to have been close neighbours for the two men sometimes travelled home together on the same train, and:

> In early times he used to call on me pretty often, sometimes with two young children, one almost a baby. His wife also used to call with the kids [5].

Leila Cooke remembers Smith as 'a lion of a man', but as she was only four years old when the Smiths left London in 1885, she must have met him with her father later.

In the first year of their existence, the Amateur Botanists, as can be gleaned from their 'Transactions', hand-written by Cooke and with a frontispiece by Smith (Fig. 7.2), held 19 meetings out of a possible 22, a most commendable average of 2 papers being read at each. Up to the end of April, 13 out of 17 were read by the President himself, but later in the year other members gained in confidence and Cooke only found it necessary to contribute 5 out of 18 papers including the Presidential Address. Most of his talks were on his favourite topic, fungi, and as might be expected in the circumstances, few were original—usually being admitted compilations from the literature. But he also spoke about a census of British plants that had taken place that year, about mosses, and about British plants found in India. Of the other contributors, Grugeon was the most frequent, and Smith spoke twice, on one occasion on the disastrous consequences to himself and his family of eating a poisonous fungus. A description of this traumatic incident was published in the *Journal of Botany* the following year.

By the end of the Society's first year Mr. Martin, on whose premises it had been born, had disappeared from the scene, and the Committee for 1864

Fig. 7.2 Frontispiece for the Proceedings of the Society of Amateur Botanists, by W. G. Smith.

lacked a Treasurer; Alfred Grugeon became Librarian and Curator, showing that the Society felt secure in its new premises, while Ketteringham was elected Secretary and Cooke remained President. One of the matters mentioned in his first Presidential Address may be noted here:

> Our members have hailed with pleasure the establishment of a *penny* [his italics] Botanical periodical, and I am sure that I give expression to their sentiments and my own when I say that they heartily wish it success, and they do so now, before any judgement of contents can influence them, mainly because it is a vast stride in the right direction. Botanists have their penny paper, at last, more shame to them, if they ever be without one again.

This remarkable little magazine was the *Botanists' Chronicle of Notes, News, Remarks, Questions and Replies*, the first number of which appeared just after the Amateur Botanists' Annual General Meeting. Its publishers, Alexander and James Irvine, booksellers of Chelsea, had no official connection with the Society at the time, but James joined in 1864, gave a talk on the botany of Merionethshire, and presented the Society with some specimens for its herbarium: thereafter the *Chronicle* published regular accounts of the Society's meetings and exhorted its readers to join. This brave venture only survived for 17 months, collapsing not through lack of subscribers, but because its founders were unable to find a publishing house to take it over. As readers were informed in the last number:

> If our residence had been in town instead of at Chelsea, we could still have issued the Chronicle, but no tradesman can be expected to send a boy or a man two or three miles for a penny periodical; hence the labour of folding, addressing and posting had to be borne by ourselves.

It is obvious that Cooke thoroughly enjoyed delivering that first Presidential Address to his new Society, and by the end the flamboyant rhetoric which we first saw in his articles for *The School and the Teacher* had taken over:

> Small beginnings often make great endings, and though we can read our own past history, our future destiny is hid from our view, we cannot even define the shadows in the mist of another year. We have a noble purpose, let it be nobly won. If the lamps in our pitchers are kept burning the little band of Gideon will conquer. We are no enthusiasts in search of a new Eldorado, no dreamers of a new commonwealth neither the votaries of a dead saint or the slaves of a living sinner, but I trust we are a band of earnest men scorning the empty foibles with which thousands waste their golden hours, with ambitions above that of roistering bibulants, seeking amusements in our youth which will bear the reflections of age, aiming to employ our faculties to their ultima thule, and not studying to forget that we are men by cultivating only the passions and appetites of animal nature. We con the unfolded page of a glorious creation, redolent with beauty, teem-

ing with life, and in its deepest recesses still revealing a deeper and a deeper mine of wonders, till the brain reels and the bewildered mind sinks in the vain effort to comprehend the hidden mysteries of life. This is our book, written in symbols, by the author of all intellect, the incarnation of all intelligence, open to all, yet ever new; never illusory but ever leading on from joy to joy.

The address concluded with a lengthy quotation from Wordsworth's 'Lines composed a few miles above Tintern Abbey' and a hideous pun on the last line:

> If solitude, or pain, or fear or grief
> Should be thy portion, with what healing thoughts
> Of tender joy wilt thou remember *us*
> *And our Association.*   (Cooke's italics)

Cooke used a much shorter extract from the same poem when speaking to the Hackney Microscopical Society in 1887, and on both occasions his selection included the words 'Nature never did betray the heart that loved her'. Leila Cooke told me that this line was indeed her father's favourite quotation, and it was for this reason that she had it engraved on his tombstone.

The Society continued to thrive throughout 1864, an increasing proportion of the membership offering talks on a more varied range of subjects. A long and commendatory notice about its activities and membership appeared in the prestigious *Journal of Botany*, approving especially of its combination of 'field club with chamber-association' [3]. In November, at the last ordinary meeting of the year, the President read a lengthy paper on 'Infinite Variety and Immutable Law', setting out his views on the origin of species, a paper which, according to the *Botanists' Chronicle*, 'was listened to throughout the entire evening with intense interest, and closed amid unanimous applause'. It will be quoted at some length elsewhere (p. 154). The year ended with the second Annual General Meeting at which the President's Address consisted mainly of an account of the year's activities and of suggestions for the future.

1865 saw Cooke firmly established as an author of books and articles both on fungi and on much broader aspects of natural history. Only his non-mycological works will be considered in this chapter and the next, his huge output of mycological publications, though written concurrently, will be discussed in a more appropriate context (Chaps. 9 & 10). In the second half of the nineteenth century it was still possible for an individual to have a good working knowledge of a wide field of biological science, for the sheer volume of information in existence, though expanding rapidly, did not remotely approach that with which we must cope today; the straight-jacket of specialisation which is now forced upon us only began to be necessary towards the end of the century. It was, therefore, quite possible for a man such as Cooke, with his

Fig. 7.3 Illustration by W. G. Smith from *Science Gossip*, reproduced by Cooke in *Our Reptiles*.

lifelong interest in, and curiosity about, living things, to write authoritative popular and semipopular books on a number of widely different groups of the English flora and fauna; many Victorian naturalists did this. Where Cooke was unusual was in that he continued to do so long after he had become an acknowledged authority on a single subject, mycology. There is no doubt that he was fascinated by the whole realm of nature and felt it his mission to enlist the enthusiasm of as many other people as possible; equally it is certain that he was a quick and facile writer. But is is probable that his large literary output was also partly a response to financial stress. For all through his life, even after he had settled into a secure job, or later had a Civil Service pension, he was to complain bitterly and frequently of shortage of money. Fortunately for him this was the age of the amateur naturalist, and the market for popular books on natural history was almost insatiable.

Our Reptiles, again in Hardwicke's 'Plain and Easy' series, appeared in 1865, and concerned 'Lizards, Snakes, Newts, Toads, Frogs and Tortoises indigenous to great Britain'. In his preface Cooke comments:

> It may cause surprise to some who are not among my most intimate friends, that my name should be attached to a volume on any other branch of Natural History than one which is in some way associated with the Vegetable Kingdom. To these it may be necessary to explain that I have only returned on this occasion to an 'old love', long deserted for the fascinating charms of 'moulds and mildews'. . . . In a foolish moment, perhaps, the old hankering to have a word or two with one's 'first Love', has come over me, and has resulted in this humble account of 'Our Reptiles'.

True to the series for which it was written it is small, unpretentious, and intended for the layman. Ebenezer was responsible for the beautifully drawn and coloured illustrations (Figs 7.4 & 7.5) of which Mordecai writes: 'My brother copied most of the reptiles from life as I had them in confinement in my conservatory.' These reptiles became legendary in the family, for one of Ebenezer's younger sons, born many years after the incident, records that some escaped and 'a blindworm was found patiently weaving its way back and forth through the bannister rails'. A delightful engraving by Worthington Smith, at

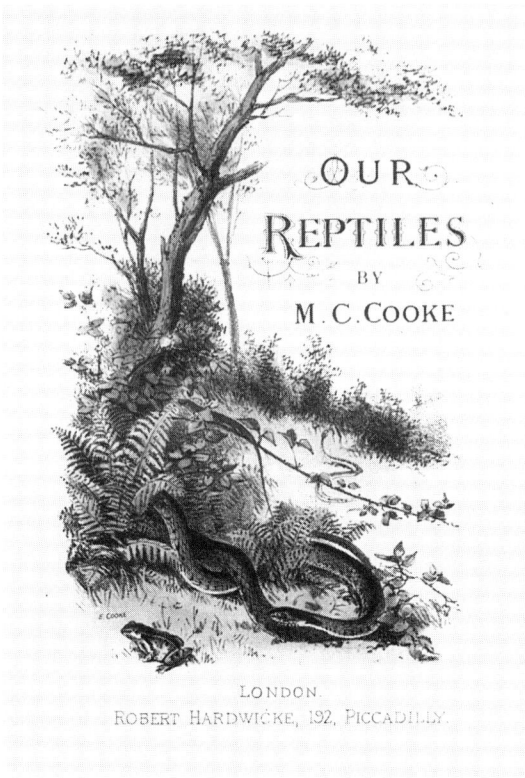

Fig. 7.4 Ebenezer Cooke's frontispiece for *Our Reptiles*. Original in colour.

the end of the chapter on toads (Fig. 7.3), shows two of these animals, each sitting on a toadstool, engaged in amiable conversation. Smith wrote later: 'I saw a *frog* sitting on the top of an agaric a few days ago, in the manner of the sketch I made...' [8] Mordecai used the drawing again in the same year in an article of his own on toadstools in *Science Gossip* [9]. No further editions of the reptile book were issued by Hardwicke but, as with *British Fungi*, the title was taken over by W. H. Allen, who produced a new and much inferior edition in 1893. Again, Ebenezer's plates were redrawn, stiffly and with little attention to detail, and were poorly coloured in nasty, yellowish shades. In particular, the frontispiece is a travesty of the original—bramble and bracken, immediately recognisable in Ebenezer's delicate drawing, are just 'greenery' in the Allen edition.

The Society of Amateur Botanists, successful though it was, only reached a tiny fraction of those Londoners who were potential recruits to the ranks of amateur naturalists, and in no way satisfied Cooke's urge to spread his gospel to all who would listen. So when, some time during 1864, the means of reaching a much wider audience occurred to him, he went straight to Hardwicke, whose 'little shop in Piccadilly' he was now visiting almost daily, with his idea:

> ... I suggested to Mr. Hardwicke and to Mr. Ketteringham that there was a good opening for a cheap monthly magazine, which should be devoted to Natural History and Microscopy, offering facilities for exchanges and copious notes and queries. This idea was eagerly entertained, and *Hardwicke's Science Gossip* was the title suggested by Mr. Hardwicke, and to this he adhered pertinaceously.

The last phrase is a little mysterious until one reads Smith's account of the same episode:

> No name was at first suggested for this periodical, but Cooke suggested as a good name *The Veil of Isis*. This made Hardwicke wild; he said 'he was not going to cry rotten fish', and suggested *Science Gossip* as a name that everyone could understand [4].

Cooke's choice of so obscure a title for a magazine specifically aimed at a broad cross-section of the population can only be explained by the wide knowledge and great love of ancient mythology instilled into him in childhood by his Uncle James.* The reference appears to be an ancient statue of the

Fig. 7.5 Illustration by Ebenezer Cooke for *Our Reptiles*. Original in colour.

---

*Cooke refers to the Veil of Isis a number of times in subsequent writings, the last occasion being in an article in the *Essex Naturalist* in 1905.

Egyptian god Isis on which was engraved the legend, 'I am all that has been, that shall be, and no mortal has lifted the veil that covers me.' Cooke wished his readers to discover for themselves some of the secrets hidden by that veil, but the only vestige of his poetical title to survive was the winged symbol of Isis which decorated the 'Avant-Propos' at the beginning of the first year's bound volume. It must be admitted that, compared with Cooke's effort, Hardwicke's title was a stroke of genius, and by the end of the first year of publication Cooke was fully committed to it:

> It has been whispered abroad that we, in our humble endeavours to 'Gossip' freely over the little extracts we collect from the book of Nature are giving offence. Not that we act as 'snappers up of unconsidered trifles', but because we give to them an undignified name. On the threshold of the Temple of Janus, with our first volume under our arm, we again announce our name, however undignified it may be, and with it gain admission to the fireside of thousands, whilst the same talisman excludes us, we hope, from the drawing-rooms of only a few. Parents seldom give to their children names which satisfy all their friends, and we cannot hope to be more successful than they [10]

Cooke, overflowing with enthusiasm and ideas as he was, was appointed Editor of the new venture. Then the question of the design of the covers for the monthly parts arose, and Hardwicke asked Smith to submit one:

> In this I introduced an eagle crouching on a rock, at the top centre; but Mr. Hardwicke would not have an eagle, because, he said, an eagle was not a British bird. 'Why not have an owl?' said he. 'Everyone knows that an owl is British.' And the owl was substituted. At last the block was drawn, approved by all, and engraved. On the publication of the first number someone wrote and pointed out that the shells at the bottom were twisted the wrong way. I had drawn them correctly on the block, but, of course, they were reversed in printing, and so became wrong. Other faults were found by other critics, who wrote to Mr. Hardwicke, and the latter went for me 'hot and strong' [5].

The first number of 'Hardwicke's *Science Gossip*. An Illustrated Medium of Interchange and Gossip for Students and LOVERS OF NATURE' (Fig. 7.6), was 24 pages long, priced at 4d, and appeared on January 1st, 1865. According to Cooke, it 'became at once a success. At the time there was no cheap competitor, and for a long time it was the only cheap journal of Natural History.' Describing the contents of each monthly part he says:

> One feature was the column for free exchanges, and another for notes and queries in which subscribers were assured of the correct naming of their objects, which soon arrived for that purpose in great numbers. I had secured the cooperation of some of the best experts of the day, amongst whom came Professor J. O. Westwood, the entire entomological staff of the British Museum, and for Botany Dr. D. Oliver and his staff at Kew, with Professor C. C. Babington* and many specialists. (Aut)

*of Cambridge

Clearly he had obtained really first rate advice for his readers, and it is a tribute to these eminent scientists that they willingly gave their time for such a purpose.

For the first 18 months each issue begins with an unsigned editorial in Cooke's unmistakable style (later these are replaced by original articles by experts on some aspect of Natural History); then follow several signed articles on a wide range of topics, often illustrated with engravings by G. W. Ruffle, and some short, usually anecdotal, contributions from readers on such matters as 'Viper swallowing its young' by 'A Norfolk clergyman', and 'The cultivation of ferns'. All through the first volume Cooke runs a series on 'Simple Objects', first describing the chosen specimen and then giving instructions for its examination under the microscope. Next, there are regular sections on Zoology, Entomology, Fish Tattle, Botany, Microscopy and Geology, all consisting of short extracts from current scientific books and periodicals—Cooke was certainly a voracious reader if he found all this material himself each month. Finally come several pages of brief correspondence, queries, replies, and notes for readers.

Naturally such a variety of contributors and subjects leads to some extra-

HARDWICKE'S

Science-Gossip:

AN ILLUSTRATED MEDIUM OF INTERCHANGE AND GOSSIP

FOR STUDENTS AND

LOVERS OF NATURE.

Edited by M. C. COOKE,

AUTHOR OF "A PLAIN AND EASY ACCOUNT OF THE BRITISH FUNGI," "MICROSCOPIC FUNGI,"
"A MANUAL OF BOTANICAL TERMS," AND OF "STRUCTURAL BOTANY,"
THE "BRITISH REPTILES," ETC. ETC.

LONDON:
ROBERT HARDWICKE, 192, PICCADILLY.
1866.

Fig. 7.6 Title page of *Hardwicke's Science Gossip*, vol. 2.

ordinary juxtapositions in the text of this lively magazine. For instance, in Volume 3 (1867) we find in succession an article on a 'Clip for a zoological trough', and one on 'Juvenile Museums', immediately followed by an original scientific paper on 'The Chignon Fungus'–an infection of the scalp hair–by W. T. Fox, a leading dermatologist of the day and Britain's first authority on the fungal diseases of man [11]. Fox entrusted the publication of his dermatological books to Hardwicke, and this is presumably how he came offer his paper to *Science Gossip* when it was of a standard which would have been perfectly acceptable to any reputable medical journal. Indeed the correspondence which it provoked appeared both in *The Lancet* and the *Journal of Botany*, indicating the exalted circles to which *Science Gossip* was penetrating. Fox's paper is followed by an essay on 'The Unity of Mankind' in which the author 'proves' to his own satisfaction that the black races are degenerate whites who had succumbed to the unhealthy tropical climates in which they found themselves. Cooke censors nothing; everyone is allowed his say, though occasionally he publishes a reply to the more outrageous old-wives' tales, such as that of the flattened and desiccated toads which come to life and hop away when water is poured over them, a popular notion of the period. It says much for the magazine that serious scientists were prepared to publish in such company, though not all approved. Bertold Seeman, the Editor of the *Journal of Botany*, remarked in an undated letter to Cooke concerning it: 'I do not think that anybody has ever attempted to combine the style of Hans Andersen's fairy tales with solid popular botanical information' [12].

Cooke's editorials vary in content from exhortations to his readers to study and enjoy Nature, through advice on specific subjects, to philosophical musings. Some, such as 'Splitters and Lumpers' (conflicting ideas on classification) concern subjects that were exercising his mind in other contexts at the time (p. 153). In March 1865 the Editorial is entitled 'Short-Commons', and heralds an interest in conservation which was to continue over the years. More and more bills granting permission to enclose 'waste lands', translated by Cooke as 'short-commons', were being pressed through Parliament at the time, and there was concern in many quarters at the consequent loss of open spaces, particularly those within easy reach of the great cities.

> It is sufficient for us to suffer the hallucination of Hampstead Heath laid out in squares, adorned with villas, cut up into terraces, and sacrificed to the demon of bricks and mortar. For us the vision of Wimbledon Common surrounded by miles of monotonous palisades, and laid out in trim parallelograms of level grass, intersected by the cleanest of gravel walks, is enough, be it only a vision, to arouse us from exploring the 'Origin of Species' to protest against the 'Origin of Parks,' and the absorption of 'the last of commons'. . . . if Hainault is gone, and Epping is going, is not the New Forest left? We are not politicians, so perhaps all this is as it should be. It may be right enough to give to 'Labour' its Saturday half-holiday, that it may go out of town and enjoy itself, and hold converse with Nature face to face, and at the same time drive Nature so far out of town,

that a half-day is too short to reach her domains. ... Let our friend the mechanic, who has for five days and a half laboured in hope that during the latter half of the sixth day he shall run down to Wimbledon to collect a few plants or insects, hear the birds sing, scramble amongst the furze, and feel the cool fresh breeze blowing the smoke out of his hair, pause awhile, and picture to himself a future. In that day there shall be no more furze, or heather, or buttercups and daisies; the bluebell and the fern must give place to asters and chrysanthemums, and the furze be uprooted that the laurel and *Aucuba* may stand in its place. ... Against the parks of our metropolis, whether patrician or plebeian, we have nought to urge; but against the appropriation of all the waste places within range of a Cockney stroll we dare to protest, because it tends to deprive those of legitimate hunting grounds, who have no other 'preserves,' and whose 'little game' are birds, beetles, butterflies, and flowers, ... the consummation of such appropriation will be–not only to deprive the operative classes of a patrimony in which they can stand erect, and feel themselves 'at home' and at ease, but to check them in the enjoyment of rational amusement whilst in the pursuit of some branch of natural history for which no other and compensatory provision is made.

In June Cooke's mood changes and he describes in a paeon of praise the beauty awaiting discovery by the owner of a microscope:

Everywhere treasures are being laid open for the microscopist, and may be had for the trouble of stooping and picking up. In the woodland the early flowers are shedding their pollen, or their leaves are spotted with unmistakable evidences of the development of microscopic fungi. On the anemone, the pilewort, the sheep's sorrel, and the violet, the cluster-cups are bursting through the cuticle, and expanding their fringed lips. ... Mosses display their fringed peristomes and liverworts are green and glorious. Every drop of stagnant water quivers with life. ... Along the sea-shore, zoophytes and sea-weeds, in variety sufficient to employ the microscope for a month, may be gathered in a day. Moths and butterflies yield their curious antennae and multiform scales; flies offer their eyes, tongues and toes ... and all the air, the earth, and the ocean are peopled with myriad forms, which only await the enquiring spirit and the earnest will to be converted into an inexhaustible source of pleasure and instruction.

By October he has returned to a favourite topic which had been much on his mind while he was formulating the activities of the Society of Amateur Botanists–'Common Things'–urging his readers not to collect simply for the sake of collecting with no more ambitious end in view, especially in the case of rarities, but to study in depth the common objects which surround them in their every-day lives.

As well as all this, Cooke answered many of his readers queries himself, and wrote articles long and short for the magazine on a wide variety of topics. They include 'A tit in moustaches' (the bearded tit); 'Under a palm tree!', a brief summary of the palm family; 'Insect moulds'; 'Toadstools'; and 'Homes without hands', on the cases built for their own protection by certain Indian insects, an article doubtless based on specimens sent to the India Museum. The periodical very soon began to attract other regular contributors, as well as occasional articles by many naturalists, including Worthington Smith and the young Ray Lankester. Cooke claims that 'at the time of its prosperity the manager told me that the advertisements alone gave a surplus of sixty pounds per month'.

> In the year 1865 a small and unpretending little work was issued under the title of *Science Gossip; Easy Guide to the Study of British Hepaticae*, which included brief descriptions of all the species known to date, illustrated by woodcuts. This was little more than an illustrated catalogue, but was the only complete work published down to the present [13].

This is how, in 1895, Cooke described his useful small book on the Hepaticae, or Liverworts, issued as a supplement to the August number of *Science Gossip*. The book included 136 species of these damp-loving, flowerless plants, closely related to the mosses, and was illustrated by 200 woodcuts made by himself; he was surely justified in claiming that, at the price of fourpence, 'this must have been the cheapest and most remarkable botanical manual ever issued.'

In his autobiographical notes Cooke claims that *Science Gossip* 'had many imitators and followers, besides "Nature" '. *Nature* began publication in 1869, and even if it was not a deliberate imitation of *Science Gossip*, the latter could have sowed the seed which finally germinated into *Nature*, for one of that journal's recent Editors has written:

> To begin with, the journal was a gossip sheet ... It was almost an accident that *Nature* stumbled into being a learned journal as well as a weekly vehicle for news and comment. After decades during which readers would write in with observations scarcely more profound than that they had heard the first cuckoo in spring, they began to use the journal for saying that they had found an electron, or a neutron...[14].

However, this interpretation of the aims and intentions of *Nature* has been questioned and it has been suggested that at least one of them

> was to mediate between increasingly diverse and sometimes antagonistic segments of late Victorian society–between scientists and artists, between professional scientists and interested amateurs, between scientific generalists and specialists, and between specialists in different fields [15].

Science Gossip was aimed specifically at an audience of amateurs and could hardly be said to be 'mediating' between any groups, although the services of professionals were freely given and a few even used its columns for original publications.

The emphasis placed by *Science Gossip* on the pleasures of microscopy ensured that, in Cooke's words:

> ... there was a very large number of patrons and contributors from the ranks of what we called at the time Amateur Microscopists, and there was always a long list of exchanges of microscopical objects, so that *Science Gossip* became the popular magazine of the microscopist.

Speaking 34 years after the events he was describing he contined:

> At this time Mr.Thomas Ketteringham had two hobbies which he pursued simultaneously; the one was his microscope, for which he was an enthusiast, and the other was his violin. To these he devoted the whole of his energies, when he had finished his daily labours in Piccadilly. I was not long in discovering that Ketteringham had a 'chum' who was also microscopical and musical, and with whom he was accustomed to spend an occasional evening, primarily with the microscope. I was soon introduced to this congenial companion, whose name was W. M. Bywater, and who was the manager of a select business house* in Hanover Square. Of course we became intimates, and it was a recognised institution that we three should meet about once a week, after eight o'clock, at Hanover Square, and go through a regular programme, which never varied. For an hour, or an hour and a half, in the counting-room at Hanover Square, we examined our objects of interest under the microscope, each bringing with him anything of special interest, which we discussed, and we discussed nothing else for the time being.
>
> At ten o'clock the microscopes were put away, and then we discussed sandwiches, cigars, and liquid refreshment, and many things besides, until eleven o'clock. Amongst other things to talk about was 'Science Gossip'.
>
> It was on the 1st of May, 1865, that a proposal from Mr. W. Gibson was published on page 116 of 'Science Gossip' to the following effect:—'It appears to me that some association amongst the amateur microscopists of London is desirable, which shall afford greater facilities for the communication of ideas and the resolution of difficulties than the present Society† affords, and which, whilst in no respect hostile to the latter, shall give amateurs the opportunity of assisting each other as members of an amateur society, with less pretensions; holding monthly meetings in some central locality, at an annual charge sufficient to cover the incidental expenses–say five shillings a year–on the plan of the Society of Amateur Botanists. By the publication of this letter the general feeling of the parties interested will be ascertained, and by this future action determined.–W. Gibson.' To which the following editorial note was appended:—'N.B. We insert our correspondent's communication, and would be glad to hear from any microscopists desirious of co-operating with him in carrying out his proposition.'–Ed. 'Science Gossip.'
>
> Of course this letter occupied the attention of the trio at their next meeting, and all of us resolved to make a strong effort for the establishment of such a society as that proposed.

*a saddlery business
†The Microscopical Society

> The reasons which appealed to us most strongly at this time were, that there were a large number of young microscopists arising in the Metropolis, who could not afford to pay the subscription to the old Society, and, if they could do so, they would not get what they most required–that is to say, sympathy and encouragement as well as practical help in their studies [2].

In fact Cooke, despite a lifetime spent in an occupation for which the microscope was his chief tool, never himself joined the Microscopical Society, a prestigious organisation founded in 1839 and soon to receive a Royal Charter. It is interesting to speculate on his reasons. The annual subscription of £1 was double that finally agreed on for the new Society, and might have influenced the ever impecunious Cooke, but it could hardly have been the whole reason. It is more likely that his feelings of social inferiority were involved, for though the membership of the older Society was very mixed, some were distinguished and most would have had considerably greater educational advantages than he. Also:

> It was frankly acknowledged that at this time the old Society was too exclusive, and self-contained, to interest itself in students, and was incapable of training the rising generation of microscopists. The feeling of monopoly had to be broken down.

Maybe Cooke was attributing to students a diffidence that was really his own. On the other hand his microscopical techniques were avowedly the simplest he could devise to produce the routine results which were all he required (p. 167), and it has been suggested that with this outlook he would hardly have been welcome in the Microscopical Society [16].

The three friends took their idea to that tireless protagonist of the cause of natural history, Robert Hardwicke, and once again, as with *Science Gossip*, it was through his encouragement and active help that the scheme came to fruition.

> Although he was not a microscopist himself he volunteered to do all that he could to assist in its realisation, by placing his office at disposal for preliminary meetings, and allowing it to be used as a provisional office for correspondence. On July 1st, it was announced in 'Science Gossip' that at a preliminary meeting, which was held on June 14th, at Mr. Hardwicke's office, it was decided to hold a general meeting for organisation at the St. Martin's National Schools, Charing Cross, on July 7th, for which purpose a provisional Committee had been appointed, consisting of the three persons already mentioned, with Mr. Hardwicke and his friend Mr. S. Highley, together with the originator of the proposal, Mr. W. Gibson.

Nothing is known of Mr. Gibson, but Samuel Highley was, like Hardwicke, a successful biological publisher. His membership of the new Society seems to have lapsed in 1872, and ten years later, following a series of financial calamities, he was to write to the Director of Kew Gardens listing his contributions to biological publishing and begging financial help to pay the bailiffs.

Cooke continued:

> Two subjects occupied most of the discussion at the July meeting–viz., the name of the Club, and the terms of subscription. Mr. Richard Beck* took an animated part in the discussions, and greatly assisted the Committee by his sympathetic attitude, aided by his experience in practical microscopical work, and his appreciation of all efforts on behalf of struggling students. I am not sure my memory is accurate, but I think it was Mr. Beck who proposed the name of the 'Quekett Microscopical Club,' and that was ultimately unanimously adopted. It was urged that the word 'Club' would better express its aims and objects than that of 'Society.'

Dr. John Thomas Quekett, after whom the new Club was named, was a brilliant microscopist who had died five years before at the early age of 45. Not only was he a renowned histologist and Professor of Histology at the Hunterian Museum, but he was a founder member of the Microscopical Society and had been its Secretary for 19 years and its President for a short while before he died. One of his books, *A Practical Treatise on the Use of the Microscope*, was widely known and would have had instant appeal to members of the Club.

The second matter discussed was the annual subscription, some advocating 5s and others 10s. The final vote in favour of the higher sum was in marked contrast to the 2s 6d subscription chosen for the Society of Amateur Botanists, and doubtless reflected the overwhelmingly middle class origins of those present. It was agreed that meetings would be held at 32, Sackville Street, very close to Hardwicke's premises, on the fourth Friday of the month, and for some years Hardwicke allowed the Club to use his address for correspondence. Finally, a provisional committee was chosen, consisting of Cooke as Chairman until such time as a President should be elected, Hardwicke as Treasurer, and Bywater as Acting Secretary. The Committee met for the first time the following month, and in September it got down to business, resolving to purchase a brass plate for the doorpost of Hardwicke's premises (which plate is still in the possession of the Club) and a ballot box. In November the need for a tin box for books was recognised, and in January 1866, a Chairman's hammer. Cooke's story continued:

> The bye-laws and regulations were drawn up by myself and Bywater, and afterwards considered and adopted by the members. I was requested also to draw up the prospectus or programme, which was endorsed by the conclave of three and afterwards officially adopted.

*of Smith & Beck, manufacturers of optical instruments.

The first and original members of the Club were twelve, who have since been, rather irreverently, styled the 'Twelve Elders,' and their names–

W. M. Bywater,	S. Highley,
M. C. Cooke,	E. Jaques,
W. Gisbon,	T. Ketteringham,
E. R. Godley,	E. Marks,
H. F. Hailes,	W. W. Reeves,
R. Hardwicke,	G. W. Ruffle.

Of these original members we hear no more of Gibson, although he remained a member for 18 years; or of Godley, whose membership had lapsed by 1868. Reeves and Jaques were members of the Amateur Botanists, who would play a prominent part in arranging excursions and social events for the Quekett, and Marks was in some way associated with Hardwicke's firm and would help with the Club's journal in the editor's absence. Only Hailes had no previous connection with Hardwicke's circle but, having attended the first meeting, remained active for a quarter of a century. Cooke paid tribute to the remarkable way in which Hardwicke inspired this group of enterprising amateurs, and also had high praise for Bywater:

> It would scarcely have been possible to have found a more thorough business man, of a more amiable disposition, prompt, punctual, methodical and unassuming, than its genial first Secretary, to whom the early success of the Club must be largely attributed.

In addition, Bywater was a skilled photographer and an excellent botanist, who would act as referee at the Club's excursions of which he was one of the most regular participants. As Assistant Secretary to the Microscopical Society, he also provided a useful link between the two organisations. Cooke's account continued:

> I was requested to allow myself to be nominated as President for the first year, but this I at once declined to do, in the event of the Committee being able to secure the services of Dr. Edwin Lankester, or some one of equal repute or influence, and this decision I have never regretted. I contended that a President who was already a Fellow of the Microscopical Society would bear evidence that the Club was *not* an opposition Society and that, with Dr. Lankester as President, the Club would at least be held in respect by kindred Societies, whereas I was myself a comparatively unknown man except within a limited circle. Fortunately, the friendship and influence of Mr. Hardwicke, added to my own request secured the services of the man whom we all regarded as the best man, under all the circumstances, to preside over the Q.M.C. during the first year of its existence.

In fact, Lankester was not only a Fellow of the Microscopical Society but the Editor of its *Quarterly Journal,* and his small book, *Half Hours with the Mic-*

roscope, was so popular that it eventually ran to 19 editions. Cooke was elected Vice-President along with Peter Le Neve Foster (1829–1879), the Secretary of the Society of Arts, who would probably have been recruited through Lankester or one of the more exalted members of Hardwicke's circle. Foster served the Club well, becoming President a few years later and remaining a member until 1876.

One other matter exercised the collective mind of the Quekett, that of members with a commercial interest in either optical instruments or biological publishing. Beck suggested that such members should not be eligible for a seat on the Committee, but Highley was much opposed to this restriction. Beck carried the day, though clearly Hardwicke, as one of the father figures of the Club, was excepted from the unwritten rule; but in later years the restriction was allowed to lapse.

Membership grew so fast that in a few months the Club had outgrown its meeting place in Sackville St., and after some negotiation through a friendly professor, permission was obtained from University College for meetings to be held in that institution's spacious Library (Fig. 7.7). The move to the new rooms was made at the eighth meeting and the arrangement continued until 1899. By the end of the first year there were 166 members, of whom two were Fellows of the Royal Society, one a clergyman, nine were medically qualified and eleven held other professional qualifications.

How did the Society of Amateur Botanists fare against such competition? Cooke had supplied its main drive and motive power, and his attention was

Fig. 7.7   General Library of University College, London, first meeting place of the Quekett Microscopical Club, seen early this century.

now fully engaged by the Quekett. Bywater, Jaques, Ketteringham and Ruffle, microscopists all, also transferred their allegiance, as did two other less active members. The young botanists Trimen, Dyer and Britten disappeared into the professional world, and Smith, who at this date was no microscopist, would soon join another field club. Cooke's old students would be unable to afford either their own microscopes or the quadrupled subscription of the Quekett, but their continued activity as Amateur Botanists would depend on the willingness of some of their number to run the Society now that the old leaders had defected. However Grugeon, the obvious candidate for leadership, seems to have been a teacher and organiser rather than a leader, so by 1866 the Society of Amateur Botanists, the undoubted progenitor of the Quekett, had died by the hand of its own offspring.

Chapter Eight

The Quekett Microscopical Club. 1865-1880

The first ordinary meeting of the Club was held on 25 August, 1865. Dr Edwin Lankester was the first President, and he took the chair. In his Inaugural Address he sketched the growth of microscopical enquiry and the gradual development of the instrument from the simple lens. He recommended the members to take up special subjects of study, rather than waste their energies on too wide a field. The microscope, he observed, was a working instrument, by the use of which the boundaries of science may be greatly extended.

The mantle of the Amateur Botanists had indeed fallen upon the Queketters, for Lankester's advice to members was exactly that given by Cooke to the Botanists three years earlier.

Cooke had the honour of reading the Club's first paper, on 'Work for the microscope'; then at successive meetings Richard Beck spoke on 'The spiracles of insects', Cooke on 'Microscopical moulds', and Cooke again on 'The application of the microscope to the discrimination of vegetable fibres'. This talk is reported in detail in *Science Gossip* and illustrated with engravings of the microscopic appearance of the fibres. Doubtless the subject came to his notice through his work at the India Museum, but the breadth of his interest in the natural world is again demonstrated by the fact that he was not only prepared to marshal the information into a paper suitable for presentation to the Club, but also to prevail upon the Committee to set up a Fibre Sub-Committee with himself as Chairman to report on the subject in detail. Only one other member of this sub-committee is known to us, the dermatologist,

Dr. W. T. Fox, and though the subject is quite extraneous to his other interests, he is said to have worked hard on it. However, there is no record that a report was ever submitted.

Right from the start a high priority was given to the Club's avowed object of helping beginners. As the first Annual Report said:

> The subject of Class instruction has been tested with the greatest success. Through the kindness of our Vice-President, Mr. P. Le Neve Foster, a room at the Society of Arts was placed at the service of a class formed under the direction of Mr. Suffolk, who has generously given much time and patience to impart to the members of it a thorough grounding in those important and fundamental principles necessary to working with the microscope, and there is little doubt that a second class, which that gentleman has signified his willingness to undertake, and for which there have been numerous applications, will be equally successful.

Suffolk's classes were, in fact, repeated on a number of occasions, and others were given too, including some on fungi by Cooke, 'illustrated by at least one hundred diagrams which were prepared especially', and on mosses by the noted bryologist, Robert Braithwaite. Study was further facilitated by the setting up of a library and slide collection based on donations from members.

The other important activity to which the Quekett Club was committed was the organisation of field excursions. Despite the example of the Amateur Botanists such excursions still seem not to have caught on in London, for the Club's Annual Report refers only to those 'long established in the North of England' as examples. It was hoped that not only would such excursions enable members to observe organisms in their natural surroundings before collecting them for study under their microscopes, but that the informal atmosphere of the events would encourage freer discussion among members. To this end, the first specialist group set up by the Club in that first year of its existence was the Excursion Committee consisting, by virtue of their previous experience with the Amateur Botanists, of Messrs. Cooke and Reeves. A modest start was made with two excursions during the summer, one to Hampstead Heath, and the other to one of Cooke's favourite collecting grounds, Darenth Wood, in Kent.

At the Annual General Meeting in July, 1866, a highly successful first year was reported. Lankester stepped down as President but remained on the Committee, Dr. Fox became one of the four Vice-Presidents, and Cooke, though he disappeared from the Committee, was not idle on the Club's behalf. On Jan. 25, 1867, he wrote to the Committee suggesting that the Club should publish either its Communications or a quarterly journal, for which he offered to make the arrangements. At the next meeting on Feb. 8, under the chairmanship of Dr. Fox, two letters were read; one from Hardwicke, who had absented himself from the meeting because his presence would have been 'improper', and the other from Cooke. Cooke offered a synopsis of the journal he had proposed in his previous letter, while Hardwicke made a counter-offer, to publish the abstracts of the Club's transactions in *Science Gossip* until such time as the Quekett could afford its own journal, which he would be prepared to publish. The discussion which followed was lengthy and apparently heated, for though Cooke's proposal was eventually accepted, Fox suggested postponing further consideration of the matter for six months, presumably to allow a cooling-off period.

However, this was not to be, for on February 15 a long memo with 18 clauses was received from Cooke proposing that he himself should edit the new journal, which would be published by Messrs Wheldon of Paternoster Row. The Committee clearly felt that Cooke's terms would allow the Club too little control over its own publication, and inserted clauses allowing it to approve papers before submission and to devote up to 40 pages to the Club's transactions. This provoked a furious response from Cooke:

> ... As one who claims to have been, perhaps more than anyone else, the founder of the Club, I have as deep an interest in it as any one of its Committee or Officers, hence it is rather uncourteous and extremely distrustful to demand what no proprietor of any existing journal would accede to, as if there were not moral responsibilities which bind me to carry out proposals on your minute book, as far as lay in my power, without appealing to the arm of the law ...
>
> Although I regret that so much trouble has been occasioned to the gentlemen of the Committee, I shall have no reason to regret altogether renouncing the design and publishing a Journal which has already been the occasion of misrepresentation and mistrust ... If the Committee feel themselves bound to press their demands, I must beg of them permission to withdraw my proposal.'

Here the Committee's minutes are terse: 'It was resolved that a Special Committee be called to consider this communication.'

On February 27, with Fox again in the chair, the Committee resolved unanimously that it could not accede to Cooke's proposals, and there the matter rested for a month, when a letter of apology was received from Cooke blaming a misunderstanding on both sides. No more is heard of the matter until September, when the Sub-Committee reported. Its suggestions were followed, with the result that the first number of the *Journal of the Quekett Microscopical Club* appeared in January, 1868, published by Hardwicke and with a Mr. W. Hislop as Editor. During the whole of the negotiations over the Quekett's *Journal* Cooke was still editing *Science Gossip*, so must have been in almost daily contact with Hardwicke. Unless some quarrel had taken place, it is difficult to understand why Cooke had passed over his old friend and benefactor in his original proposal for a publisher for the new *Journal*. The first paper was by Cooke and was the text of one he had read to the Club on March 3, 1866, 'On Universal Microscopic Admeasurement', a topic of vital importance to members. It will be further quoted elsewhere (p. 169), but here it can be said that Cooke deserves the gratitude of all British microscopists for insisting, at this early date, on the use of metric measurements in this country as on the Continent.

Meanwhile, back in 1867, the Club had held its first Soirée at University College. These Soirées, which were held annually for ten years, were extremely popular events and were the only one of the Club's activities to which the ladies were invited. They must have been considerable social occasions, as well as offering the opportunity to all members to show off their skills, and to the trade to display the latest instruments and gadgets. At the height of their popularity 1400 people attended, and the Club's minutes record that so many demonstrations were offered that it had proved impossible to obtain a complete list. The Soirées were discontinued, not because of lack of support, but because they became too much of a drain on the Club's funds.

Simultaneously with editing *Science Gossip* and with his active participation in the life of the Quekett Club, Cooke had been carrying on with his mycological activities with ever increasing intensity (Chap. 9), but nevertheless he still found time to bring out *A Fern Book for Everybody* in 1867. In writing about fungi and liverworts for the lay public he had been breaking new ground; there had been no popular presentation of these groups before, and his books sold well. But with the *Fern Book* it was different; indeed it clearly represents an attempt, either by Cooke himself or by his publisher, Warne's, to take advantage of the huge, and apparently assured, market which had been opened up by the Victorians' extraordinary craze for these plants. Hardwicke had already published a 'Plain and Easy' book on ferns, written by Mrs. Lankester, which is presumably why Cooke moved to Warne's.

The fern craze had begun back in the 1830s [1] when Nathaniel Bagshaw Ward, a doctor in one of the grimier parts of London and a devoted amateur naturalist, showed that plants could be grown for almost indefinite periods in closed glass cases, if they were supplied with moisture and light. The containers became known as Wardian cases and were the progenitors of the modern botttle garden. They were, of course, ideal for growing fastidious exotic and water-loving plants, including many species of ferns. Botanists, later joined by horticulturalists, finding they now had a simple and reliable way of keeping ferns in cultivation, set about hunting for them in earnest, especially in their damp hideouts in western Britain. Books were published about them, and the craze was on among the wealthier and more leisured classes. After a while so many wild plants had been torn up by the roots that there was a real danger of extinction of the rarer species.

But the nineteenth century was the heyday of the natural history society and field club, and these organisations were by no means the prerogative of the rich and learned. With them, all classes flocked into the countryside on Sundays and holidays to learn from nature, and the most popular way of doing this was to collect and press the plants found. By the 1850s the field clubs had followed in the wake of the professionals, and members had taken to the ferns in a big way, encouraged by an ever increasing flow of popular books on the subject put out by publishers hoping to cash in on the craze. The peak years for the publication of fern books were 1865 to 1867; by 1869 the excitement was virtually over. Cooke's little book, published in the last of the peak years, and containing chapters on the structure and classification of ferns, their cultivation indoors and out, and descriptions of the British species illustrated in colour by the author himself (Fig. 8.1), was not out of the ordinary, never ran to a second edition, and was probably of little financial profit to him.

In April of the year in which the *Fern Book* was published there appeared in *Science Gossip* and unsigned article, illustrated by woodcuts, on 'Hardy foreign ferns'. Now the *Fern Book* itself ends with an annotated list of 'Exotic Ferns' which includes those mentioned in the article together with others which would only flourish in a stove (greenhouse). As in a few places the wording in both book and article is identical, there can be little doubt that Cooke was the author of the latter and that it was probably originally intended for inclusion in the book.

At this time Cooke was indulging in yet another interest–entomology–of which the only forewarning we have is a remark in one of his popular books [2]: 'When we were boys and insect hunters . . .'. In 1867 he put together an 8-volume, leather-bound scrapbook on mites entitled 'Monographia Acarinae', together with a ninth volume on 'Arachnidae. Chelifers', (spiders) all of which are now in the library of the Quekett Microscopical Club. Each

Fig. 8.1 Illustration from *A Fern Book for Everybody*. Original in colour.

volume consists of text and illustrations cut from published works and pasted on to the blank sheets of the scrapbooks, the text, frequently in German, on the left-hand page and the illustration on the right. Each page is headed in Cooke's hand-writing with the name of the family to which the insect belongs. In Volume 3 he has listed the publications he used, almost all of them German, and Volume 8 contains a hand-written index to all the preceeding volumes. How his attention turned to mites, and what the object of the scrapbooks was it is difficult to know, unless it was a basis for the eventual publication of a text-book. But such a work would have been of no use in the field, as it included large numbers of exotic species, and there is no record of the Quekett Club ever having had a special interest in mites.

With spiders it is different; there is evidence that Cooke had for a time a quite considerable interest in this group. Not only are the pages of Volume 9 of the scrapbook quite copiously annotated in his own hand-writing, but at a meeting of the Quekett Club in February 1871, four years after the completion of the scrapbook, he reported that he had been out collecting spiders with a friend, had found them fascinating, and would commend them to members as a subject for study. Nearly all the material, both text and figures, was taken from Koch's *Deutschlands Crustaceen, Myriopoda und Arachniden*, which was published between 1835 and 1844 in the form of 40 packets of loose sheets, one sheet per species, which the purchaser could bind according to his own requirements. Cooke obviously intended his scrapbook to be as complete as possible, for not only had he filled in missing species with material from other authors, but he left a number of gaps for those for which he could not at the time obtain suitable illustrations.

His only considerable published work on insects appeared in the *Pharmaceutical Journal* of 1872, in the middle of his series of articles on Indian medicinal plants. It is a most surprising run of 17 articles on "Vesicating Insects", in which the author describes in detail beetles of the genera Mylabridae, Cantharidae and Meloïdae from all over the world, sometimes with drawings. All these Blister Beetles produce the drug cantharidin, which raises blisters when applied to the skin in very minute quantities and was much used in medicine at the time as a so-called counter-irritant in diseases such as rheumatism. The descriptions are so detailed that Cooke must have obtained them from an authoritative source, though he does not acknowledge one.

At the Quekett Club during the year 1866–1867, Cooke read four short papers on microscopical technique, and Dr. Fox gave an account of the subject he had made so much his own, 'Human vegetable parasites', by which he meant the fungi pathogenic for man. Though much work had been done on ringworm and thrush on the Continent, the aetiology of these diseases had been almost entirely neglected in England, and Fox was the first Englishman to study them seriously. He wrote extensively about them and their causal fungi, adding his own original observations to distillations of translations by Edwin Lankester of authoritative German sources [3]. It is strange that though he and Cooke must have known each other well through the Quekett Club, and especially through its Fibre Committee, neither mentions the other in any of their publications, Fox obtaining all his mycological advice from Cooke's mentor the Rev. M. J. Berkeley.

By the time of the second Annual General Meeting, held in July, 1867, the Quekett Club was truly flourishing. A Librarian, Mr. Jaques, a Curator, Mr. Ruffle, and an exchange of Slides Committee had been appointed, and another new post, that of Honorary Secretary for Foreign Correspondence, had been filled by Cooke. The Committee had taken the precaution, when creating the post at its June meeting, of emphasising that the holder would *not* (Committee's italics) be an *ex officio* Committee member, which was perhaps just as well, for Cooke held the post for 14 years. Most of the foreign correspondence emanated from the U.S.A. and originated in Cooke's mycological contacts there as well as elsewhere in the world. One of the first and most regular of the letter-writers was the Rev. E. C. Bolles, Vice-President and Record Secretary of the Portland Society for Natural History, Maine (p. 173) who was to do much to facilitate Cooke's contacts with American mycologists. Occasionally the correspondents, while visiting England, would attend a meeting of the Club.

One of Cooke's more unexpected contributions to the transactions of the Quekett was a lengthy paper on 'The hairs of Indian bats' [4]. The subject apparently suggested itself to him after he had come across a description of such a hair by Quekett himself, the identity of which had been questioned by other workers. Cooke, having easy access to Indian artefacts, examined and drew the hairs of 60 species of bat and, while finding none identical with Quekett's, saw no reason to doubt that the latter's came from a related species. A version of this paper also appeared in *Science Gossip*.

The year 1868 saw field excursions firmly established as an important part of the Club's activities. Eleven were undertaken that summer, and reports of the finds and of the localities where they were collected were printed in the *Journal*. The June outing, to Leatherhead, was followed for the first time by an Excursionist's Dinner at the Swan Hotel, an occasion which proved such a success that it became an annual event. The Swan*, in Leatherhead's main street, was a seventeenth century coaching inn (Fig. 8.2) which was still functioning as such in 1905, despite the fact that the railway had come to the town in the 1850s. In the 1860s, however, the inn was famous for more than its 'post-horses and carriages; superior lock-up coach houses': a handbill issued by its proprietor, William Moore, proclaimed, under an engraving of the building,

> The situation of this beautiful village is too well known to require much comment, as it is presumed to stand unrivalled for its views and delightful rural retirement. . . . To the

---

*Demolished in 1938 to make way for a parade of shops.

admirer of Nature, geologist, or disciples of Isaac Walton, Letherhead [sic] and its immediate neighbourhood hold out many inducements, and the quick transit which may be made by Railway to the Metropolis, must render it most desirable . . .

An excellently supplied Larder.
Choice Wines and Piquant Liqueurs.'

An account written about ten years later confirms this:

It would be too provoking perhaps, if I were to enumerate all I can recall of the stacks of varied eatables and drinkables which were heaped upon the tables . . .'

Obviously the Excursion Committee knew exactly what it was doing when it picked on the Swan for its first Dinner, and it is easy to imagine the gentlemen, after a good day in the field, and untrammelled by the presence of the fairer sex, relaxing under the influence of the Excellent Larder and Choice Wines until the evening was drawn to a close with one of those sing-songs so beloved of the Victorians, and in which Cooke, with his good tenor voice and propensity for writing verse, would play an important part. A collection of the songs he wrote for these dinners was published in 1878 by Keating & Co.,

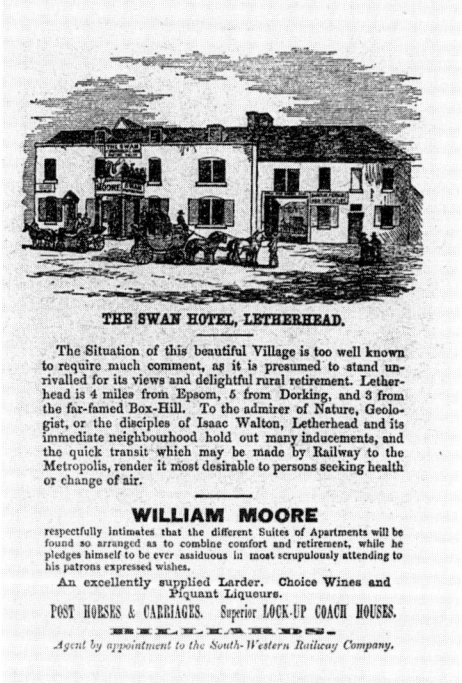

Fig. 8.2   The Swan Hotel, Leatherhead.
An advertisement of about 1860.

Steam Printers [5]. Nearly all have at least a reference to the pleasures of alcoholic refreshment of which Cooke was clearly, by now, a devotee. He had come a long way from his teetotal upbringing and his days in the Temperance Band in Norwich, but he has left no record of the circumstances which led to this radical change in his convictions.

The first poem, 'The History of a Dinner', (to be sung to the tune of 'Guy Fawkes') tells of the origin of these events:

> Four and twenty Queketters
> Went out to catch a beetle
> They tried their best, without success,
> The Cara-bides to wheedle.
> Says GAY* to REEVES, I think the game
> Might be a little quicker;
> I feel an aching void within,
> Suppose we 'do a liquor.'
>
> . . . . . .
>
> Now while these sages on a stump
> In conclave close were sitting,
> The rest of the excursionists
> Around a bog were flitting.
>
> . . . . . .
>
> Oh! tell it not in Gower Street,
> Nor hint it at 'the College'
> They found at length a wayside 'pub'
> To some, a tree of knowledge.
> Where RUFFLE, HAINWORTH, JAQUES & WHITE
> Sucked wisdom from the 'pewter',
> and gallons followed gallons, as
> The suited follows suitor.
>
> . . . . . .
>
> The hours spend on, and one by one,
> The 'Swan' beneath her pinions,
> Collected all *sound* Queketters
> Within the Queen's dominions.

---

*F. W. Gay, a Fellow of the Microscopical Society, had joined Reeves on the Excursion Committee.

> The dinner came, and dinner went,
> With sherry all were sated,
> Each subject was injected well
> And well illuminated;
> . . . . . .
> And when the Editor* should go,
> And all the rest were ready,
> Mixed with the wine he drank his notes,
> And both were rather heady.

A few years later, in a prelude to a longer verse saga, Cooke was in more lyrical mood:

> Down in a green and shady bed,
> Not very far from Leatherhead,
>   A modest violet grew;
> Its stalk was bent, it hung its head,
> And thither sixty Cockneys fled,
>   To take a private view.
>
> This lowly flower in secret bower,
>   Invisible did dwell;
> They found it out, 'tis past a doubt,
>   Discovered by the smell.
>
> They sniffed the odour with delight,
> They saw the sun set out of sight,
>   They saw the moonlight pale;
> They heard the 'Swan' of Leatherhead
> Flutter her pinions overhead, . . .

'The Days of Olden Time', a new song written for 1871, tells of changes wrought by then recent advances in biological science. (Tune–'A Fine Old English Gentleman').

> . . . . . .
> I'll sing of days when cholera and fever used to come,
> As 'providential visitors' to wicked peoples Homes;
> But now we're choked with sewerage and sanitary law,
> And suck in germs and molecules with every breath we draw.

---

*The 'History' was written for the Dinner of 1869, so this reference must be to Cooke himself (see below).

I'll sing of days when pretty girls were angels in our eyes,
And niggers scarcely dared to think they were brothers in disguise;
But now we're Darwinised to dream, that every smiling lass
Is first cousin to a monkey, or half sister to an ass.

I'll sing of days when chalk was *chalk*, and had no other name,
And blood a sanguinary fluid circling through our human frame;
But now folks talk of corpuscles, and leucocytes, and then
We're bored to death with *protoplasm* by very learned men.
. . . . . .

So the social life of the Quekett Club flourished, but its more serious objectives were by no means forgotten either by the Committee or by Cooke himself. During 1868–1869, when he was one of the Vice-Presidents, he gave two talks on microscopic fungi, and dedicated a small book on microscopy to members. The vogue for this hobby was not, of course, confined to the Quekett–it was being taken up in many circles where there was sufficient affluence for the purchase of a microscope. The huge sales of Lankester's *Half Hours with the Microscope* were sufficient measure of the popularity of the hobby. Cooke too decided to cash in on the market, and his *One Thousand Objects for the Microscope*, dedicated 'To the President and Members of the Quekett Club', appeared in 1869 in the same series as the *Fern Book for Everybody*. In his Dedication he says:

> I have endeavoured to produce 'a guide to the cabinet', which will be of service to the microscopist of smallest pretensions; and I claim to associate it with a Club eminently 'popular' in its constitution – to the establishment of which I had the honour of being chiefly instrumental–in the hope that it may aid in rendering the use of the microscope still more popular . . . If these pages should induce but a few readers to appeal to works of greater pretensions, my labour will not have been altogether in vain.

The objects were drawn from both plant and animal kingdoms, and were chosen to avoid the necessity for complex preparation such as section-cutting: his own or borrowed woodcuts illustrated 500 of them. Despite the competition, the book was approved by the public and ran to several large editions before the publisher, Frederick Warne, suggested that a few introductory chapters on the workings and use of the instrument might enhance its value. These were duly written for the last edition, which appeared in 1900.

At the Club, extra winter meetings were started to which members were invited to bring their microscopes and demonstrate and discuss their work informally, events which proved so popular that the following year they became part of the regular programme, alternating with the well-established monthly paper-reading meetings. The Committee's Annual Report describes them thus:

> ... Your Committee desire especially to call the attention of the members to the many pleasant advantages derivable from the meetings on the gossip nights; for on these occasions, difficulties in Microscopical manipulation and the various methods of mounting Microscopic Objects are communicated and discussed, and while an interchange of thoughts and plans is freely given, friendships are formed that are destined to be life-long.

These meetings became known as Gossip Nights, and are still a feature of the life of the Club today.

At about this time too it was realised that much useful scientific information which might have been obtained from the summer excursions had findings been pooled, was being lost because each member collected for himself alone. An Annual Report reads:

> These excursions afford opportunities for research and investigation such as few other Societies in London can supply. Your Committee hope to receive at the termination of this season more copious records of the objects collected and their habitats, as noted by members, than have hitherto been furnished. It is intended that the results of the excursions shall form a chapter in the *Journal*.

In January, 1869, after a year of peace in the Committee, there were once again problems with the *Journal*; Mr. Hislop suddenly, and without explanation, sent in his resignation from the editorial chair. This led to 'an animated discussion' in the Committee, and it was finally decided that the journal should, in future, consist solely of a record of the Club's transactions, and that the single editor should be replaced by an Editorial Committee consisting of the President, the Secretary, the Reporter (a newly created post) and the Secretary for Foreign Correspondence. Cooke was in business at last! In fact, it is clear that he was the *de facto* Editor, the others presumably being there to keep a restraining hand on him if necessary, for two years later the Committee thanks him for his services 'in conducting the *Journal*', and requests him to accept an honorarium of 20 guineas. It was noted with pride, at about this time, that the huge success of the Quekett Club had resulted in the founding of new societies modelled on it in Liverpool and Chicago, and in fact these proved to be the first of many imitators.

To the very varied papers given during the year Cooke contributed an illustrated talk on 'Microscopic moulds', and a Mr. Lowne discussed 'So-called spontaneous generation', coming to the conclusion that microscopic life could arise *de novo* by a process of 'aggregation', and that this in no way conflicted with Mr. Darwin's theories. In reply Mr. Cooke remarked that:

> ... as for agglutinated atoms forming fungus spores, he could only say that this was so extraordinary that he should be very glad to see the fungus spores so formed.

1870 saw the resignation of the Club's first Secretary, Witham Bywater, by common consent one of those who had contributed most to its success. A committee, including Cooke, was set up to 'make arrangements for a testimonial' and Bywater was presented with 'a service of plate consisting of a salver, silver coffee-pot, teapot, cream jug and sugar basin'.

> Mr. Bywater expressed his deep sense of obligation to the members for this valuable mark of their kindly feelings and esteem, and he hoped that he should be pardoned, if his words failed to express all he felt up on the occasion.

The highly successful year 1869–1870 had been marked by the Presidency of that wise man, Peter Le Neve Foster, one of the Club's first Vice-Presidents. But its very success, particularly in drawing in the professionals, must have been diverting it from its original purpose of providing a training ground for beginners. For Foster's Presidential Address was deliberately designed to pull the Club up in its tracks and force members to consider seriously which way they were going. After pointing out that the astonishnig growth in membership, though laudable, was but one mark of success, he continued: 'The Quekett is . . . a society of students', but–

> Many of the papers have quite sufficient integral merit to have been read at any society. They touch on the usual topics in the usual manner. Of real students' papers there are very few, and I would appeal urgently to students for more. . . .
>
> Again, few remember how important, how essential, are those two powers, to dissect and to draw; nothing can replace or compensate for them, and yet how many who profess and call themselves microscopists can do neither.

Then he goes on to suggest collaborative research as a means of widening the horizons of every member:

> There are many microscopists with the means at their disposal, and magnificent instruments, who have no leisure for collection, nor skill in drawing and dissection; there are young and active members in the Quekett with leisure and skill and small means . . . how valuable to us would be their combined labours and what a characteristic volume would our transactions become.

Again, some members could devote time to:

> . . . the abstraction and reduction of the vast mass of British and foreign literature bearing on our subject . . . the historical enumeration of previous labours often supplies a valuable stimulus to further investigations.

And finally, it would benefit everyone to make a thorough study of the optical sciences which are the foundation of all their work, and so enable themselves to attain optimal results from their studies.

The immediate result of this address was an appeal from an eminent member for more contributions to the paper-reading meetings, with the speaker emphasizing and re-emphasizing that *startling* results are not required, for the Club is not a scientific society: he asks for *short* papers on *any* subject, especially such as show the value of *systematic* work and are accompanied by *faithful* drawings.

Ever since 1868, *Land and Water* had been one of several periodicals presented regularly to the library of the Quekett Club, the donor being its founder and editor, Frank Buckland (1826–1880). Buckland was a quite extraordinary character [6], a naturalist and zoologist known to every nature lover in the country, of high or low estate, through his articles, books and eccentric behaviour. The son of the much respected geologist, William Buckland, Dean of Westminster, he started his career as an army surgeon, but from childhood he had a passion for animals, and it was to them that he devoted most of his life. Anyone interested in them, from the rat-catcher to the Superintendent of London Zoo, was his friend. He kept as pets every sort of animal, from a lion cub to fish and frogs, often carrying them about on his person, and he wrote copiously about them, in a highly entertaining style, in popular books and articles. He was also a zoophagist and, being greatly concerned about the country's food supplies, he considered it his duty to test the palatability of any species of animal he could lay his hands on, not excluding field-mice and young rats, which he found very tasty.

In 1863 he began to devote his attention to fish culture in all its aspects, and particularly to improving Britain's fisheries, to developing artificial methods of hatching fish, and to educating the public about the fishing industry, and a few years later he accepted the post of Her Majesty's Inspector of Salmon Fisheries. It must have been at about this time that Cooke came to know him personally. He mentions Buckland briefly and quite inconsequentially in his autobiographical notes, in the middle of describing the writing of *Our Reptiles*, saying simply:

> One of my most intimate acquaintances, and whom I often met for a gossip was Frank Buckland, who died lamented by all good naturalists in 1880.

It was as though the subject of reptiles reminded him of his friend, and considering Buckland's wide interests, this may well have been so.

Buckland had started his magazine, *Land and Water*, in 1866. In some ways it resembled *The Field*, but it was directed more towards the common man than the landed gentry, as was the latter. The founder himself edited the section on 'Practical Natural History', including in it accounts of his own exploits and a brisk correspondence column. An article by Cooke in 1871 on 'How to examine mushrooms' must have been solicited by Buckland who, in a previous number, had recommended Cooke's *British Fungi* to a reader. Buckland, with his encyclopaedic knowledge of natural history, found little difficulty in answering all his reader's queries himself and, soon after the appearance of Cooke's article, had identified as *Marasmius oreades* (the fairy ring champignon) a toadstool sent to him as being poisonous. However, the reader was not entirely happy at being told that it was, in fact, edible, so Buckland sent the specimen to Cooke (this species dries well so would not have decayed in transit) who not only confirmed the identification, but announced his intention of proving the point by making a meal of the specimen.

In 1871 Cooke saw his way to helping both the Quekett *Journal* and his brother Ebenezer. It will be remembered that when Ebenezer finished his apprenticeship in about 1862 he set up in business on his own as a lithographer, though he also continued to teach drawing in Ruskin's class at the Working Men's College. This must still have been his situation in 1871 when he moved house from Camden New Town back to Kenish Town, to 8, Falkland Road, only a little way round the corner from the house he had shared with Mordecai's family ten years earlier. Though the houses in Falkland Road were not large, and Ebenezer now had five children, he was once again sharing with another family, whereas Mordecai had recently moved to Grosvenor Villas which his family had to itself: apparently, despite the elder brother's perpetual shortage of money, the younger was even worse off. But Mordecai was now in a position to put work in his brother's way. The *Journal of the Quekett Microscopical Club*, which had been in existence for three years, was generous to its contributors in the matter of illustrations, so it was necessary that it should employ a lithographer. At first a number of firms seem to have been used indiscriminately, but between 1871 and 1880 Ebenezer Cooke was responsible for almost all the plates, and the Committee's minutes record regularly the payment to him of sums of between £5 and £9.

However, it was during this period that he began to realise that, of the two occupations in which he was engaged, lithography and teaching, it was the latter which was his true calling. As well as Ruskin, he was meeting other influential educationalists at the College and, inspired by them, was beginning to develop his own methods of teaching drawing and nature study to both chilren and adults, based on the Pestalozzian ideals which had so influenced him as a schoolboy. Gradually he was becoming known in progressive teaching circles, and as a result, in 1875 he obtained two part-time teaching posts, one at an experimental school in Caterham, and the other at Miss Buss's Camden High School for Girls. Though having less and less time for his lithography business, he must have continued to run it, presumably with paid employees, until at least 1880, when the Quekett paid him for the last time.

Towards the end of 1871 the already strained relations between Cooke and Hardwicke seem to have come to a head, though which of them precipitated the quarrel it is impossible to discover from the two brief and conflicting accounts which remain. In his autobiographical notes Cooke writes:

> I remained Editor [of *Science Gossip*] for six years, when the publishers discovered a cheaper editor who lived in the country and could answer all the correspondents himself and so, the change was made, and the journal gradually declined until it sank almost into oblivion.

It was, however, at this moment that Cooke was about to publish with Macmillan, rather than with Hardwicke, his first major mycological work, the *Handbook of British Fungi* (p. 175), and Worthington Smith, in his obituary of Cooke, writes:

> Mr. Hardwicke turned sulky, because, he said, Mr. Cooke ought to have allowed him to be the publisher. This dispute soon resulted in Cooke retiring from the editorship of the journal [*Science Gossip*] he himself had made [7].

There may well have been another reason for Cooke's departure from the editorial chair, for no sooner had he resigned than he started his own mycological journal, *Grevillea*, with Williams and Norgate as his publishers (p. 178). He must have been planning this for some time, and even he would have realised that he could hardly continue editing *Science Gossip*, the Quekett *Journal*, and his new periodical, while carrying on with his other multifarious activities. Whatever the true facts of the case, there is no doubt that *Science Gossip*, although it was its publisher's especial pride and joy–his 'pet journal', according to the magazine's obituary notice for him–declined in value after the change of editor, although it survived for another 30 years. Cooke never again published under Hardwicke's imprint.

Simultaneously with these events, trouble had once more arisen over the Quekett Club's *Journal*, though again it is impossible to discover the extent to which Cooke was personally involved. At the Committee meeting on November 24th, 1871, which Hardwicke did not attend, it was minuted that:

> ... the Journal Committee [of which, of course, Cooke was a member] consider it advisable that the Journal should for the future be published at some office where there is no other journal of a similar character.'

No reason is given, but this proposal bears an extraordinary resemblence to that put forward by Cooke in 1866, and eventually turned down by the Committee. On December 2nd Hardwicke, understandably, asked for the reasons for the decision, and the Committee postponed replying until the following month. On February 23rd, 1872, it was carried unanimously

> that in the absence of any complaints as to the conduct of the *Journal* and the general satisfaction which has been felt ... as to its conduct during the past year, no change to be made as regards the Publisher or Editor at present.

On April 26th Cooke's position as *de facto* Editor was regularised when the Committee appointed him as Editor at a salary of £20 a year.

But matters were only quiescent. A year later, on May 23rd, 1873, at a meeting which was not attended by Cooke, for he had just left the country for the Vienna International Exhibition, the Committee passed a motion that the *Journal* be reorganised; that the present version should be discontinued and a new one started under a publication Committee consisting of the President, the Secretary, and three members of the general Committee, which excluded Cooke unless he were specifically elected to that Committee. In August of that year he received his last payment as Editor, presumably in his absence as the Vienna Exhibition was still open. One wonders whether this prolonged absence was the reason for his loss of the editorship, or whether there had been more trouble between himself and the Committee. He himself may have suggested a change, or the Committee may, with some justification, have felt that in the circumstances he was unable to pull his weight. In any case, the reorganisation meant the final severing of Cooke's connection with Hardwicke's firm. Despite the new arrangements, misunderstandings and disagreements over the *Journal* continued, but Cooke, no longer on the Committee, was not involved.

By 1874 Cooke was well established as one of Britain's leading mycologists (Chap. 10), and not only this, but his pioneering work for field studies in London would also be widely known; with the result that from then on he was invited to the meetings of more and more field clubs all over the country as their expert on fungi. Many nationally known naturalists stemmed from his home county, Norfolk, where the tradition of natural history societies had first begun in the previous century, and it would have been strange if, during his frequent visits to his family (p. 163), Cooke did not sometimes take the opportunity to attend local club meetings and get to know their members. In 1870, in addition to the older societies, a new one based on Norwich had been started, the Norfolk and Norwich Naturalists (it had as a founder member a Mr. W. A. Cubitt of Neatishead) and it was this society which elected Cooke, a distinguished son of the county, to honorary membership in 1874. However, he never contributed to the Society's transactions, and does not seem to have been a very active member. In contrast, he had, for some years, been exchanging specimens with a well-known mycologist from King's Lynn, Dr. C. B. Plowright (1848–1910), who was Medical Officer of Health for the district for 32 years. A number of Plowright's collections appear in Cooke's *Exsiccati* (p. 161), and Cooke illustrated one of Plowright's articles on micro-fungi in the *Transactions of the Woolhope Club*.

In the same year that he was honoured by his home county, Cooke was made an honorary member of another provincial society on the other side of the country, the Woolhope Naturalists' Field Club. It took its name from a small village near Hereford and, unlike the Norfolk club, had been established as far back as 1851 'for the practical study in all its branches of the natural history of Herefordshire'. Worthington Smith had been a very active honorary member since 1869 and Cooke began to attend its field meetings perhaps a year or so later, continuing to do so for many years, for these meetings were of particular importance in the history of British mycology.

The membership of a Club based in a small town deep in the countryside of the Welsh borders of necessity differed greatly from that of a similar club in London. There was no pretence at working class membership here. 'About a third of its original members were clergymen and another third professional people from the county. It has a small number of non-local members' [7]. Excursions took place on week-days, for Saturdays were inconvenient for most members, and the clergy could not attend on Sundays. Travel arrangements to excursion sites were also very different from those in London. Only if the collecting area were some distance away and there happened to be a station nearby would excursionists go by rail. Usually it was not necessary to travel far in that beautiful countryside, and carriages would be hired to pick up the gentlemen at a fixed rendezvous, complete with their picnic baskets, and transport them to the chosen locality packed tightly inside and outside the vehicles, in their black suits and hard hats (Figs. 8.3, 13.2). The distance walked during one of these outings was usually about eight miles, and often a local squire would play host to the Woolhopeans during the day, while afterwards there might be a convivial meal at the local inn or hotel.

It is to one of the leading members of the Club, the local physician, Dr. Henry Graves Bull (*c.* 1818–1885) (Fig. 8.4), that the Woolhope owes its mycological importance. Bull was an outstanding naturalist who had different interests at different times, but towards the end of the 1860s he began to make a special study of the larger fungi. During a meeting in 1867 a paper was read on an edible toadstool and Bull, after making a few comments:

> remarked that as the fungi of the neighbourhood seemed excellent he thought the Woolhope Club should make them their special study . . . The following year the Royal Horticultural Society, at the instigation of the Rev. M. J. Berkeley offered prizes for the best collection of edible fungi. The Woolhope Club gained first prize and Mr. W. G. Smith the second. (JR)

That same autumn a notice was circulated to the effect that on Friday Oct. 9th there would be:

> AN EXTRA FIELD MEETING in the form of a 'FORAY AMONGST THE FUNGUSES'; that 'the carriages will leave the Green Dragon Hotel [Hereford] at 10 o'clock–to arrive back again a little before three o'clock, to examine the collection of Funguses.
>
> The dinner will take place at the Mitre Hotel, at 4 o'clock p.m., when several of the edible Funguses, cooked from the Club receipts, will be served. Tickets 4/- each [8]. (Fig. 8.5).

Fig. 8.3　En route to a Woolhope excursion. (From a cartoon by W. G. Smith, *Pictorial World*, 1877).

Fig. 8.4　Dr. Henry Graves Bull of the Woolhope Naturalists' Field Club.

The excursion was to a local park, Holme Lacey, by kind permission of its owner, and to a hill nearby. Twenty-one people sat down to dinner after this first fungus foray, and Smith was one of the referees for their finds. So began the forays which are still an integral part of the programme of mycological and natural history societies to-day, not only in Britain, but all over the world. Later, the Woolhope forays were extended to several days and mycologists from outside the county, of whom there were many, put up at the Green Dragon, where there might be as many as 60 gentlemen sitting down to the annual dinner. Years later Cooke described the care which Bull lavished on the organisation of the proceedings:

> Practically, the date was always the first week in October ... For at least six weeks beforehand the director was considering the most favourable localities, and personally going over the ground to see what were the prospects for the future. When the programme was determined, and during the week before the commencement of operations, the same indefatigable feet were treading the scene of the projected excursions, and deciding on the course to be pursued; ... Nothing but the weather was left to chance,
> ...
> Another feature must be alluded to, as contributing not a little to the success of a Foray, and that is the willing consent of the excursionists to submit to reasonable control ... It is well known that the most satisfactory results are obtained, not by scattering all

over the ground, but by unison in direction, within the limit of a whistle call. Who does not remember that whenever a tendency towards lagging, or scattering, was observed at a Woolhope excursion, the well known cry of 'For-ward'! brought all together again, so that, at any time, the whole party was within the range and under the control of the conductor [9].

Participants in a modern foray are less amenable to discipline!

At that time every British mycologist was a Woolhopean, and many Continental workers came over to join in the forays, which soon widened their scope to include all types of fungi, not only the agarics. Smith's services became quite indispensible, for not only was his knowledge of the larger fungi extensive, but it was usually he who was responsible for reporting the results of the foray at the Club's next meeting, and he also took it upon himself to immortalise the proceedings in a series of highly entertaining cartoons which decorated the menus for the Green Dragon dinners and were even reproduced in the national press. In recognition of his services to the Club [10], Dr. Bull organised the presentation to him, in the same year in which Cooke became a member, of a set of 48 items of silver, mostly forks and spoons, on the oaken case of which was inscribed: '... in pleasant memory of fungus forays, assisted by his experience, illustrated by his pencil, and chronicled by his pen.' The handle of each fork and spoon was engraved with a different fungus, all copied from his own drawings.

Cooke seems to have been in his element in the Woolhope Club. Here, for a few days at least, he could share the company of most of Britain's leading mycologists, making up a little for the desperate loneliness he felt in London, isolated for months on end from personal contact with his peers (p. 181). And again, out in the woods everyone was informal and relaxed; at the dinners wine flowed freely and all was friendly and jovial. So he could, for a while, forget his constant awareness of social inferiority and allow his natural geniality and sense

Woolhope Naturalists' Field Club.
1868.
President—DR. M'CULLOUGH.

AN EXTRA FIELD MEETING
WILL BE HELD AT HEREFORD, FOR A
Foray amongst the Funguses,
ON FRIDAY, OCTOBER 9.

Fig. 8.5 The announcement of the first fungus foray. (Reproduced in the *Transactions of the Woolhope Naturalists' Field Club*, 1877).

Fig. 8.6 The Woolhope Club setting out on a fungus foray in the rain. (From a cartoon by W. G. Smith, *Graphic*, 1875).

of fun free rein. While he did not neglect to read serious papers from time to time at the Club's meetings, he was also well known for his 'humourous postprandial addresses' at the foray dinners. Even Smith at his most sour allowed that Cooke was 'an excellent after dinner speaker, never at a loss for words.'

The Quekett Club continued to flourish after the dispute over its *Journal* had been settled, but on March 8th, 1875, both it and the natural history world suffered a grievous and irreparable loss; Robert Hardwicke died suddenly at the age of 52.

> While travelling up to London by train from his country house at Sydenham, he was siezed with an attack of paralysis. On arriving at the Victoria Station he was found speechless, and removed to his place of business, where he had the best medical attendance. Although he rallied to some extent, death ensued early on Monday morning [11].

From this description he seems to have died of a stroke. At the next meeting of the Quekett Club the President, Dr. Matthews, F.R.M.S., is quoted as having told members that Hardwicke:

> was one of the founders of their Society, and was always ready to promote its objects, and his great tact and talent in doing so would be acknowledged by all. As their Treasurer, he discharged his duties with uprightness, and in a manner which gave offence to none. As a member of the Committee he was never absent from his place except from unavoidable cause, and greatly would they miss from his accustomed seat his pleasant face and portly form. As their publisher, his services to the Society were very great, for in all times of difficulty his valuable technical experience was readily placed at their service, and his loss to them in this respect could hardly be over-rated. As a man of intellect his acquirements were far above the average, and it was ever his pleasure to collect around him men of science, by whom it might fairly be said he was held in high appreciation. As a publisher of scientific works he showed great discrimination, and to his judgement many valuable and popular books of this class owe their existence . . . He

had placed his office at all times at their [the Club's] disposal, received their letters ... his rooms were always at the service of the Society when special meetings or special business required. ... Several of the leading members of the Club, Mr. Bywater, Mr. Cooke, and the Secretary, together with himself, had attended Mr. Hardwicke's funeral, and upon no occasion could it be said that there gathered around a grave more real mourners.

It was his ability to draw round himself a lively circle of active and respected scientists that made Hardwicke unique as a publisher. Many another aspiring young naturalist must, like Cooke, have been given a start in his scientific career by being drawn into that circle and enjoying the opportunity of informal contact with established men that it made possible. Hardwicke's business was carried on by David Bogue, who, after publishing under the name 'Hardwicke and Bogue' for a few years, continued as 'Bogue' until 1882 when, like Hardwicke's publisher friend, Samuel Highely, and in the same year, he went bankrupt, and a unique publishing house sank into oblivion.

As to the Quekett Club, H. W. Gay replaced Hardwicke as Treasurer while retaining his seat on the Excursion Committee, and Bogue joined the Club, though he never seems to have taken any active part in its affairs. He continued to publish the *Journal* until his firm's collapse. Members settled back to their now well established routine of paper-reading meetings, gossip nights, excursions and soirées, and for some years Cooke contributed only the occasional mycological paper, and made a few comments in discussions. Once, when it was suggested that Club members who put on courses for their fellows should be paid for their trouble, he spoke vehemently against the idea, although he was one of those who gave such courses: the suggestion was dropped. Then in 1878 a great honour was bestowed upon the Club; Professor T. H. Huxley, probably through the good offices of Edwin Lankester, was prevailed upon to become its President. The Minutes of the Committee Meeting for May 24th read:

> It being necessary that Professor Huxley should be elected as an *ordinary* member ... previously to his being nominated as President, it was ordered that Professor Huxley should not *at any time* be called upon to pay any subscription.'

The Quekett Club was just the sort of popular educational organisation for which Huxley would wish to show support; his problem was to spare the time to do it justice, and before agreeing to take office he warned that he would only be able to attend a few meetings.

That year Cooke made two contributions to paper-reading meetings, the first being quite unprepared. At the session on August 24th:

> the Chairman said it appeared that they were without any formal paper to be read that evening. ... Perhaps while others were preparing to say something Dr. Cooke would be

kind enough to offer some remarks upon some subject which from him would be sure to be interesting. ... Dr. M. C. Cooke. ... said that when one had ever so many ideas stored up in one's head, it was sometimes rather difficult to pick and choose amongst them at a moment's notice; he had, however, had so much to do with one group of organisms, that it would be strange if he could not find something to talk about in connection with them. It was that old *mouldy* subject which always cropped up whenever he got upon his feet. One of the commonest things in the whole group was the common blue mould ... They had doubtless heard something of a vinegar plant, a film of ropy substance which was often cultivated domestically for the purpose of producing vinegar by letting it grow in any sweet mixture such as sugar and water. In most cottages in the country they would find it to be well-known, and most cottagers were very proud of it. ... If this vinegar plant was exposed to the air, and its means of sustenance was allowed to get dry, it would no longer increase in its former manner, but it turned about to seek some other means of development. ... The vinegar plant was naturally vegetative only; year after year cells were added to cells,* and filaments to filaments, but there was nothing to be found about it of the character of a seed. So long as sweet water was supplied to it, so long it kept growing in this manner; but if a check took place and it could no longer get what it wanted to feed upon, directly it began to starve, it shot up a number of filaments, and it became covered with a beautiful crop of blue mould,

that is, a *Penicillium*. Now the vinegar plant, or Mother of Vinegar, is in fact a culture of the bacterium, *Acetobacter*; Pasteur had described it some ten years earlier, but the Quekett Club does not seem to have been aware of this. Inevitably, as such a culture is grown for months on end with no aseptic precautions, it becomes contaminated by airborne spores, and any that can tolerate the acidic substrate produced by *Acetobacter* will germinate when conditions become suitable. *Penicillium* species and yeasts are some of these, and *Penicillium* will only sporulate in aerobic conditions, i.e. as the liquid dries out. Hence the persistent myth that the vinegar plant consisted of yeast and *Penicillium*, two stages of the same organism, while the importance of the bacteria was completely ignored. Huxley had fallen into the same trap eight years before his Presidency of the Quekett Club (Fig. 8.7), in one of his rare excursions into mycology, [12] but he was not present at the meeting to enter into discussion with Cooke.

This situation could arise in 1878 because pure culture techniques were then in their infancy. Though sterilisation by boiling, both of liquid media and of natural substrates such as slices of potato, had been practiced for some time, it was more difficult to keep inoculated media aerated yet free from contamination over long periods. The sealing of culture vessels with cotton wool plugs had only been introduced four years earlier, and R. J. Petri would not invent his 'dish' for another 10 years or so. Also, until the use of solid media was introduced into mycology, it remained very difficult to obtain a pure,

---

*i.e. a yeast.

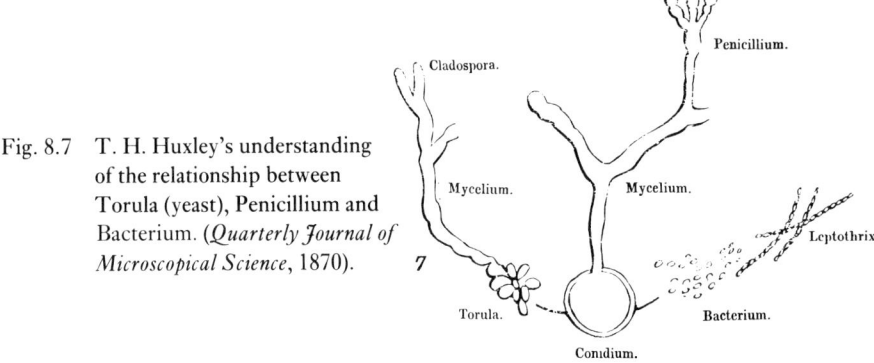

Fig. 8.7 T. H. Huxley's understanding of the relationship between Torula (yeast), Penicillium and Bacterium. (*Quarterly Journal of Microscopical Science*, 1870).

uncontaminated isolate of a fungus with which to start a culture, and though bacteriologists had been solidifying their media with gelatin for 20 years, Brefeld, in 1872, was the first to use the technique for fungi. Gelatin was not superseded by agar agar until the 1880s. It was also in this same period that Brefeld, Koch and others first elaborated methods which are still in use today for making single spore cultures.

In view of these developments it is perhaps surprising that Cooke could make a similar error five years later with the 'so-called Ginger-beer Plant, which somewhat resembles large, irregular grains of boiled sago', but in fact consists of a mixture of a bacterium and a yeast and is used in the home brewing of ginger beer. A specimen of the 'plant' was being demonstrated at a meeting of the Essex Field Club at which both Cooke and Smith were present. Cooke is reported as saying that he had frequently examined specimens but:

> could only determine it to be a cellular vegetable body very similar to some other low forms of Fungi such as the Yeast and Vinegar plants, but apparently distinct from them. He regards it as an immature form of some species which [in a sugar and water mixture] is under unnatural conditions and cannot develope [sic] its normal characters' [13].

He makes no mention of the possibility of using pure cultures to solve the problem, but simply suggests growing the plant 'comparatively dry' so as to encourage it to 'assume its proper from'. Worthington Smith, on the other hand, was very close to the truth when he concluded 'from a cursory examination' that the 'plant' was a sedimentary yeast mixed with a pin-mould and various bacilli.

The explanation for Cooke's failure to move with the times would seem to lie in the fact that he himself very rarely made use of cultures. Though he prided himself on his method of studying the habit of microfungi growing undisturbed on their natural substrate without damaging them under a coverslip (p. 167), he never made similar studies on cultures. Indeed, on only one occasion in all his writings does he describe an observation made by himself on cultured material (p. 177). He thus had no incentive to keep up to date with new culture techniques and their applications.

The Quekett Club had indeed a devoted and enthusiastic membership: it held an Ordinary Meeting on December 27th, 1878, at which 50 persons were present, half the usual attendance, it is true, but nevertheless extraordinary considering the season. At that meeting Mr. Charles Stewart, Vice-President of the Club, surgeon at St. Thomas's Hospital, and Fellow of the Linnean and Microscopical Societies, gave a talk on a subject much under discussion at the time, the nature of lichens (Fig. 8.8). For years there had been confusion among botanists on this subject. The plants resembled fungi in so many ways that it was tempting to regard them as such; and yet they differed fundamentally in other ways, notably in their pigment and need for light–no other fungus contained chlorophyll. Mycologists tended to regard them as a special group, distinct from fungi, but in 1867 the Swiss botanist, S. Schwenender, after meticulous and detailed studies, propounded the then astonishing theory that lichens were dual organisms consisting of a fungus parasitic upon an alga. It was a report on this theory that had come to Stewart's notice and that he now put forward to the Club with great enthusiasm. After the talk the Chairman 'called upon Dr. M. C. Cooke to say something upon' the paper. Dr. Cooke, after some complementary remarks on Mr. Stewart's delivery, said that:

> although this theory was widely extended and widely entertained, it was by no means to be concluded that it was therefore true. He thought that it was quite fitting that, at a time of year when romance was held to be in fashion, this, one of the prettiest little bits of scientific imagination, should be brought before them. Nevertheless he regarded it as being from beginning to end, root and branch, totally and entirely false, and without a single good fact to stand upon. . . . He had, during his time, some little experience as to fungi; there was hardly, in fact, a night on which he did not examine something like fifty of them, so that he might be said to know something about them; but directly he put his knife into a lichen he knew what he had got, even if his eyes were closed, such an entire difference was there in the substance; and although there might be some sort of apparent resemblance, when they got them under the microscope it was evident directly what they were.

And more in this vein, ending by congratulating Stewart 'on having made the best that he could of a very bad job'. Nothing daunted, another member then got up and listed some eminent botanists, including Sir Joseph Hooker, who supported the theory, and the discussion concluded with a few additional observations from Stewart, holding firmly to his previous opinions, but making no attempt to argue with Cooke's forthright attack.

However, Cooke was determined to kill this pernicious theory before it could gain further ground, and in his opposition to it, it must be said, he

Fig. 8.8 Two lichens.

would be supported for several years yet by a considerable number of his fellow botanists. He mounted his attack at a meeting on March 28th, 1879, delivering a paper on 'The Dual-Lichen Hypothesis' which was not only abstracted very fully in the Club's *Journal*, but printed in full in his own mycological publication, *Grevillea*. In it he went into his objections to the theory in considerable detail, but broadly speaking they can be reduced to two: that it was precluded by common sense–it was inherently impossible: and that the relationship between the organisms could not be that of parasite and host, as claimed by Schwenender, as the supposed alga was not damaged by the fungus. In fact, the idea of mutual interdependence between organisms, as opposed to parasitism, was a very new one at that date, the word 'symbiosis' having been introduced only two years earlier as a direct result of the realisation that 'parasitism' did not describe the fungus-alga relationship proposed for lichens. Though this development had not yet reached the members of the Quekett Club, it is obvious from the lively discussion which followed Cooke's paper that members interested in other groups of organisms had realised the need for some new term for species which live in close contact with one another without causing each other damage. Cooke, however, was never able, to his dying day, to accept the 'dual-lichen hypothesis', even though, by the end of the century, the great majority of botanists had done so. He was still inveighing bitterly against it in print in two books, one popular and one scientific, in 1893.

In July, 1879, Huxley completed his term as the Club's President with the customary Address. First he lamented the amount of time that the professional scientist must spend reading the current literature.

> The truth is that the increase of knowledge has its drawbacks ... during the last quarter century science has taken to grow in the most portentous and astonishing fashion. I remember the time when it was quite easy for the ordinary industrious man to keep up with the scientific literature of the period ... but, at the present day, to keep up with all

that goes on ... he must do an amount of reading which is singularly demoralising to his habits of thought.

How familiar is this cry of one hundred years ago to today's scientists! Huxley continues:

> But you, members of this Club, are in this respect vastly better off, because you can give your attention to any one point which you want to get to the bottom of, and you are not likely to be pulled up by some student in the lecture room who has read the latest thing published, and who expresses surprise that you do not know all about it too. Consequently you can give your attention to your own subject as exclusively as you may desire.
> ...
> Several amongst your number have asked me to indicate those courses of enquiry which may best be commended to members of such a society as this. ... It is exactly in that field–the following up of details, tracing out minutiae of structure, in occupying themselves with such questions as are only to be solved by long and patient devotion of time and dexterity, and a thorough knowledge of instrumental manipulation–it is exactly there that men of science find their difficulties, because the amount of time consumed is so great.

He ends, like Cooke before him, by recommending that such studies be carried out on the commonest of organisms.

Cooke meanwhile, once again at the request of a publisher, had returned to the writing of popular books. At a meeting of the General Literature Committee of the Society for the Promotion of Christian Knowledge, held in April, 1878, a new series of popular books on natural history had been proposed, and members had set about drawing up a list of possible authors. Among the names put forward at the May meeting was that of Cooke, known for his association with *Science Gossip*. He was asked to write a volume on *Woodlands*, he accepted, and it was published the next year, to be followed in 1880 by *Ponds and Ditches* in the same series, 'Natural History Rambles'.

From the evidence of the minutes of the General Literature Committee it is certain that no more than 19 months could have elapsed from the time that Cooke first accepted the commission for *Woodlands* to the time it appeared on the shelves of the bookshops, yet it was the first book on general natural history that he had written, and it covered all the denizens of British woodlands, from trees to fungi and from mammals, through insects, to slugs and snails, in 17 chapters and nearly 300 pages. In the circumstances he could hardly have prepared the material before hand and so been ready to start writing immediately the proposition was made. We can only suppose that he already had an encyclopaedic, if sometimes superficial, knowledge of all branches of natural history, and rapid access, perhaps partly in his own library, to the appropriate works of reference; and that all this was meticulously indexed. Perhaps he had accumulated such an index while scanning the literature for his various columns in *Science Gossip*. For in both books, though there are lively passages based on his own observations, large sections consist of judiciously chosen quotations from numerous authors recounting stories, lore and gossip about the organisms described. Both books are also freely sprinkled with extracts from the poets and references to the Almighty, the latter reminiscent of his articles in *The School and the Teacher*, and doubtless deliberately included as a gesture to the religious leanings of his publishers.

The destruction of the British countryside was going on at an alarming rate in Victorian times as it is today, and by some of the same means, including urban sprawl and the drainage of the Fens and other marshlands. At that time also the enclosure of common land and the laying of new railway tracks were a menace, and all provoked much opposition from an active conservationist lobby striving to save what it could of our national heritage. Cooke, as we have seen, was a supporter of this lobby, and the first chapter of *Woodlands*, entitled 'Forests and their Uses' might, apart from its style, have been written today. It describes the dreadful effects on soil, rivers and climate, of the destruction of forests all over the world, and particularly draws attention to the wholesale extinction of plant and animal species dependent upon those forests. On a smaller scale, he commends those who have succeeded in stamping out the 'barbarous practice' once common among country school boys of vying with one another as to which could collect the largest number of birds' eggs.

It is interesting that in describing the life of the Elm Bark Beetle, the vector of the condition now known as Dutch Elm Disease, but which at that time took only a minimum toll, Cooke attributes the death of the tree directly to the ravages of the insect; it had yet to be discovered that the actual killer is the fungus, *Ceratocystis ulmi*, which is carried into the tree by the beetle.

Ponds and Ditches, which appeared a year after *Woodlands*, is concerned mainly with the microscopic forms of life to be found in still water, and is a much livelier and more readable book than its companion volume because, to a very large extent, Cooke is describing his own observations rather than quoting the work of others. Time and again, his own huge enjoyment of the antics of the 'low-life' he watches under his microscope is passed on to the reader. Of *Coleps hirtus* he says:

> We have often been greatly amused in watching the movements of this little creature. Its form is not unlike a barrel. . . . As it moves it continues to roll over and over on its longitudinal axis, the opaque spheres rolling in its interior. . . . A dozen of these organisms rolling about at the same time across the field of the microscope form a very amusing object. Now and again a specimen will be seen with a part of the upper portion broken away, as if gnawed by a mouse, and yet the animal rolls on as happily as if no accident had befallen it.'

And of *Stylonichia*:

> ... an active vigorous Infusorian ... now going a little way in one direction, then turning sharply in another, and anon, rushing among the filaments of Algae, never resting an instant, and meanwhile causing a strong vortex in the water round it.'

The style of these two books was to form the pattern for all those on natural history which would follow. One can only speculate as to whether any of them would have been written if the S.P.C.K. had not initially given him the idea. When one considers that their preparation was proceeding while their author was already occupied with a full-time job, more than full-time 'leisure' activities, and a nerve-racking private life, one can understand how the legend arose that, at his desk, he always had three sheets of paper before him, one on which he was writing and two on which the ink was drying.

It was in 1880, the year that Frank Buckland died, that Cooke completed a task which he doubtless owed to the recommendation of his friend–a report to the Fishery Board on the disease of salmon which was decimating the Scottish fisheries at the time [14]. A great many investigations had already been made into the problem, and theories as to its cause abounded, but no-one had been able to suggest any means of control or cure. All agreed that the water mould, *Saprolegnia ferax*, was present in the necrotic lesions of the fish's head and tail which characterised the disease, but opinion was divided as to whether it caused them, or invaded them after they had already appeared. Though Cooke had certainly handled *S. ferax* and knew its life history well, there is no evidence that he undertook experimental work with the host; his 'Observations on the Salmon Disease', published by H.M.S.O. as the first part of the Inspector's (Buckland's) report on the subject, seems to be based entirely on a wide and critical reading of the literature and, given the state of biological knowledge at the time, the conclusions to which he came are very reasonable.

He is in no doubt that the fungus is the cause of the disease, but is equally certain that it cannot attack a healthy fish with unblemished skin, but only one that has previously been bruised or damaged in some way, perhaps while fighting its way up-river to spawn. The problem was that, as the motile zoospores of the fungus are common, why was the disease confined to certain rivers only, and why had it never manifested itself before? He is under the impression, by analogy with the fungal diseases of plants, that the stock may have been in some way weakened, but cannot explain how. However, he is sure that water pollution is part of the cause, combined with the low level of the rivers due to lack of maintenance, and any other environmental factors which might lead to exhaustion of the fish. Therefore the only 'cure' which he can suggest is general remedial attention to the rivers, the improvement of which should allow a build-up of a more healthy stock of fish, for:

> if the annual recurrence of the disease is destroying the constitution of the salmon, it

must as a consequence follow that the disease will increase rather than diminish, unless some means of invigorating the salmon can be found. If the parents are becoming debilitated, how can a healthy offspring, or a robust future generation be anticipated from the same stock?

As E. C. Large has pointed out, Cooke was a good Darwinian! [15] But though his 'cure' was correct, his reasons for suggesting attention to the rivers may not have been entirely so. He had overlooked one piece of evidence:

> Another alleged cause is the presence of bacteria in diseased spots, upon which the fungus is supposed afterwards to locate itself. I fear this view cannot be maintained, as the presence of bacteria is the consequence, and not the cause, of morbid tissues.

It is now considered by some that it is indeed these bacteria which first gain entry to a damaged spot on the skin of the fish, so preparing the way for the *Saprolegnia*, but the matter is still undecided, and in 1880 the idea of the bacterial causation of disease was in its infancy.

With two books, two papers to the Quekett Club, and the report on the salmon disease completed, not to mention the crescendo of mycological work described below, Cooke took on two further responsibilities in 1880. The absence of field clubs in and around London that had led to the outstanding success of the Amateur Botanists and the Quekett was now rapidly being made good, and a number of new organisations had sprung up over the years. One of the more recent was the Hackney Microscopical and Natural History Society, founded only two years before, and this group now made Cooke an Honorary Member, preparatory to electing him President in 1881. He was to remain closely associated with the Society for many years (p. 253).

Back at the Quekett Club, Cooke's name was one of four put forward by the Committee that year for President. Its first choice was unable to accept nomination, so it held a ballot among the remaining three, the vote went to a surgeon whose hobby was entomology, and Cooke was then overwhelmingly elected Vice-President over a list of three others.

Chapter Nine

The Making of a Mycologist. 1862-1870

Already in this story of Mordecai Cooke it has been necessary to return once to the year 1862, when he took up his post at the India Museum, to trace the separate threads in the tangled skein of his life–the paid employment by which he supported himself, his family affairs, and the unceasing drive to publicise the pleasures of the study of Nature. It might well be thought that these preoccupations would more than fill the waking hours of one man, even in the era which gave birth to the 'work ethic', and especially before communications were eased by the introduction of motor transport, the telephone, or even the typewriter. But this was not so. These same crowded years also saw the great flowering of his mycological career, and we must now retrace our steps to 1862 for the second time to examine his growing influence in that field.

No sooner was Cooke sure of his appointment in the Civil Service than he set about fulfilling his mycological aspirations in earnest. But first he took a brief busman's holiday: 'Before I commenced my daily occupation at the India Museum I resolved to undertake my long proposed pedestrian tour through North Wales'. He was accompanied by Alfred Grugeon, and together they set out 'with drying paper and other necessaries strapped to our backs, Grugeon in search of wild flowers and myself of parasitic fungi'. By now Cooke's interest had turned to microfungi, the tiny species that are the cause of spots, blisters and 'mildew' on the leaves, stems and fruits of plants, and it was these which he was bent on collecting.

> We passed from Corwen to Caernarvon and back by another route over Snowdon to

> Capel Curig ... We laid out [our finds] each evening in our drying papers, and made an average of thirty miles walk each day for a week, passing from the Beddgellert side up to Snowdon to encounter a thick mist in which we lost our way, which might have been serious, had we not met with a miner who for a gratuity put us in the right track and down on the Capel Curig side.
>
> On our way down we encountered a large cluster of the filmy fern *Hymenophyllum unilaterale* in a small cascade of which I carried away a tuft, and this when I reached home was tied to a half a brick, placed in a bowl of water and stowed in a dark cupboard, where it survived for a twelvemonth, and then succumbed to the uncongenial surroundings. In this excursion we found two or three fungus parasites which I never afterwards encountered elsewhere. (Aut.)

This was the first of several Welsh walking tours undertaken by the two men, and it must have left a deep impression on Cooke, for alone of all the hundreds of fungus forays and tours in which he took part during his life, it occupies one whole page of the brief 25 pages of his autobiographical notes. One cannot help wondering if some slight exaggeration of the distances walked has not crept in during the fifty years which elapsed between the event and the recording of it, for one would not see many microfungi while walking 30 miles a day. However, to have covered anything like that distance in mountainous terrain they must have been very fit, despite being city dwellers.

By now Cooke would have been a competent mycologist with a wide knowledge of his subject, but he was entirely self-taught. To progress further he desperately needed contact with others more experienced than himself with whom he could exchange opinions and from whose experience he could learn. The Rev. Miles Joseph Berkeley (1803–1889) (Fig. 9.1), to whom Cooke turned, was a greatly respected and much-loved figure among the botanists of his day, and is universally accepted as the Father of British Mycology [1]. He was educated at Rugby and at Cambridge, where he took Holy Orders, and spent his whole working life as a country parson, first at Apethorpe, near King's Cliffe, Northamptonshire, then from 1868 until his death, at Sibbertoft, near Market Harborough, Leicestershire. He was a good and consciencious priest, but his stipend alone was inadequate for the support of his large family, so for a time he took on the duties of a schoolmaster as well, and also coached private pupils.

Berkeley had been interested in natural history since his youth, having been attracted first to the molluscs, and a little later to botany and the algae, but always studying fungi as well. In 1836, at the request of Sir William Hooker, the Director of Kew Gardens, he wrote the supplement on fungi to the *English Flora*. Mycology at that time was an almost totally neglected subject and, as his obituarist, G. M., writes:

> The gathering of the material, the sifting of it, and putting the result into shape, represented much labour, and called for the highest excellence in judgement. Berkeley was

Fig. 9.1 The Rev. Miles Joseph Berkeley, 1803–1889.

equal to the task, and in the opinion of most mycologists he never surpassed his first great work on Fungi—nor have any who have come after him.

From then on Berkeley published innumerable articles in various journals describing not only British fungi, but specimens sent to him from all over the world, including those brought back by Darwin from his voyage in the Beagle. Some of the countless replies he wrote to the senders are still preserved.

In 1845 the potato murrain struck in Ireland. The potato plants on which the peasants depended for their basic diet rotted in the fields, and the lifted crop turned into stinking, sodden masses in storage. Famine set in. The cause of the disaster was unknown and many suggestions were put forward, the most popular being the appallingly cold, wet weather of that summer. Many workers, including Berkeley, had noticed that a mould could always be found on the blighted parts of the plants, but it was generally considered that the organism was merely colonising tissue which had already been killed by a disease of unknown origin. Berkeley, however, after collating all the available observations and the experimental evidence, stated unequivocally, in an article in 1846, that the fungus, now named *Phytophthora infestans*, was the cause of the blight of the potatoes. His revolutionary idea antedated Pasteur's germ theory of disease by nearly a quarter of a century, and put the science of plant pathology well ahead of that of human or animal pathology. Thanks to Berkeley, Cooke was to have no difficulty later in accepting the concept of pathogenicity in fungi.

From that time on Berkeley devoted much of his effort to plant pathology, and between 1854 and 1857 he published in the *Gardener's Chronicle* a series of 172 articles under the general title of 'Vegetable Pathology' in which the plant in both health and disease was studied. A popular magazine such as the *Gardener's Chronicle* would have been easily accessible to Cooke in his school-teaching days, as the scientific journals in which Berkeley usually published would not, so it is almost certainly this series which Cooke had collected so avidly and treasured so much at that time and which was to set him on the path to his future wide-ranging studies of the microfungi.

Cooke had become a knowledgeable mycologist himself before he presumed to approach directly so great an authority as Berkeley, asking for his opinion on a toadstool on which he was working. The letter making the request, dated Nov. 3rd, 1862, together with many others which followed as the two men entered into frequent correspondence, is filed with Berkeley's papers at the British Museum (Natural History), but unfortunately Berkeley's replies are lost. Of these letters Cooke said:

> ... I came into regular correspondence with Berkeley which continued until his death and was of immense service to me, and helped to the solution of many a difficulty, and he recommended me to extend my study of fungi ... as much as possible all over the world, to which he assisted me by the loan of specimens from his herbarium. (Aut.)

Cooke's first letter is that of a grateful student to a revered teacher, but while he always retained his profound respect for Berkeley, both as a mycologist and as a man, subsequent correspondence shows a gradual change in the relationship of the two, with Cooke beginning to discuss Berkeley's ideas, and putting forward his own, as an equal. Berkeley disliked controversy, though he was quite capable of holding his own in an argument if necessary, as over potato blight. He and Cooke clearly did not see eye to eye on a number of subjects over the years, but this in no way altered Cooke's feeling for the older man. He repeatedly referred to him in his writings as 'the Prince of Mycologists', and in the obituary he wrote for him in his own journal, *Grevillea* [2], he says:

> With a kind and genial disposition, a warm heart, and a benevolent presence, he was beloved by his family, in his parish, in the various societies of which he was a member, and, indeed, by all with whom he came in contact.

Berkeley had a close associate, Christopher Edmund Broome (1812–1886) (Fig. 9.2) of Elmhurst, near Bath, who for many years collected fungi with and for him, and helped in the determination and description of the hundreds of others sent to him by collectors in Britain and abroad. Indeed, so closely are the names of the two men associated that it is almost impossible for a modern mycologist to think of one without the other. Like Berkeley, Broome was a Cambridge graduate. He lived a quiet and uneventful life which he devoted

Fig. 9.2 Christopher Edmund Broome, 1812–1886.

entirely to fungi, and left his large herbarium and much of his mycological correspondence to the British Museum (Natural History). Though he had none of Berkeley's brilliance and originality he made an important contribution to the mycology of his day by sheer hard work.

Cooke first entered into correspondence with Broome two years after he started writing to Berkeley, and it is evident from Broome's letters to his friend that Cooke was sometimes a sore trial to him. One argument started as early as 1866, but before quoting it a digression is necessary concerning 'splitters' and 'lumpers' in the field of biology. The classification of living things into species, based on their similarities to and differences from one another, would seem on first consideration to be a comparatively simple task; and so it may be with organisms showing marked and unvarying divergence between themselves and their nearest relatives. But as Darwin himself remarks: 'No one definition [of a species] has yet satisfied all naturalists; yet every naturalist knows vaguely what he means when he speaks of a species'[3], and the situation is only a little different today. When the organisms to be classified vary with their geographical origin, their environmental conditions, or the substrate on which they grow, and when a range of individuals can be found with characteristics intermediate between those of two supposed species, the scope for individual preference in classification is enormous. In circumstances such as this the problem of the delimitation of species engenders much heat today, and there was even more scope for disagreement in Cooke's time. As a result, taxonomists have themselves been classified into those who like to break up a large and convergent group of organisms into many separate species based on very small differences, who are known as 'splitters', and those who prefer to keep together the whole group, with all the variations, as a single species, these being termed 'lumpers'. Their controversies add greatly to the spice of biological life.

Cooke himself was a lumper by inclination, and he justified his conviction on the ground of his own interpretation of Charles Darwin's then novel theory of the origin of species. In 1864, five years after the publication of the latter's controversial book of that name, Cooke gave a talk to the Society of Amateur Botanists entitled 'Infinite Variety and Immutable Law' which sets out his views very clearly [4]. He starts with a résumé of Darwin's ideas on change and chance but, accepting that change takes place, he denies that chance plays any part in the evolution of species.

> *Individually* [his ital.] the organic world is a world of order, eternally progressing, retrogressing, or changing and exhibiting its phenomena as the results of the working of certain immutable laws.

Variability between individuals is due 'much but not altogether' to the parents of the individuals, progeny partaking of the characters of both, but never being identical because of the law of variation. He does not define the law, but the statement was made long before Mendel carried out his work on inheritance. Cooke continues:

> Accepting genera and species as we find them, we must admit that they are useful, and that there could be no science without them, but it is at the same time our belief that they are artificial, and that there is really no such thing in nature ... The whole scheme of the universe is unity. Rather that all organised beings in the present and in the past and in the future are linked together in one unbroken chain. That they form a wheel within a wheel, a vast circle with God at its centre and a circumference extending from the beginning to the end of time.

It is clear from this that even if he had broken with organised religion he was still very far from being an atheist. Change, he says, should be studied. All change has a cause which botanists should discover: 'another cause of variability is the power of *adaptability* in the organs of plants to the circumstances in which they may be placed'. It is not clear whether he is talking here about inheritable variability. He defines the limits of species as two plants which, whatever they were like in the past, are:

> now sufficiently distinct as to afford distinct species capable of description, and which appear to be *relatively* permanent, because *positive* permanence, like *positive perfection*, is just one of the things we permit ourselves to doubt.' Individuals, he continues, in philosophical mood are linked together 'not like the stones of a temple dedicated to his [God's] honour, because they have too separate an individuality, but like the atoms in

the stone itself. Then we form a higher conception of the Creator of a united work, proclaimed as *one* through all its members, one in spirit, one in design, one in execution, and admit a variety that is truly infinite, a unity that is complete.

Only a year later we find him giving practical advice to amateurs based on these beliefs:

We think that were a young student to ask our advice, we should recommend him to imitate the most inveterate splitter that he could imagine, because such a course would, of necessity, compel him to a rigid scrutiny and comparison, which might thenceforth grow into a habit of close observation. On the other hand, we should feel disposed to caution the mature naturalist against new species, recommending him to cultivate a 'conservative' spirit, and to regard all allied forms as the same species until he recognised unmistakeable evidences that they could not have diverged from each other, to that extent, under any ordinary conditions to which animal or plant life could be submitted [5].

It was Cooke's propensity for lumping that led, in 1866, to the argument with Broome. This concerned the identity of a little fungus called *Diatrype sordida*, so named by Berkeley and Broome, with a very similar species, *Valsa taleola*, named by an eminent Swedish mycologist some years earlier. Both fungi belong to a large group whose members form minute, flask-shaped fruiting structures embedded just under the surface of the twig or other object on which they grow. After quoting Cooke's reasons for regarding the two fungi as identical, a letter from Broome to Berkeley continues:

... all this is Mr. Cooke, and he certainly has spun a web which is enough to try Job's patience ... Mr. Cooke says he is about to publish a volume on British Sphaeriae [the group concerned] and is anxious not to cause greater confusion than already exists, which I conceive he will probably do. I send such drawings as I can for the paper and hope we shall not incur again Tulasne's castigation [an eminent French mycologist] or further letters from Mr. Cooke, I know not which is worse [6].

The argument rumbled on for four years, Broome writing to Berkeley in 1870 that 'Mr. Cooke is at us again', and a little later: 'I have answered Mr. Cooke as best I can about his Diatrypes'. Cooke prevailed in the end, for in his major mycological work, *The Handbook of British Fungi*, published the following year, *D. sordida* is given as synonymous with *V. taleola* (Fig. 9.3).

Very soon after beginning his correspondence with Berkeley, Cooke took another important step along the road to mycological eminence–he began to publish in botanical, rather than in popular, journals. His first paper, 'Rare or new British hymenomycetal fungi' (i.e. toadstools), appeared in 1863 in the first number of the *Journal of Botany*, often known as *Seeman's Journal of Botany* from the name of its founder and first editor. Cooke would have had foreknowledge of the birth of this journal, for the first of its many publishers

Gen. **351.**   **VALSA**, Fr.

Perithecia carbonaceous, perfect, circinating, elongated into converging necks; ostiola erumpent, joined together, or ending in a common disc.—*Fr. S.M. Berk. Outl. p.* 389. (*Fig.* 388.)

Fig. 9.3   Description of the genus Valsa, from Cooke's *Handbook of British Fungi*.

was Robert Hardwicke. Altogether, in the four years to 1866, he contributed four long articles on the classification of various groups of fungi, and a number of shorter ones describing newly discovered species. E. C. Large [7] has pointed out that one of these new species, found on dead apple leaves and called by Cooke *Sphaerella inaequalis* (Fig. 9.4), was none other than the sexual state of that notorious orchard scourge the apple scab fungus, now known as *Venturia inaequalis*. It would be 21 years before a German mycologist, R. Goethe, realised that the fungi were different states of one and the same organism, and that Cooke's *Sphaerella* was the means whereby the scab fungus overwintered, to infect the tree's new growth in the spring.

Nowadays taxonomy, the study of organisms with a view to their classification, is but one aspect, albeit an important one, of the large and varied discipline of mycology. To-day by far the greatest number of mycologists are experimentalists–physiologists, plant pathologists, geneticists, ecologists, etc. But this was not so in the nineteenth century, when the few who professed to be mycologists, were, especially in England, almost all taxonomists. Because

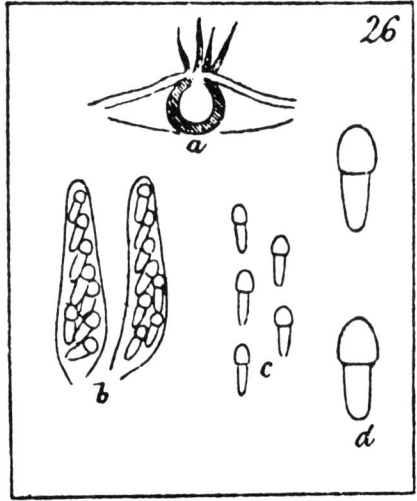

Fig. 9.4   Original drawing by M. C. Cooke of *Sphaerella inaequalis*, now known to be the sexual state of the apple scab fungus, *Venturia inaequalis*.

fungi had until then been a sorely neglected group, the scope for taxonomy, at that time a purely descriptive science, was enormous, and there was little impetus towards experimental work. Discovery and description of new species was going on apace all over Europe and America, especially when the price of microscopes became low enough to bring them within the reach of many a gifted amateur, but there was no accepted overall framework of classification. Indeed there could not be, because nothing was then known of the sexual reproduction of the fungi. Until the 1860s fungi were believed to be sexless, despite the fact that in the 1820s C. G. Ehrenberg of Berlin had illustrated, in a little known pin mould, *Syzygites megalocarpus*, the conjugation of two hyphae to form a sexual spore, the zygospore. It was not until the 1860s that another German, Anton de Bary (1831–1888), one of the giants of mycology, together with the Russian, M. S. Woronin, confirmed and extended Ehrenberg's work and established sexuality in the Zygomycetes as a fact.

Then, in the same decade, de Bary himself, using *Pyronema confluens*, the little cup fungus found on burnt ground, and *Erysiphe cichoraceum*, one of the powdery mildews, showed that their sac-like asci containing ascospores (Figs 9.4, 10.8) which had been described in the previous century, were also the result of sexual fusion between hyphae, thus establishing sexuality in the second great group of fungi, the Ascomycetes.

There remained a third group, which we now know as the Basidiomycetes, of which the most conspicuous members are the toadstools and their allies. These were clearly distinct from the Zygomycetes and the Ascomycetes, and were characterised by the club-shaped basidia, each bearing (usually) four spores, which covered their gills or other spore bearing surfaces, but in which no method of sexual reproduction had been found. As early as the 1830s the Swede, Elias Fries (1794–1878), had put forward an important taxonomic study of the group, and suggested a classification based on the colour of the spores which was still in use one hundred years later; and during the 1860s and early 1870s many mycologists claimed to have discovered the sexual organs of the agarics, though all were eventually discredited. It was left to another German, Oscar Brefeld (1839–1925), in 1876 to state his personal conviction that the agaric fruit body and its basidia were not the result of a sexual process. The complex story of sexuality in this group, and in other Basidiomycetes such as the plant pathogens, the rusts and the smuts, would not be unravelled until this century, when much more had become known of fungal cytology and genetics.

Cooke was working in an exciting but difficult era, when new ideas were flooding in from Europe (British workers had little to contribute to these vital and fundamental studies) and the whole subject was in a state of flux. However, the morphology of all organisms (their form as seen with the naked eye and under the microscope), not only fungi, was still the basis of taxonomy; the experimental approaches in use to-day, such as developmental and biochemical studies, were as yet unheard of. The main controversy was between those using the old 'artificial' classifications, i.e. number of stamens, colour of spores, etc., and those attempting to use 'natural' systems based on supposed evolutionary relationships, a particularly difficult matter in fungi. Cooke, as we could guess from the passages quoted, belonged to the latter group, and this was apt to bring him into conflict with certain other mycologists.

That he was a born collector and classifier there can be no doubt; it is obvious from his initiative in starting the small museum in his schoolroom in the 1850s, and later the Scholastic Museum; from his skill as a cataloguer; and from his long tenure of his post at the India Museum. In fact, he was almost certainly better suited temperamentally to the profession of taxonomy than to the experimental work he sometimes regretted he never had time to undertake. The only book he published in 1863, the *Index Fungorum Britannicorum*, was aimed at helping those with similar tastes to his own, for it was 'A Complete List of Fungi Found in the British Islands to the Present Date, arranged so as to be applicable either as a check-list or as herbarium labels', and it contains the names of 3079 fungi grouped according to Family, Order and Genus. Reviewers hailed it as a useful addition to the scanty mycological literature.

An essential working tool of any botanical taxonomist is his herbarium, or collection of dried specimens of the groups which he is studying. The herbarium specimen, if properly prepared, classified and annotated, is a lasting and absolute record with which both the owner and other workers can compare future finds. Though the herbarium of a novice is mainly of use to himself, that of an expert, who may have been collecting for years in many parts of the world, will be of assistance in reassessments of the group concerned and in controversies about it, not only in his own lifetime, but for generations afterwards. Herbaria which contain 'type specimens' have an additional importance, for such a specimen consists of material from the original collection of a newly named species, and any future collections given the same name must conform to the type.

Different workers may employ slightly different methods in setting up their herbaria, especially when the organisms they are studying are of such diverse nature and texture as the fungi. Cooke described the method he used for leaf parasites as follows. After pressing the diseased leaves between sheets of blotting paper, a sheet of white cartridge paper, foolscap size,

> receives within its folds the specimens of a single species; these are affixed to the right-hand page, when the sheet is open, and a small envelope is attached by its face to the same page at the bottom, in which loose specimens are kept for minute and special examination, or as duplicates. When the sheet is folded, the specific name is written at the right hand lower corner, or, what is better, a strip containing that name and its number is cut from a copy of the *Index Fungorum* [Cooke loses no opportunity of advertising his works] kept for the purpose, and gummed in its place. The remainder of the page . . . is occupied with memorand a relating to the species enclosed, sketches of the spores, synonyms, references to descriptions, etc. All the species papers of each genus are placed together

within a sheet of brown paper, half an inch larger in each direction, with the name of the genus written at the left hand corner. A piece of millboard, the size of the covers when folded, separates each order [8].

In dealing with agarics, a notoriously difficult group to preserve, he advocates making a coloured drawing of the whole toadstool and of its section while it is still fresh, then drying it in three parts, the stem, half the cap with the gills removed, and a thin section of the whole fungus cut from another specimen. He added to his own herbarium throughout his working life, until when he finally sold it to Kew in 1885 it contained 43,500 specimens.

The efforts of Victorian educationalists to promote an appreciation of science among the lay public led not only, as we have seen, to a great flowering of museum displays of all types, but also to the appearance of numerous popular magazines, some of them extremely cheap, on all aspects of science. One of these was the *Popular Science Review*, another Hardwicke publication, which first appeared in 1862 and was to flourish for 15 years. Hardwicke's authors, including Dr. and Mrs. Lankester and Peter Simmonds, were much in evidence as contributors to the new magazine. Each number included information about the state of science teaching, both in the schools and as organised by the Department of Science and Art, and the book reviews included favourable notices of the *Manual of Structural Botany* and the 'Plain and Easy' *British Fungi*.

It is characteristic of Cooke that his first articles on parasitic fungi should appear in the *Popular Science Review* rather than in the *Journal of Botany* or other learned periodical. 'Microscopic Fungi Parasitic on Living Plants' came out in 1864, in four parts, illustrated in colour with six plates by the well-known botanical artist John Edward Sowerby (1825–1870), a grandson of the famous author of *English Botany*, James Sowerby (1751–1822). Berkeley's articles of the previous decade, so avidly collected by Cooke, had begun to bear fruit. Cooke gave as his reasons for writing a popular account of these fungi at a time when no scientific study had appeared for 30 years, the following:

> In these latter days, when everyone who possesses a love of the marvellous, or desires a knowledge of some of the minute mysteries of nature has, or ought to have, a microscope, a want is occasionally felt which we have essayed to supply. This want consists in a guide to some systematic botanic study, in which the microscope can be rendered available, and in which there is ample field for discovery, and ample opportunity for the elucidation of facts only partly revealed. Fungi, especially the more minute epiphyllous species, present just such an opportunity as many an ardent student would take advantage of [8].

As always, Cooke was championing the amateur and helping to widen his field of study. But there may have been another reason for his choice. Writing in

1892, he says of the year 1865:

> This was a sort of 'middle ages' for mycology. Those who pursued it were persecuted, and pestered with the enquiry of 'What good is it?' 'Will it put money in your purse?' The only possible answer was, that it enabled a person to distinguish good from bad, as esculents, and that as a means of acquiring knowledge it would ultimately secure its own reward, in addition to the pleasure it gave to all lovers of nature to explore her mysteries, without regard to whether they were profitable or not [9].

Perhaps he felt that it was only among amateurs that such altruistic 'lovers of nature' could be found.

Because of the interest aroused by the articles in 'these little pests of field and garden', Hardwicke asked Cooke to revise and expand them into book form. Further illustrations were added (Fig. 9.5) and the scientific level of the work was raised by the addition in an appendix of a brief description of the appearance and habitat of each fungus. A number of the species included in the book were new to science, having recently been discovered either by Berkeley or by Cooke himself, and others were recorded for the first time in this country. In fact, the book is a strange blend of the popular with the scientific, for it was not usual then, and would never be done to-day, to publish new species in any but a scientific work.

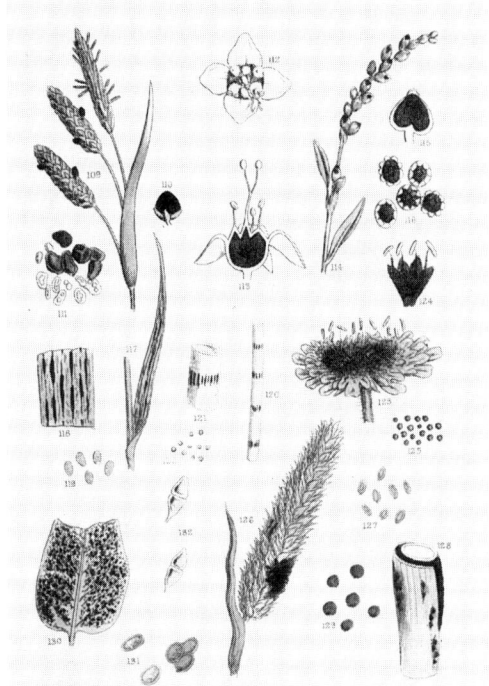

Fig. 9.5 Illustration by J. E. Sowerby of infections by smut fungi of various hosts. From Cooke's *Rust, Smut, Mildew and Mould*. Original in colour.

THE MAKING OF A MYCOLOGIST. 1862–1870 161

Cooke evidently intended his book to be of use to a wide range of practical people, for he ends his Preface thus:

> I trust that what has been done will be found of interest and utility to agriculturalist, horticulturalist, microscopist, and, in fact, to every student and lover of nature.

It certainly achieved its purpose, for by 1869 it was being used as a text-book by T. J. Burrill, Professor of Natural History at the University of Illinois, for classes in botany and plant pathology which are acknowledged to be the first ever given in America to include practical work [10]. If Cooke knew of this, it must have given him great satisfaction.

Rust, Smut, Mildew and Mould, which appeared in 1865, was universally well received and ran to six editions, the last by W. H. Allen in 1902. In the final three editions the appendix was completely revised to bring the names and descriptions of the fungi into line with the latest taxonomic ideas.

At the beginning of the first edition is a 'Notice' to the effect that:

> A collection of about one hundred specimens of minute Fungi, illustrative of this work, is in course of preparation. It will be published at one guinea, and is expected to be ready shortly.

This heralds the first of numerous 'exsiccati', or collections of dried fungi, named and numbered, which Cooke would be issuing over the coming years. In the nineteenth century it was a common and valuable practice, which has largely died out to-day, for competent botanists to issue such sets, which might be presented to herbaria for reference purposes, or sold more widely for the personal use of amateurs and professionals alike. A set of 100 specimens was known as a 'century', a smaller collection as a 'fascicle', and a mere ten species as a 'decade'. The perpetually impecunious Cooke undoubtedly regarded his exsiccati as a source of income, but even so he presented sets to Berkeley and other respected figures as gifts. The preparation of exsiccati was time-consuming work, for not only had sufficient material to be collected in the field to make a worthwhile series, but the identity of the fungi had to be carefully checked, they had to be prepared, mounted and annotated, the sheets had to be suitably bound together, the finished volumes advertised, wrapped and posted, and the payments collected from the purchasers–all of which Cooke undertook himself.

In 1977 sets of his First (undated, presumably 1865), Second (1866), Third (undated, presumably 1867) and Sixth (1872) Centuries were discovered in the Botany Department of the University of Bristol. There is no record of their provenance (Figs 9.6 & 9.7). The 50 loose sheets in each volume, on each of which two fungi are mounted, are held between dark green cloth covers with ties in the front, and on the spine in gold lettering are the words 'Cooke. Fungi Britannici'. The covers were produced by Hardwicke, doubtless to Cooke's

Fig. 9.6 A volume of Cooke's exsiccati.

Fig. 9.7 Pages from Cooke's exsiccati.

design. Most of the specimens, which are glued to the sheets, are still in reasonable condition 120 years later, though some have crumbled. Each specimen is named, and the date and place of collection is given, but in no case is the host plant identified, a strange omission. Presumably the first three series, at least, were the collections refered to in the 'Notice' in *Rust, Smut, Mildew and Mould*.

A practical problem in issuing exsiccati was whether they could be considered to be books, and therefore qualify for a cheap rate of postage. Cooke evidently carried on a long-lasting and bitter war with the Post Office on this matter, for he refers to it a number of times in his correspondence. On one occasion Broome had been surcharged for some exsiccati Cooke had sent by book post, for in an indignant letter Cooke told him that the Post Office had agreed to the cheaper rate, and commented: 'It is certain that [it] was never conducted so bad [sic] as at present.' He had even worse trouble with postage abroad. In 1878 he wrote to an American mycologist: 'I do not think that our Post Office will take them by mail, except at letter rate, which will be monstrous'[11]. This batch of exsiccati was eventually despatched through the Canadian Express Co.

In order to collect material both for sale to other collectors and to make up his exsiccati, Cooke spent as much time as possible on field excursions and country walks. Already in 1864 he was selling material to the Botany Department of the British Museum [12], for it is recorded that in 12 months he received from the Director a sum of £9.9.0 for three collections. Doubtless the proceeds of one trip helped to finance the next. The details which accompany each specimen in the Bristol exsiccati provide an unexpected and unique record of his movements in the years 1865–1867, and the extent of his travels, despite the fact that he had a full-time job, is astonishing. It was, of course, quite easy to reach the country on foot from Kentish Town. Not only was the suburb close to Hampstead Heath but, though growing fast, it was still very near to green fields, while neighbouring Highgate was in the country. In addition, the rapidly growing network of railways made it ever easier to take day trips into the home counties, and longer excursions, for a weekend or a week, as far afield as Norfolk and Wales. The exsiccati show that Cooke did all these things.

In the September of each of the three years he returned to Norfolk, making frequent collections at Neatishead and also visiting other parts of the county. In 1865 and 1866 he was again collecting in North Wales, on the first occasion at least in company with Alfred Grugeon, who was somewhat shocked at the discovery that his friend's main object was to make enough money from his collections to cover his expenses. Grugeon, although only a working man, had no such materialistic ideas, but went purely for the joy of botanizing among his beloved wild flowers. Cooke went often, too, to Darenth and Swanscombe, in Kent, where his younger brother Zenas lived and brought up a large family, and where a cousin was the proprietor of a cement works: and occasionally he travelled to Chichester where William, who had left his uncongenial job in Norwich, was employed as works manager at the firm of a manufacturing iron-monger. William became well-known as a local botanist, even to giving evening classes in the subject, and helped to compile the botanical section of the *Victoria County History* of West Sussex. At Shere, in Surrey, Cooke collected with the local general practitioner, Dr. Edward Capron, a keen botanist and entomologist who was to be of great assistance to him a few years later. Apart from the fact that he was a Fellow of the Entomological Society, we know little of Dr. Capron, for he consistently failed to communicate with the compilers of the *Medical Directory* throughout the 1870s and 1880s, and supplied the minimum of information at any other time. Most of Cooke's other collections were made in Hampstead, Highgate, and districts which are now part of the inner suburbs of London, but had not at that time been encroached on by the builders. He obtained one specimen from outside the India Museum.

By 1872 he was including in his exsiccati large numbers of specimens collected by others, both local amateurs and established mycologists, all duly acknowledged. He actively sought such specimens; for instance, in a letter to Broome in 1865 he wrote:

> I may be excused for sending you by this post a marked list of my desiderata, so that at any time, should specimens fall in your way, you will not forget a smoke-dried citizen [his addiction to his pipe must, even in these early days, have been his hallmark] whose opportunities for finding them himself are limited.

This common practice enabled sets to be completed far more quickly than would otherwise have been possible; doubtless Cooke's finds were included in the exsiccati of other workers.

It was probably in about 1870 that Cooke bought the microscope he was to use for the rest of his working life, and which he sold to the Royal Horticultural Society the year before his death. The Society still keeps the instrument at its Wisley Gardens, and a special study has been made of it by Mr. G. L'E. Turner [13] of the Museum of the History of Science, Oxford. I am indebted to his report for the information which follows.

There can be little doubt that Cooke was describing his own instrument when he wrote in his book *One Thousand Objects for the Microscope*:

> The reader may inquire what we consider to be the essentials of a good practical working instrument: and here again we must fall back on personal experience. We should require a firm stand with a good spreading base not easily overset or given to vibration.... Then we have a strong predilection for a large-sized tube; hence we had ours made specially, on the model of a large microscope. The movements up and down must be perfectly smooth, with coarse and fine adjustments; the stage as clear of encumbrances as possible, with no elaborate machinery to move the object in all directions, but to leave that, as

much as possible, to the fingers of the operator. Underneath should be the facility for a simple condenser, movable in all directions, with a ball-and-socket joint: and then, what more?

The glasses. One good eye-piece, not too deep ... with a slot cut for the micrometer– and the objectives may be confined to two, namely, one inch and a quarter inch, to which others may be added as found requisite ...

This description fits Cooke's instrument exactly. In a note that he sent with it to the Royal Horticultural Society he explains that it is a 'Student's Portable Microscope specially constructed to order by Moginie with large body and rack movement' (Figs 9.8, 9.9 & 9.10). Clearly he had progressed far beyond those 'young and foolish' days when he had impulsively speculated £50 on his first, ready-made microscope. Now he knew exactly what he wanted from his most important working tool, and to whom to turn to have an instrument constructed to his own requirements. Mr. Turner was able to discover that the 'source of Cooke's microscope was C. Baker of 224, High Holborn' and that the design was introduced in 1867. Cooke's instrument, however, was modified in several ways and would have been more expensive than the standard model, about £12 to £14 with accessories.

The relationship of its designer, William Moginie (1829–1881), to the firm of Baker is uncertain, as he does not seem to have been an employee. He was not only a competent designer of microscopes; he was also an inventor of gadgets for field naturalists, such as the 'New Collecting Stick' for retrieving spec-

Fig. 9.8   Cooke's microscope and accessories, now in the possession of the Royal Horticultural Society.

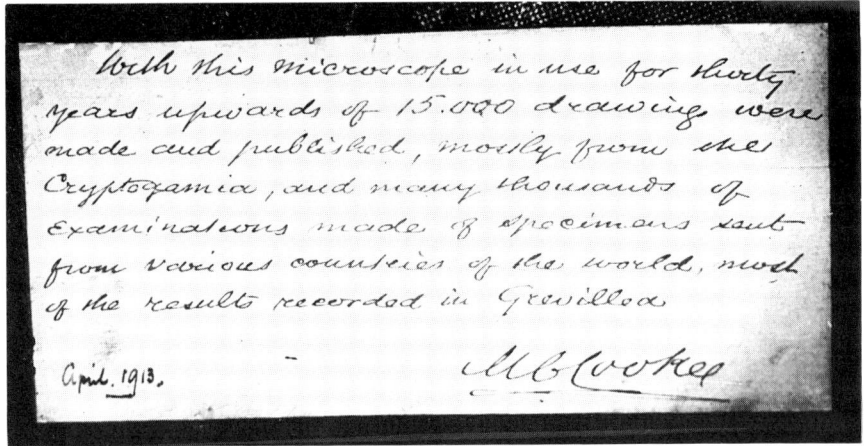

Fig. 9.9   Note accompanying Cooke's microscope to the Royal Horticultural Society.

Fig. 9.10   M. C. Cooke, aged about 85 and almost blind, with his microscope. Photograph taken by his daughter, Leila Cooke.

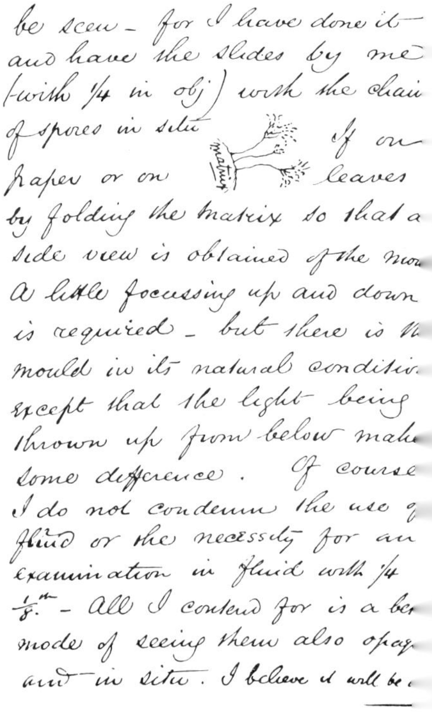

Fig. 9.11   Part of a letter from M. C. Cooke to Broome, describing his method for microscopic examination of living f— Dated June, 1870.

imens from ponds, which was also marketed by Baker's. Perhaps he was an enterprising amateur whose designs were taken up by the firm. Cooke would have met Moginie at least as early as 1866, for it was in that year that the latter joined the Quekett Microscopical Club. Moginie was an active member, exhibiting microscopes and gadgets at its meetings. After his untimely death at the age of 53 his wife presented to the Club twelve of his micro-photographs in the form of lantern slides.

Cooke's microscope was designed as a travelling instrument, and it seems very probable that he used it as such, for he not only attended field meetings at which the evenings would be spent examining the day's finds, but he would also have needed an instrument to take to the Quekett Club's Gossip Nights.

Though he recommended a cheap and simple microscope body, Cooke claims that he used 'the best glasses that could be procured'. Of the two objectives that he sold to the Royal Horticultural Society with his microscope, the quarter inch is of excellent quality, but the higher magnification, $\frac{1}{8}$th inch, is less impressive. There is no one inch objective at Wisley, though Cooke certainly used one. As to oil immersion objectives, he states specifically that they are unnecessary for beginners, who can 'afford to leave them in more experienced hands', and is very scathing about 'those who, nowadays, spend days and nights in endeavouring to resolve the striae of diatoms, to see more than can be seen'. The implication here is that the writer himself did not use one, despite being a 'more experienced hand'. Certainly the speed at which he worked would have precluded the regular use of such an objective, and the criteria employed for the classification of fungi at that time was such that a $\frac{1}{8}$th inch objective would have given sufficient magnification for most purposes. He would have had little need of an oil immersion for his own work.

Cooke was much concerned about the best technique for the microscopical examination of fungi, and describes his experiences in a letter to Broome in 1870:

> I feel confident that much is still to be done with seeing moulds *as they are*. What is required is no obstruction to light in the objective–and that the objective should work as far away from the object as possible ... and finally as much penetration as can be secured. ... Of course moulds, in water or spirit–compressed and seen as transparent objects do not give all that is wanted, especially the mode of grouping of the spores, which immediately float away on coming in contact with water. I find compressing (dry) carefully a portion of the mould–not moving the glass–but fixing the cover at once and then by viewing as a transparent object–I can make out the structure, attachment of spores better than in water with $\frac{1}{4}$ inch. A *Penicillium* or *Aspergillus* may thus be seen–for I have done it ... with the chains of spores in situ. If on paper or on leaves by folding the matrix so that a side view is obtained of the mould, a little focussing up and down is required–but there is the mould in its natural condition except that the light being thrown up from below makes some difference. Of course, I do not condemn the use of

fluid or the necessity for an examination in fluid with $\frac{1}{4}$ and $\frac{1}{8}''$–All I contend for is a better mode of seeing them also opaque and in situ.' (Fig. 9.11).

His determination to see the habit of a fungus undisturbed by mounting is much to be commended, for it was immensely difficult before slide cultures (the cultivation of fungi on a drop of nutrient on the slide itself) came into general use. However, his claim to be able to see the attachment of spores under the conditions described may account for the glaring errors to be found in a few of his drawings, notably one [14] in which the phialides, or asexual, spore-bearing cells, of an *Aspergillus*, though in fact they only bear one spore chain each, are furnished with four projections at their tips like the basidia, or sexual, spore-bearing cells of a toadstool, which bear four spores each (Fig. 9.12). Two years after his letter to Broome, and following continuing efforts to improve his technique, Cooke wrote again:

I have had a $\frac{1}{5}$ objective constructed with a long focus and the mount bevelled off to take the front lens–thus–so that the bull's eye will throw the light well upon the object.'

It is not known how satisfactory he found this arrangement.

In *One Thousand Objects for the Microscope* Cooke gives a very precise description of his microscopical techniques, including the preparation and mounting of specimens with the utmost neatness for use with both direct and transmitted light (the light source at the time would have been paraffin and, later, gas). He regarded correct lighting as absolutely critical. From the

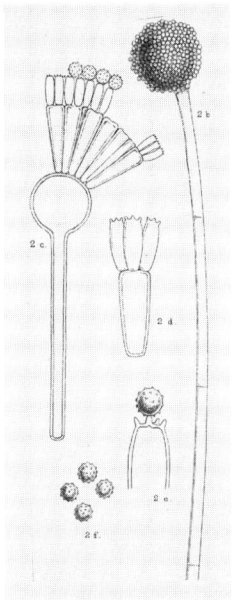

Fig. 9.12 Cooke's erroneous drawing of the sporing head of a species of *Aspergillus*.

THE MAKING OF A MYCOLOGIST. 1862–1870

instructions to others given in this book, and from the immense amount of microscopic work he got through himself, it must be assumed that he was an extremely neat, precise and orderly worker.

For one thing all English-speaking microscopists must be eternally grateful to Cooke. As we have already noted, his inaugural address in 1866 to the newly formed Quekett Microscopical Club was entitled 'On Universal Microscopic Admeasurement'. It was an impassioned plea, in his most declamatory style, for the world-wide adoption of an agreed unit of microscopic measurement, for at that time the English-speaking world used the inch while the millimetre was in use on the Continent. First he defines the microscope-using world of his day:

> We may fairly confine ourselves to Europe and America, without fear lest Asia and Africa should protest against being left out, because all the students in these regions will be 'exports' from Europe and America. Australia is not much troubled with microscopical students. Her sons have not yet found time to stand for hours at one end of a microscope.
>
> In Europe we may particularise Germany, in its widest sense, including all who speak or write the German Language, with whom could be included the few Scandinavians who pursue the study, and thus the German District would be held to mean all Northern Europe, from the shores of the German Ocean to the confines of Russia. Then France may be alluded to as including also Switzerland, and wherever the French tongue is employed. The South of Europe is represented only by Italy, for Austria belongs to Germany, the Turks are too idle, the Greeks too miserable, the Spaniards too intriguing, and the Portuguese too illiterate, to produce any contributions to microscopical literature. Of America we speak as restricted to North America, not including Mexico and Labrador, but the United States and Canada, where the English tongue prevails.

Then he gives the calculation necessary to convert '0.33 mm' to '.0129921 of an inch', and enumerates the reasons for preferring the millimetre to the inch, including the commonsense one that it is simply easier to imagine say, $\frac{1}{25}$ of a small object than $\frac{1}{625}$ of a larger one. He continues:

> The machinery and labour which it would cost to introduce our pet inch as the standard all over the world, would be enormous. The power to introduce the European standard into all countries in which English is spoken, lies in the hands of a few: it rests with the microscopists of London. If they adopt it, those in the provinces will follow the example, and, in self-defence, our colonists will do the same; then if America does not think fit to fall in, she will become as isolated in microscopy as if she spoke and wrote in an unknown tongue. . . . The adoption of the millimetre as the standard is, therefore, incumbent upon British microscopists, if they would advance, and not obstruct, the cause which they are presumed to have at heart.

Finally he declares his own intention of using the 'French measurements' in all his published works from then on. The article ends with a stirring call to the Quekett Microscopical Club:

> 'It is for you to decide whether the Q.M.C. shall bind the laurel wreath about its young but sturdy brow. It is for you to determine whether the honour of appealing on a common object to the microscopists of Europe shall be yours.
>
> It is for you to determine if the crowning effort of a year of glorious success shall be to prepare the way for the more glorious success of future years, by repudiating the use of figures which, to thousands of fellow-workers, from the Seine to the Danube, have no meaning; and to whom they are almost as barbarous as the hieroglyphics of Egypt, or the 'pot-hooks and hangers' of Nineveh. It is your good fortune . . . that you should have the credit of breaking through the sullen and selfish moroseness of Englishmen, and take the initiative of spreading abroad your arms from Sweden to Italy, and from Paris to St. Petersburg, to shake all fellow-workers by the hand, whether they date their ancestry to Maximilian or Charlemagne, to Julius Caesar or to Peter the Great.

Chapter Ten

'A maze of Correspondence'. 1868-1880

In the New World, as in Britain, the nineteenth century was a time of great upsurge of interest in natural history. In the first half of the century American naturalists were in constant communication with their British counterparts, comparing the huge numbers of a new species of plants and animals then being discovered with their better known relatives in the Old World, classifying them and naming them. In the field of mycology the Rev. M. J. Berkeley had examined and described hundreds of new species of fungi, most of which had been sent to him by M. A. Curtis of Carolina. The material had included not only Curtis's own finds, but some interesting collections by a fellow naturalist, Henry W. Ravenel (Fig. 10.1), a landed gentleman from the same state; and when, in 1848, Ravenel had asked to be put into direct touch with a competent European mycologist, Berkeley, who could not himself undertake any further work, suggested his friend Broome, though a few years later Ravenel was, in fact, writing to Berkeley in person. In 1860 the American Civil War broke out and transatlantic collaboration was severely disrupted, but Ravenel had just managed to complete a series of five centuries (p. 161) of American fungi, the first ever to be published. He suffered badly in the Civil War; all his property was confiscated and he was left penniless. Soon after hostilities ceased in 1865 he wrote to Berkeley telling him of his plight, saying that he hoped to support himself by selling collections of dried plants to naturalists, and asking for Berkeley's help. The kindly clergyman, of course, did all he could by putting Ravenel in touch in touch with British naturalists, and Cooke, who had not yet made his name as a mycologist when Ravenel had been active before the

Fig. 10.1 Henry W. Ravenel of South Carolina, U.S.A.

War, must have been one of the names he suggested. Cooke did his best, as a letter of July, 1868, to the Department of Botany at the British Museum shows:

> I have received from H. W. Ravenel of South Carolina two sets of specimens of the fungi of South Carolina for sale on his behalf–as he is nearly ruined by the War. Can you become the purchasers of one set. It is the most complete set I believe that he has yet sent to Europe. I have no interest whatever in their sale except the desire to help Mr. Ravenel.'

The first intimation that we have that Cooke was in contact with American mycologists is in the autumn of 1867, when he wrote to Ravenel: 'I duly received the specimens of Carolina fungi through the Rev. E. C. Bolles–for which I am obliged'[1]. It must have been an unpleasant parcel, for he continues:

> I am sorry to have found all the Agarici [fleshy fungi] irrevocably mouldy and the Clavariae [club fungi] and some of the Polyporae [pored fungi] devoured by Acari [mites]. Several of the specimens, I could only throw the remains into the fire.'

Most of the letter concerns Ravenel's offer to sell him a large collection of American fungi at $3 per 100 plus carriage, which Cooke refuses; understandably in view of the condition of the parcel he had just received, and of the fact that the offer is based on the number of specimens sent, with no guarantee that they will be of different species. Cooke ends his letter: 'I am very sorry to learn of the uncomfortable position in which events have placed you, and hope you will soon realise the dawn of a brighter future.' Not long after this Ravenel obtained a post in the Government service and though his correspondence with Cooke may have continued, no more is heard of him for nearly ten years.

A year after this exchange it is recorded that Cooke 'had received from Maine some Canada balsam'[2] a then new mounting medium for microscopic objects which later came into universal use. The balsam had undoubtedly been sent by the Rev. E. C. Bolles, of Portland, Maine, Ravenel's intermediary for the despatch of his exsiccati and Cooke's first American correspondent (p. 125). This early contact with Bolles must account for the fact that by 1869 Cooke was already a Corresponding Member of the Portland Society and had published his first paper on American fungi, 'Decades of Maine Fungi', in its *Proceedings*. The introductory paragraph of this paper reads:

> As far as I am aware, no attempt has been made to secure a List of the fungi of any state except South Carolina and Ohio, and certainly no effort has hitherto been made in favor of the state of Maine. Through the kindness of the Rev. E. C. Bolles, I hope gradually to obtain specimens, at least of the more minute, and less perishable, fungi [Cooke has learned his lesson from Ravenel's parcel], from the neighbourhood of Portland, so that in the course of time a respectable list may be constructed of the Fungi of Maine.

Cooke, Bolles, and the Editor of the *Proceedings* all had a fine disregard for geography, for although all the fungi described were undoubtedly collected by Bolles, they were by no means all from Maine, some having their origin in very distant parts of the United States.

The Rev. Edwin C. Bolles (1836–1920) Fig. 10.2), though he himself left no great mark on mycology, must have been extremely useful to Cooke, and seems to have been the inspiration for one strange incident in the latter's career. Bolles had been a minister in the Universalist Church ever since he was 19, and during the Civil War had acted as a chaplain in the 1st Maine Regiment; then shortly after the War was over, probably in 1866, he took over the ministry of the church in Portland. He must have been a keen naturalist and have joined the town's Society for Natural History almost immediately, for he was Vice-President and Record Secretary by 1867 when he was sending his own and Ravenel's collections to Cooke. He was indeed a man of parts, for simultaneously with his pastoral duties he was at different times Professor of Microscopy at St. Lawrence University, Canton, New York, and Professor of English and American History at Tuft's College. St. Lawrence is a small institution, not a University in the British sense, and at that time many of its staff came from Maine, travelling over to give their lectures. It's Theological School is strongly in the Universalist tradition, so it is not surprising that Bolles was in close touch with it, and that it was here that in 1870 he had conferred on him the Honorary Degree of Ph. D. Perhaps it was given in anticipation of his taking up the University's Chair of Microscopy the following year. The syllabus of his ten very comprehensive lectures is still in existence and shows his excellent grasp

Fig. 10.2 The Rev. E. C. Bolles of the Portland Society for Natural History, U.S.A.

both of the principles underlying microscopy and the microscope, and of the animal and vegetable worlds as viewed through the instrument.

It was in the same year that Bolles received his Degree that Cooke claims that the University conferred on him the Honorary Degree of M.A. But St. Lawrence, which has always kept meticulous records of the recipients of its honorary degrees, has only one listed for 1870, that of Bolles; nor does Cooke's name appear in preceeding or succeeding years. What are we to make of this? Did Cooke misunderstand a letter from Bolles concerning his own Degree, or perhaps a suggestion that Cooke might like to put his own name forward for a similar honour? or might he, on hearing the details of Bolles' Degree, have decided that here at last was a way of achieving the status and recognition with which a university education had automatically endowed his more fortunate colleagues, and deliberately have conferred a St. Lawrence degree upon himself? The answer can never be known. Though he continued to use the degree after his name to the end of his life his claim never seems to have been challenged by the University, so it is probable that Bolles, who alone was in a position to question it, believed it to be genuine.

In 1868, two years before this episode, Cooke was elected to the Lyceum of Natural History of New York* as a Corresponding Member. He had not yet

*Now the New York Acadamy of Sciences.

published on American fungi, and no citation for his election survives, so we can only guess that the proposal must have come about through the good offices of E. C. Bolles. Cooke was to make use of the Lyceum's *Annals* for publication purposes only twice, one paper appearing in each of the years 1877 and 1878.

By 1869 Cooke had a very wide experience of British fungi, and sufficient standing in the mycological world to enable him to embark on a major scientific work, as distinct from the more elementary and popular books he had written up to that time. This work was the *Handbook of British Fungi*, which appeared in 1871 and which contains descriptions of each of the 2810 species then known. It was the first complete fungal flora of Britain to appear since Berkeley had described 1390 species in his Supplement to the *English Flora*, 35 years earlier, and also the last one ever to be attempted.

As he explains in the Foreword, the circumstances in which he undertook the job were not easy:

> Pursuing the study of Fungi as a recreation in the intervals of the daily business of life, it was no easy task to prepare and arrange the descriptions of nearly three thousand plants, compare specimens and figures, and measure their spores.

Throughout 1869 and into 1870 he was appealing to colleagues for help with the *Handbook*; to Broome:

> As to the proposed Handbook–I shall be glad if you can bring it under the notice of any Mycologists so that I may hope to obtain sufficient subscribers to enable me to publish, as I must do at my own risk.
>
> Allow me to hope for your co-operation. . . . I have no ambitions to make money out of it. What I most require is descriptions and measurements of the Discomycetes [a group of fungi characterised by approximately disc-shaped fruit bodies] . . . I shall accept them gladly and will give you credit for any such aid honestly. It can scarcely be supposed with my other preoccupations, that I can revise everything. Hence for every kind of help I shall be grateful.

Among others whose assistance is acknowledged are the Rev. M. J. Berkeley, Dr. Capron and Ravenel, while '. . . the arrangement of the Hymenomycetes (the group to which the toadstools belong) is based upon the latest views of the illustrious Fries, with such additions as were recommended by Mr. Worthington Smith.' Smith was also responsible for the measurements of the spores of the toadstools, which relieved Cooke of a very time-consuming job, and for the woodcuts which illustrate each genus.

The book's finances are described in Cooke's autobiographical notes:

> It being a general complaint that there was no Handbook of British Fungi in existence I offered to supply the deficiency if sufficient number of subscribers at 10/- per copy were

found to guarantee me against loss.... When half the copy was in type I discovered that it was already sufficient to form one volume, therefore it was put into paper covers and issued to subscribers, with a promise that the residue should follow soon. When this was done I balanced accounts, and found that each copy had cost me twelve shillings and six pence so that I was a loser by half a crown per copy. The extra copies, remaining after the subscribers were supplied, were bound in cloth, two volumes, and offered for sale at one guinea each. When these were all sold, and the woodcut blocks disposed of, profit and loss was nearly balanced, and the loss, if any, was but small. There was only *one* defaulter, and I forgave him as he was not an Englishman.'

In the Foreword he remarks: 'Subscribers will not regret that, instead of 600 pages they will receive more than 900; and instead of 200 figures upwards of 400.' At the end of April, 1871, he reports to Broome that the book is nearly finished (Figs 10.3 & 10.4).

True to his strongly held beliefs, Cooke makes it quite clear that, as far as classification is concerned, he is having no truck with the vogue for splitting that was then sweeping Europe:

... in the face of the bewildering chaos of new genera which have of late been proposed on the Continent, especially for the Ascomycetous forms, I have endeavoured to avoid,

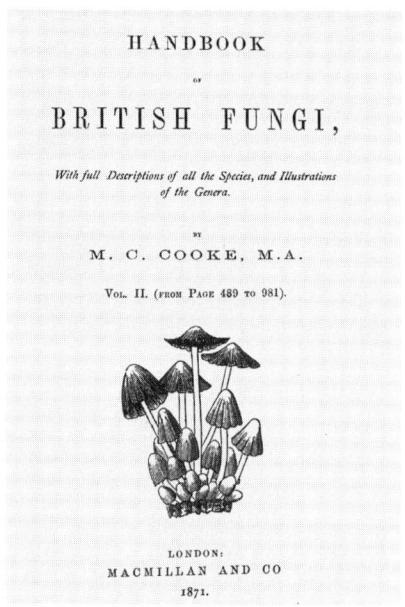

Fig. 10.3 Title page of the *Handbook of British Fungi*, Vol. 2.

Fig. 10.4 Frontispiece of the *Handbook of British Fungi*, Vol. 2, illustrating *Botryosporum pulchrum*. Original with a touch of colour.

as much as possible, encumbering these pages with a nomenclature often fanciful, seldom necessary, and which may, at best, be regarded as transitional.

The *Handbook* was very well received, even Broome commenting that though it was too cheap at 10s 6d, here at last was a reliable text within the reach of all. Among the features he particularly commends are the woodcuts, and the fact that spore sizes are given in metric measurements as well as in inches, as would, of course, be expected of Cooke. As we have seen, he entrusted the publication of the book to Macmillan and Co. rather than to Hardwicke. No second edition appeared 'for lack of the necessary leisure', though a 'second and revised edition' of the Hymenomycetes alone was published as an appendix to *Grevillea* in 1883–1891. Some ten years after publication the *Handbook* was sold out, and Cooke was unable to satisfy the Italian mycologist, Saccardo's (see below) request for a second copy.

As Cooke made clear in a talk he gave a few years later, the publication of the *Handbook* did not have at all the result he had hoped for:

If I may judge from my own experience, it is often the case that the individual has little power to control the direction of his studies after he has once started on his course. My own predilections would be in favour of more exclusively physiological investigations, but finding no systematic arrangement of the British Fungi in existence ten years ago, as the basis of operations, I set myself at first to what I conceived the most essential work, and issued the 'Handbook.' Then, I thought, surely it would be permitted me to pursue some course of investigation; but, on the contrary, the publication of the book increased the number and interest of the workers, and gave such an impetus to British Mycology, that at once I became drawn into and involved in a maze of correspondence, not only in the British Islands, but over Europe and America, with regard to the species contained in this book, and the two thousand soon became three thousand; so that in self-preservation and in self-defence I was driven to the study of allied species in other countries, and now I am so committed, by the study of certain groups of which I have consequently acquired a large experience, that I have no alternative but onwards, or to relinquish the study altogether ... There is still another reason why I am precluded from absolute physiological studies. Having espoused what might be termed strong conservative views on mycological matters, I feel it my duty to science to resist, not only by words, but by work, the innovations of a modern school of Radicalism which threatened to sweep away all old landmarks, and, by wholesale manufacture of new genera and species on illogical, shifting, and unstable bases, to bring science, in so far as Mycology is concerned, into contempt [3].

This 'duty' Cooke had already been carrying out assiduously for five years when he wrote the paragraph quoted, and was to continue to do for the remaining thirty years of his intensely active mycological life. He has only left one record of a behavioural observation made by himself. In a short paper in 1883 entitled 'Circumnutation in Fungi' [4] he describes how he grew the spores of several species in drops of liquid on microscope slides, incubated them in a moist atmosphere, and watched the spiral growth of the germ tubes. He suggests that further observations might yield interesting results.

Cooke's *Handbook* appeared in the same year in which Annie left him to become Mrs. Cubitt, and in which her daughter Mabel was born, so it would have been entirely understandable if emotional distress and domestic upheaval had combined to slow down his leisure activities: but this was not the case. Although he published no more books for four years, and his pioneering efforts on behalf of amateur naturalists diminished, new mycological ventures proceeded apace. Either he was genuinely untouched by his family problems, which seems unlikely, or he was one of those people who, in a crisis, bury themselves in work in order to stave off unbearable introspection.

His reputation as a mycologist had already reached Scotland, for as far back as 1865 he had been elected a non-resident Fellow of the Edinburgh Botanical Society, and had presented copies of his books up to that date to its library. What contact he had had with the Society since then we do not know, but now he was approached by Professor John Balfour and asked to arrange in correct botanical fashion the fungal herbarium of the Edinburgh Botanic Garden, to which end the collection was sent down to him in London. There were considerable gaps in it, and the story still circulates among to-day's mycologists that, during its stay south of the border, some of the specimens found their way into Cooke's own herbarium. The tale would seem to be entirely without foundation in the light of a letter he wrote to Broome in September, 1871:

> I am arranging the fungi of the Herbarium at the Botanic Gardens, Edinburgh, which has been sent me for that purpose by Professor Balfour. It is very deficient and I should be glad if you could send me any duplicates that you can spare to augment it. He will gladly acknowledge them. As it is the National Scotch herbarium I hope you will do your best for it.

In March the following year Cooke was acknowledging Broome's contribution.

We have seem that at some time in 1871 Cooke left the editorial chair of *Science Gossip* and so found himself with some 'spare' time on his hands. His next major venture followed swiftly; in his own words:

> I had long discovered the misfortune that we had no journal in this country which was devoted to Cryptogamic Botany, and the existing Botanical journals evidently gave the cold shoulder to Cryptogamic communications. After consulting M.J.B., [Berkeley], Dr. Robert Braithwaite [p. 120] and other cryptogamists, I resolved on starting a journal myself, and in 1872 appeared the first number of 'Grevillea, a journal of cryptogamic literature', for the first year *monthly*, and afterwards to the end a *quarterly* journal with a subscription of five shillings per year, and it continued and flourished for twenty years in my hands. It was named in honour of R. K. Greville, the author of Greville's Scottish Cryptogamic Flora

It is doubtful whether Cooke had ever met the great Scottish botanist–of whom the *Dictionary of National Biography* says, 'few men have done as much for descriptive cryptogamic botany in Britain'–for he died in 1866; but his herbarium had been bought by Edinburgh Botanic Garden and may have formed part of the collection Cooke was then arranging. In any case, Scottish mycology must have been in the forefront of his mind at the time.

The purpose of the new journal is set out as follows:

> The intention of its projectors is to furnish, month by month, descriptions in English of new species discovered in the British Islands, and to record habitats of rare and interesting forms, ... to furnish a record of the literature; and as far as space permits, descriptions of Exotic species, especially those of the British Colonies and dependencies and the United States of America, or wherever the English language is spoken.

Grevillea, with its numerous full-page plates of which some were in colour, proved its usefulness immediately, for:

> ... throughout the first volume [in fact, the first four volumes] Berkeley contributed a series of 'Notices of North American Fungi', being the descriptions of new species, and others worthy of note, which had remained unpublished from the numerous specimens which had from time to time been communicated to him from the United States.' (Aut)

From now on Cooke was to publish most of his own taxonomic studies in his journal, the later volumes consisting almost entirely of his articles. Contributions were received from other well-known mycologists and lichenologists, but other branches of cryptogamic botany were little represented (Figs. 10.5 & 10.6). It is of interest that at least by 1881 [5] Cooke was insisting that his authors describe new species of exotic fungi in Latin as well as English, a custom already established in the *Annals of Natural History*, but which would not become a requirement under the International Botanical Code until 1935.

The circulation of so specialised a periodical would of necessity have been limited, although after two years it had extended

> ... itself at home and abroad so as to attain a circulation exceeding that of any botanical magazine yet emanating from this country, ... for twenty years the cryptogamic botanists supported [it] without pecuniary loss to the proprietors ... an event which has no parallel in the history of this country [6].

Nevertheless, the time taken up by editorial duties must have been prodigious, especially as Cooke kept in his own hands, though assisted by Annie, the accounts, the records of subscribers, and the despatch of parts. Some idea of the pressure under which he worked, and of its effect on him, can be gleaned from a letter written in 1885, when *Grevillea* had been appearing for

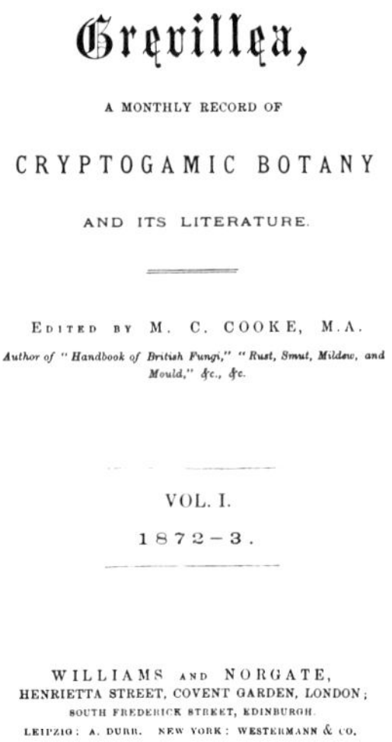

Grevillea,

A MONTHLY RECORD OF

CRYPTOGAMIC BOTANY

AND ITS LITERATURE.

EDITED BY M. C. COOKE, M.A.

Author of "Handbook of British Fungi," "Rust, Smut, Mildew, and Mould," &c., &c.

VOL. I.

1872–3.

WILLIAMS AND NORGATE,
HENRIETTA STREET, COVENT GARDEN, LONDON;
SOUTH FREDERICK STREET, EDINBURGH.
LEIPZIG: A. DURR. NEW YORK: WESTERMANN & CO.

Fig. 10.5 Title page of *Grevillea*, Vol. 1.

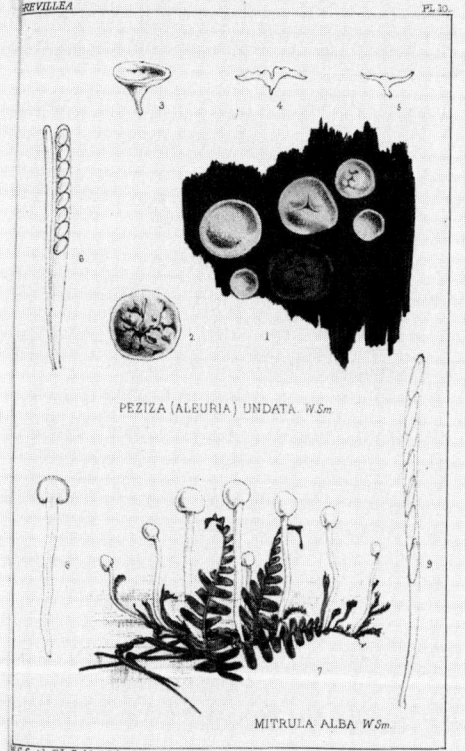

Fig. 10.6 Illustration from an article by W. G. Smith in *Grevillea*, Vol. 1. Original in colour.

13 years, to the Librarian of the Linnean Society, who had evidently been chiding him for some omission:

> I do not forget, but–it is not a part of my business to concern myself with the difficulties of my friends and bear their burdens. No doubt you always do your best for everybody and think it very unkind when they do not estimate it all at its proper value, but such is life!
>
> I have myself to drive everything along at high pressure and bear in mind as little as I can all things that can be got out of it. I cannot afford to keep a clerk or secretary and hence to keep accounts is a bore and memoranda soon get mislaid, Hence I am compelled to be sharp and 'Now or never' for my watch-word–pass on to something else.

In a letter to Broome at about the time that the first issue of *Grevillea* appeared, we obtain a rare insight into a matter which greatly troubled Cooke the mycologist:

> I must confess to considerable disappointment in not meeting you at my house this evening–it was a prospect of an agreeable gossip on subjects of interest to both on which I had long counted. Although in London I work alone and month after month I never see or speak to anyone interested in my own speciality so that I was anticipating more a little gossip than any investigation of knotty points or obscure species. . . . Are you sure that you must return home so soon and that there is no chance of a spare hour *this time*? It is quite useless attempting to enter on explanations upon paper but I was desirous of explaining my prospects and views in respect to the new journal 'Grevillea'! . . .

This letter is more than regretful–it is pleading. Did Cooke feel he was being cold-shouldered by the older man, who was, after all, a gentleman of leisure, and who, to judge from his letters to Berkeley, looked at times on his younger colleague as a considerable nuisance? There were very few mycologists in London at the time, and Cooke's only opportunity to meet such as there were would be outside working hours; so that it was indeed remarkable that he should, with so little opportunity for face-to-face discussion, have been able to establish himself as an acknowledged authority on taxonomic mycology. Soon, when he began to attend more frequently field meetings entirely devoted to fungi, this terrible isolation would be largely overcome, but may it not have been these solitary early years which shaped the profound mycological conservatism for which he is now so well known? The personal discussions he longed for were in no way replaced even by the 'maze of correspondence' which resulted from the publication of the *Handbook*, and which was further augmented by *Grevillea*, with the result that

> parcels of dried fungi came to me periodically from New York, Carolina, New Jersey, California etc., etc. to [be] examined and named or described in *Grevillea* and were soon numbered by hundreds. Afterwards from the Cape, Natal, Brazil and the Australian col-

onies, until they numbered by thousands, and had all to be submitted to the microscope, good, bad, or indifferent.

Meanwhile, he had gained genuine recognition in the United States. Whatever the truth about the degree he claimed from little known St. Lawrence University, now that the publication of the *Handbook* and the first issues of *Grevillea* had established his reputation, the prestigious University of Yale did, without any shadow of doubt, confer on him, in 1872, its own Honorary Master of Arts Degree, though unfortunately no record survives of the circumstances of the award. Cooke was very proud of the honour, and kept the framed certificate hanging over the mantlepiece in his study.

Cooke had spent a considerable part of the year 1873 on duty at the International Exhibition in Vienna. While he was there not only was his son Harry killed, but another event took place which is best described in his own words:

> ... an invitation came to me from the visitor of the Colleges in New York that I had been conferred the honour of LLD [sic] by a College in New York but, on account of the cross postage and absence for a year from England the diploma never reached me and the original letter appears to have been consumed in the bonfire which was kept up for a week on my leaving Junction Road for smaller and cheaper residence. (Aut)

As with the St. Lawrence degree, it is impossible to know what to make of this claim. There are numerous 'colleges' in New York, and more in New York state, many of which would not come near to the standards of a British university. I have, either myself or through the kindness of Prof. D. Rogers (University of Illinois), approached all I could discover, but none have any record of a degree conferred on Cooke. It therefore seems unlikely that the college was one of any standing; possibly it was a small establishment that has lost its older records; possibly it is no longer in existence. Whatever the solution, Cooke was convinced of the authenticity of the honour and henceforward took the title of Dr. Cooke. Berkeley may have been voicing a more general curiosity when, early in 1875, he asked Cooke about his degrees, and the latter replied:

> As to your query.... I am indebted to my U.S. friends connected with Universities for the Honorary Degrees of which three have been awarded to me during the past three or four years–although I have done little to deserve them.

There must have been considerable ill-feeling about them among colleagues, although apparently no doubts, for Leila Cooke told me that her father was often hurt by jibes about their worthlessness.

It has been noted that as early as 1864 Cooke had been selling specimens to the Department of Botany at the British Museum, but the first record we have of his actually working there is a letter dated July 6th, 1872, from Henry Tri-

Fig. 10.7 Undated portrait of M. C. Cooke.

men (lately of the Society of Amateur Botanists and by then an Assistant at the Department) to its absent Keeper, William Carruthers, in which he says: 'I send ... No. 1 of *Grevillea* which looks very well: Cooke is at work here again.' This was presumably the period during which he was arranging the Department's herbarium. The staff of the Botany Department did not include a cryptogamic botanist, the Keeper himself being a palaeontologist, and his two assistants being chiefly interested in flowering plants; so it was probably for this reason that Cooke, having arranged the Edinburgh Herbarium, was asked to undertake the same task for the British Museum. He must have found considerable gaps in the collection, for at Carruthers' request he sold to the Museum '100 special specimens of fungi–from my sparse duplicates of rare new species' for which he received £4.13.0, but found that he really could not part with his own collection and still continue to work on fungi.

That magnificent Victorian shrine to science, the Natural History Museum at South Kensington, was not completed until 1880, so in the 1870s the natural history collections were still housed at the grossly overcrowded British Museum at Bloomsbury, where conditions for the small staff of the Botany Department were almost intolerable. It consisted of:

> ... what was practically one long gallery divided into two portions: the first, where the assistants worked, contained the arranged portion of the herbarium; in the second, out of which opened a small public gallery were the attendants–in this the unarranged collections were stored in large cupboards ... the Keeper's room adjoined the herbarium and was entered from it [7].

But the Department was lucky in its newly appointed Keeper, William Carruthers, a tough but kindly Scot, of whom his obituarist writes:

> With his assistants ... his relations were always most friendly: during my twenty-four years' association with him, I can remember no occasion on which any friction arose between us. ... An assistant from another Department ... once referred to the botanists as 'a happy family', and the phrase not inaptly expressed the prevailing atmosphere [7].

To those outside his Department with whom he did not see eye to eye, however, he could be less amiable, and if he had made up his mind of the justice of his cause, nothing would move him.

For many years past, a battle had been raging at the highest levels between the Department of Botany at the British Museum and the Royal Botanic Gardens at Kew, over the very existence of the former. The Museum owned some extremely valuable herbaria collected by eminent botanists of the past, and Kew, led by Sir Joseph Hooker, argued that they would be of much more use if housed together with its own extensive collections and in close proximity to its Gardens. The transfer would, of course, have crippled or extinguished the Museum's Department, something to which the Government would not have been averse, as it saw little point in contributing to the support of two botanical establishments. However, there were no buildings at Kew suitable for housing all the collections, and in any case they would not have been open to the public as they were at the Museum, two arguments which the Museum was able to use to good effect. In 1871 the dispute came before the Royal Commission on Scientific Instruction and the Advancement of Science, and much lobbying for a number of different solutions went on in the scientific community. It fell to Carruthers, only a few months after his appointment as Keeper, to defend the Museum's case before the Commission, and this he did to such good effect that in January 1873 the tables were turned, and a memorial was addressed to the First Lord of the Treasury opposing a proposal to move the Kew collection to the British Museum. It was signed by 50 eminent botanists, headed by Berkeley in his capacity of Botanical Director to the Royal Horticultural Society, and including Charles Darwin and Cooke. This was one of the rare occasions in which Cooke became even marginally involved in public affairs. In the event, of course, no change was made at the time, and both collections were still intact in their original locations until very recently, when the Museum's fungi were transferred to Kew.

In Nov. 1872, presumably when the arrangement of the fungal Herbarium was nearing completion, Cooke proposed to Carruthers [8] that he should:

> prepare a Catalogue of the Fungi in the Museum Herbarium on the terms of £6 per 8vo sheet of 16 pages–the drawings for any plates made and supplied by him to be allowed for at the rate of one 8vo page plate to be equal to two pages of letter press.

On Dec. 14th the Museum's Trustees approved this arrangement and:

> directed that Mr. Carruthers secure the manuscript of so much of the Catalogue as could be produced on the terms mentioned for £150 the sum which Mr. Carruthers had been authorized to spend during the current financial year.

Cooke began with the Discomycetes, a group of Ascomycetes characterised by their disc- or cup-shaped fruit-bodies, and such was the speed with which he worked that he had completed the first part of the Catalogue, and been paid for it at the agreed rate, before leaving for the Vienna Exhibition in March. On his return from Vienna, probably some time in November, he set to work again, and was paid for the second part and some illustrations on March 14th, 1874. In all he received the sum of £156, £6 more than had actually been authorised by the Trustees.

It was seemingly not until October that he realised that he had seriously underestimated the time and labour that such a work would entail and, apparently after having discussed the matter with Carruthers without receiving satisfaction, he wrote him a long and detailed letter explaining his reasons for seeking an increased fee. The letter gives an excellent idea of the protracted and painstaking work involved in the preparation of a herbarium catalogue.

> Suppose the collection contains 50,000 or 60,000 specimens of fungi. These have all to be gone over and grouped first of all in their Natural Orders, for this elementary work had not even been accomplished; . . . Once grouped into Orders, the special work of the Catalogue commences and all the specimens have to be classed again under the different genera, and laid down for mounting by the assistants. When this is done each species has to be reviewed, containing sometimes 20 or 30 specimens, and *this* in both British and Exotic sections of the Herbarium. Upwards of 2000 species are included in the first volume of the Catalogue now ready. You are aware that in the Ascomycetous Fungi the specific characters depend primarily on the structure of the fruit. In *all*, this fruit is enclosed in the hymenium. Before any specimen can be determined from its fruit a very small fragment of the disc must be placed in water and remain there until sufficiently softened, and the peridia are expanded by endosmosis enough to furnish the necessary elements of diagnosis. In the Discomycetes two hours soaking may do this, but in more coriaceous forms six hours is necessary . . . the most expert and experienced of mycologists could not discriminate one from the other by any means except that above described, so alike are they in size, colour, and all external features, even when magnified. No other course being available, this labour must be encountered. This is independant of measuring the sporidia by micrometer which you wished to have tested in every instance, but which the rate of remuneration renders impossible. When the species are all determined and mounted, the specific name has to be inscribed on every sheet, and the whole grouped to their affinities and ultimately numbered.
>
> The literary portion of the work has also to be done, since all the foregoing relates to the herbarium each species has to be described, all synonyms tested, hunted up, com-

pared, references given to figures, and geographical distribution ascertained. Then, follows a catalogue of specimens in the National Collection. Including of course a thorough exhaustion of all sources of information bearing on individual species. Finally the proof sheets will have to be carefully read over and corrected as they pass through the press.

And what of those fungi which have three or four different spore forms, each of which must be put through all the stages of the examination separately, though all will be grouped under one specific description? He goes on to point out that journals such as the *Popular Science Review* pay their authors ten shillings per page for an article which has entailed no previous laboratory work. Finally he writes:

> Assuming that I am not overpaid in my own department [the India Museum] to which I have been thirteen years attached, and am paid at a *per diem* rate of one guinea and a half for five hours work, it may be fair to compare this with the Catalogue work. I have calculated that the time occupied by me upon the first volume averages for the whole a payment of not exceeding five shillings a day, and I contend that this is not a fair remuneration for scientific work. Even assuming that one guinea per day is to be accepted as a respectable minimum rate, what are we to say to one fourth of that sum, except that it is unreasonable to suppose that anyone who values health, or can turn his leisure to profitable account in other directions could continue to be satisfied with a less payment for scientific work than is given to a journeyman carpenter or locksmith.
>
> I do not ask that my statement should be accepted on my own word, but that someone who has had experience, and knows all the practical difficulties connected with fungi, such as the Rev. M. J. Berkeley, should be consulted as to what is a fair remuneration for such work in its entirety as I have had to perform, and, if the case is thus fairly stated I am willing to accept his adjudication.

Carruthers' reply to this lengthy letter has not been preserved, but by November 18th a fresh storm had broken–Carruthers was not satisfied with the completed manuscript and had apparently threatened to withold it from publication. Cooke writes:

> I regret being compelled to recognise the fact that you are dissatisifed with the Catalogue of *Discomycetes*. If you will permit me I will go over the remainder of the MSS succeeding that returned to you, and add in their places the few additional species and some references since obtained. This will according to my view complete the Catalogue up to date agreeably to the terms of my agreement. I shall *then* have no hesitation in relieving you, on your part, from the arrangement in so far as regards the residue of the Fungi. I have no desire to devote myself to work in which I can given no satisfaction, and which affords no commensurate remuneration.
>
> Having thus completed the MSS–as I suggest, it remains with you to employ your own discretion as to publication. I shall not however feel disposed afterwards to go over the MSS and make additions and corrections which may be rendered necessary by delay

in publication, except by a special arrangement for remuneration for time so employed. As to the work already done, I contend that it has been honestly and conscientiously performed and fulfils all the conditions of the agreement.

He again offers to submit his work to a mutually acceptable referee, ending his letter:

> It is only just that I should elect to be judged by an expert and this I claim, because I am not ashamed of my work, which concerns my reputation quite as much as yours, and you must concede that the only equitable course to adopt is that which I have suggested.
>
> I am quite conscious of your zeal for the honor of your Department, and am sure you will recognise no necessity for dimming its glory by even the semblance of injustice.

On November 24th Carruthers replied that he would now accept the first part of the Catalogue, as Cooke had admitted his fault by 'completely remodelling' the manuscript, a statement which Cooke vehemently denied in a letter to Berkeley the following day, insisting that he had made only very minor alterations. Carruthers then goes on to tear to shreds the second part, which Cooke had by then submitted, accusing the author of simply reproducing descriptions already published by himself and others, especially in one genus (*Morchella*) where:

> I cannot find . . . a single indication of new and independent work for the monograph.
>
> I submit whether this can be considered an honest and conscientious fulfilment of the work you undertook.
>
> As the MSS. on which the foregoing statements are based is in your view 'complete', I beg to say emphatically that it is impossible for me to publish it.
>
> I have a right to expect an original monograph exhibiting honest bona fide work; and I have to request that you will as soon as possible fulfil this engagement.

The next day Cooke sent Carruthers' letter to Berkeley, commenting on each criticism raised and asking for Berkeley's judgement. On the accusation of plagiarism he writes: 'I do not deny this–I contend that I could do no better,' and claims that there was no occasion for independent work in the genus *Morchella*.

> Do you think from this letter that I have done wrong as far as you can judge–or do you think that Mr. Carruthers is at all competent to judge on the details of a Mycological monograph?

Berkeley's reply has not survived, but the fact that Cooke was prepared to submit to the arbitration of the highest mycological authority in the land implies that right was on his side.

In the event, the Catalogue was never published, despite Carruthers' final acceptance of the first part. Needless to say, Cooke received none of the extra payment he had sought. It seems that Carruthers was in the habit of driving hard bargains with the casual employees of his Department, for Worthington Smith fared just as badly. In May, 1871, only a few months after Carruthers' appointment as Keeper, Smith, who had close connections with the Museum, wrote to Broome as follows:

> Carruthers has now made this [new] offer to me ... He says if I can give them a few hundred dried specimens he will get £4 a hundred paid me–if I will give four days of my time for each 100. This is better than nothing, and I can easily spare some 300 duplicates. This will give me a chance of giving twelve clear days to arrange the plates.

Smith's income for twelve days' full-time work would, therefore, have been £12. At £1 a day, from which he must subtract the value of the 300 dried specimens he was losing to the Museum, the rate was little better than that originally agreed by Cooke, though Smith was only handling Agarics, which would have been less troublesome than the Ascomycetes with which Cooke was dealing.

Though Cooke continued to use the Department's collections, as every member of the public was entitled to do, he was never again employed there and, at least for a while, his opinion of it was low, as is shown by his comment to an American correspondent in 1876:

> I never received from you any communication addressed to the British Museum–and I may add that I have never heard of the authorities of the Botanical Dept. of the British Museum sending on any communication addressed to persons not at the time under actual engagement–in which this Department has much to learn from similar departments in all parts of the civilised world. It is proverbially one of the most uncourteous branches of our scientific institutions [9].

Cooke's labours on the Discomycetes were not wasted. Early in 1875 he wrote to Berkeley:

> It seems an endless task to reconcile the species published by the different authorities –hence I think it would be a good piece of work if I could publish coloured figures of all the species possible of *Peziza* [one of the genera covered by the Catalogue] clearing up as much as I can and giving authorities for all the specimens figured.

Though the illustrations for such a publication would have to be freshly drawn, much of the taxonomic spadework would already have been done, and the first part of the new work, called *Mycographia, seu Icones Fungorum. Figures of Fungi from all Parts of the World*, appeared later the same year. Once again Cooke undertook the publication at his own financial risk, backed only by subscribers he had himself solicited. The introductory 'Notice' begins:

> It would be more than could be expected of one man, within the limits of a natural life, to publish coloured figures of every known species of Fungus . . . Nevertheless one person may contribute to this desirable result, hence I have commenced the present work, intending for the present to confine it to the *Discomycetes*.

Six parts eventually appeared, the last in 1879, when: 'it was discovered to have been attempted in too expensive a style, and did not proceed further than the first volume . . . with 406 coloured figures' made up into 113 plates. Each illustration consists not only of several aspects of the fruit-body, but also drawings of the microscopical appearance of the asci with their contained ascospores (called 'sporidia' by Cooke), which line the spore-bearing surface of the cup, and of the paraphyses (hairs) among which they are embedded.

Soon after the last part of *Mycographia* appeared, Cooke wrote a long article in *Grevillea* [10] setting out what he considered to be the relative importance of the various features of *Peziza*, (Fig. 10.8) on which the delineation of its species might be based, his object, as usual, being to reduce the number of species recognised by placing the greatest emphasis on the

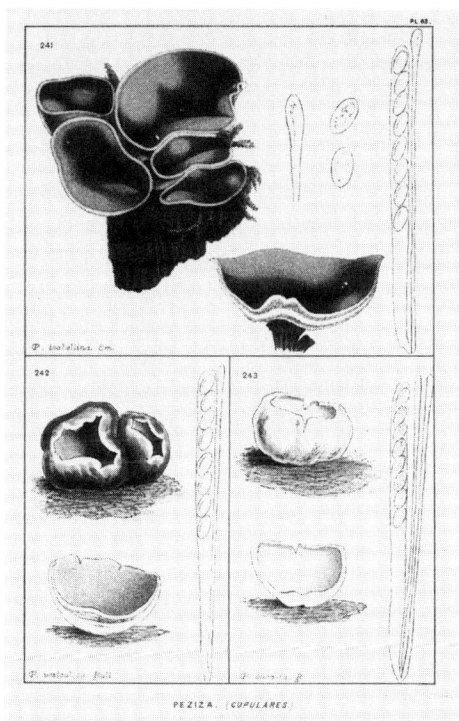

Fig. 10.8 Illustration from *Mycographia*, showing cup fungi of the genus *Peziza*. Original in colour.

most constant characters. Inevitably he comes to the conclusion that features visible to the naked eye—colour, size, habitat, etc.—on which species distinctions were usually based, are in fact the least reliable, and he contrasts these with the remarkable constancy of the morphology of the sporidia (ascospores) of which he says:

> However much we deprecate a system so artificial as the establishment of genera based on the form and septation of sporidia, we accord place of honour to sporidia, in the series of features to be taken into account, in the diagnosis of a species.

Second only in importance to the sporidia he places the paraphyses; other characters, though they must obviously be taken into account, are of less moment.

In the opinion of the eminent contemporary French mycologist, J. L. E. Boudier (1828–1920) the *Mycographia*, despite the inaccuracy of many of the figures due to their being prepared from dried, herbarium specimens, was of undoubted importance because in preparing it Cooke had had access to type specimens in many herbaria, including that of Berkeley. As so much of the work, including those genera most strongly criticised by Carruthers in the Catalogue, must have been based on that prepared for that Catalogue, Boudier's commendation of the *Mycographia* is an indirect vindication of the earlier work.

It was in 1875 that Cooke first began to correspond with the great Italian mycologist P. A. Saccardo (1845–1920) (Fig. 10.10), Professor of Botany at the University of Padua, a correspondence which was to continue for 20 years. Saccardo was working towards the publication of the first volume of his monumental *Sylloge fungorum hucusque cognitorum*, an attempt to catalogue and describe every known species of fungus. But the *Sylloge* was more than merely a compilation of the work of others, for Saccardo made his own, first hand, observations on many of the fungi, and often used these observations in his critical reclassifications. He had apparently written to Cooke early in April, 1875, seeking specimens and descriptions for this work, for on April 30th Cooke replied, sending him two volumes of *Grevillea* and a volume of his exsiccati, and accepting an offer of some of Saccardo's material in exchange. Frequent postcards, and sometimes letters, followed, usually discussing taxonomic matters, but occasionally including an outburst concerning the shortcomings of the postal system.

Meanwhile his work on American fungi had been continuing, and the number of transatlantic mycologists with whom he was in contact was steadily increasing. One of the foremost was Charles H. Peck (1833–1917) (Fig. 10.9), a graduate in Botany who, in 1867, had been appointed by the State of New York to 'fill the Herbarium with specimens representing the plant life of the State'. In 1883 he was given the newly created post of State Botanist, and every year until he retired he submitted a report of his work which, together with his

Fig. 10.9 Charles H. Peck of the New York State Herbarium.

Fig. 10.10 P. A. Saccardo of the University of Padua, Italy.

Fig. 10.11 Job B. Ellis of New Jersey, U.S.A.

scientific papers, still provide a mine of information on the fungal flora of the State, for it is said that he described 2700 new species. He corresponded with a number of European mycologists, and in 1872 had collaborated with Cooke to write two papers on American fungi in the *Journal of Botany* and one in *Grevillea*

Cooke did not publish on this subject again until 1875, when two more papers appeared, this time in the American press, but his name had been associated with only six papers in all when he was elected a Corresponding Member of the Academy of Natural Sciences of Philadelphia, probably at the instigation of an existing member, Henry Ravenel. Cooke's letter of acknowledgement ran as follows:

> Dear Sir, permit me to thank you for the honor [sic] which the Academy of Natural Sciences of Philadelphia have done me in electing me a Correspondent of the Society. The father of American Mycology the illustrious Schweinitz [1730–1834] communicated the results of his labors through your Academy. I trust that a humble disciple may do something if health and life be spared him in the way of supplement to Schweinitz's work. So many claims are made on my time and not the least my efforts to collect materials for a revision of the Fungi Flora of the United States that I fear I shall be but a bad Correspondent. Nevertheless I appreciate the honor done me, and hope at some not very remote period to produce something which may find a corner in your 'Proceedings'

This he succeeded in doing only once when, two years later, he described nearly 100 species of American Ascomycetes in the pages of the journal. Apparently a further paper was submitted the following year, but this never appeared.

At this time too, he was receiving specimens from another American, Job B. Ellis (1829–1905) (Fig. 10.11), a classics teacher from New Jersey, who dedicated all his spare time to mycology and eventually published a large book on the Pyrenomycetes (another group of Ascomycetes) and numerous exsiccati. He was a shy and gentle man, and a very hard worker, and seems to have been loved by all who knew him. Early in his mycological career he was in correspondence with Ravenel, a correspondence which continued until the latter's death and was probably the foundation on which Ellis's knowledge of fungi was built. It may also have been through Ravenel that Ellis came into correspondence with Cooke. Their series of ten joint papers appeared in *Grevillea* from 1876 to 1880 and included the descriptions of nearly 200 new species.

A little later Cooke was in correspondence with yet another American, H. W. Harkness (1821–1901) of San Francisco. Harkness had practiced medicine for about 20 years when, in about 1870, he retired and devoted the rest of his life to fungi, becoming the first mycologist on America's west cost. He was in touch with colleagues in that continent, and it was doubtless through them that he began sending specimens to Cooke. This collaboration resulted in seven papers in *Grevillea* and one in the first volume of the *Bulletin of the*

Californian Academy of Sciences, all between 1880 and 1885, in which over 200 new species were described. Altogether during his career Cooke wrote, alone and with others, 42 papers on American fungi, in which they described nearly 900 new species. When Cooke's conservative outlook on this matter is taken into account, the enormous scope for taxonomic mycology in the United States at the time can be fully appreciated.

It seems that Cooke always felt more at ease with his Scots colleagues than with the English botanical establishment. He seems to have had as little as possible to do with the latter, even through the medium of learned societies, and most of those with whom he did come into contact were at best irritated by him and at worst angered. Perhaps the Scots were more prepared to ignore the social origins about which he was so sensitive and to accept him as an equal for his contributions to science. If so he might not feel the need, in their company, to be continually on the lookout for condescending or slighting behaviour, and would in consequence be a less prickly and abrasive companion. He had obviously kept up his Scottish connections since arranging the Edinburgh Herbarium, for in September 1875 he was a guest of honour at the dinner in Perth which marked the foundation of the Cryptogamic Society of Scotland. 'There were about seventy gentlemen present ... chair Sir Thomas Moncrieffe of Moncrieffe Bart ... supported on his right by Dr. M. C. Cooke, London.' In proposing the toast to the new society Cooke said:

> I do it with pleasure because I am not a Scotchman; I do it with pleasure because I have a great many friends who are Scotchmen; and some of the finest and best of the botanical friends I have had are Scotchmen. I remember with pleasure that there is scarcely a Professor of Botany in the whole of Scotland that I cannot claim as a personal friend–(applause)–but I cannot say that I am acquainted with every person who holds a Botanical chair on the other side of the Tweed.

Cooke was, of course, made an Honorary Member, and for seven years the Society's meetings were regularly noticed in *Grevillea*. He always claimed that the Society did for mycology in Scotland what the Woolhope Club, with its field meetings devoted to fungi, did for the subject in England.

The last few years had not been confined exclusively to taxonomic studies; Cooke had also found time to write a book on the wider aspects of mycology, *Fungi, their Nature, Influence and Uses*, which appeared in the spring of 1875 in the International Scientific Series. Its publishers, H. S. King and Co. (later editions by Kegan Paul, Trench) had originally asked Berkeley to undertake the work, but owing to 'a multiplicity of engagements' and 'uncertain health' Berkeley would agree only to act as editor, asking Cooke to write the book. The contemporary verdict on it is summed up by the *Popular Science Review* [11]

> We certainly did not think so much could be said on the subject of this volume as the

book contains; but we have been enlightened, and that, too, in the most forcible manner. It must be admitted that the work which Mr. Cooke has completed is a most valuable one; and, in addition, is most interesting reading. He has written a full account of fungi, their nature, structure, classification, uses, notable phenomena; their spores, germination and growth, sexual reproduction, polymorphism, influence and effects, habits, cultivation, geographical distribution, and lastly their collection and preservation. It will be seen from this list of contents that it is essentially a work on the natural history of the fungi, one which deals with their manners and habits, and not one which has to do in the slightest way with species. It is therefore a book which everyone who loves biology must procure, [the word 'biology' is now coming into use] and it is one too which any person who has an ordinary English education may read with the greatest advantage to himself. That it must have a large sale is unquestionable; for it is *the only book of the kind we know of, and it is written by the very highest authority in the country save one* [my italics] and that one (Mr. Berkeley) has gone over the proofs . . .

Indeed, the emphasis placed on all aspects of the life of the fungus and on its interrelationships with its surroundings is a deliberate departure from the Victorian emphasis on discrete life histories and anticipates the much broader modern outlook. An unsigned review in *Nature*, while praising the book as a whole, remarks that it will be of no use to students simply because of the absence of life histories!

The cross-roads at which the biological sciences were hovering at the time is clearly demonstrated by the need Cooke found, in the first few pages, to rebut the theory of spontaneous generation, while emphasising the importance, though extreme difficulty, of ensuring that experimental work was carried out with pure cultures. This is despite the fact that Pasteur's classic paper on the subject, 'Memoire sur les corpuscules organisés qui éxistent dans l'atmosphère. Examin de la doctrine de géneration spontanées', had appeared as long ago as 1861.

One jarring note in the chorus of approval with which the book was received was struck by the ill-tempered and petty review by Worthington Smith [12] in the *Gardener's Chronicle*. After a very long essay in which he damned the work with faint praise, he ended:

> We must confess that after many years' study of fungi our feelings are very much the same as those expressed by Goethe's *Faust* in his opening soliloquy—that the reward is in no way commensurable with the labour of the study . . . we confess that we consider the knowledge of the plants themselves . . . to be little more than a difficult mental exercise.

Writing to Berkeley soon after this review appeared Cooke comments:

> Such a sneer at microscopical species or at a study which brings no pecuniary reward comes ill from the Gardener's Chronicle. However I resolved to have nothing to do with a controversy on the subject to gratify Smith's desire to use his pen . . . you will

estimate his sneer at Doctors and Museums and Clergy when you remember that he adopted the study of fungi under the impression that he could make plenty of money out of it and finds himself disappointed–as he has confessed to me. He believed, he said, that you and me made plenty of money out of fungi–whatever you may be able to do–I never could find the receipts to equal the expenditure on books, instruments etc. not to mention labour, and if I followed mycology for the sake of its substantial remuneration I fear that I should very soon starve.

With this book, however, Cooke was for once to have a financial success. It ran to many editions,

as well as an edition in French, published in Paris, and another edition published by D. Appleton and Co of New York, to which royalty was loyally paid to the very last. I may be permitted to intimate that this was the most pecuniarily successful of all my books, having realized nearly 300 pounds in royalties.

As *Fungi, their Nature, Influence and Uses* had shown, Cooke had an extraordinarily wide and detailed mastery of all aspects of mycology, even though so much of his own energy was expended on taxonomy. From his early interest in the microfungi infecting leaves, which he had collected during field excursions and country rambles, it was but a short step to the study of plant pathology–fungal diseases of plants–for which, in any case, his mentor and friend, the Rev. M. J. Berkeley, already had a world-wide reputation. Berkeley's knowledge of the flowering plants was nearly as encyclopaedic as of the fungi and plant disease, and for many years he had been an active and much respected member of the Royal Horticultural Society, being the co-founder, in 1868, of its Scientific Committee, of which he was now Secretary. But he was over 70 years old in 1876, and it may well have been he who, with an eye to the future, suggested that Cooke should be invited to join the Committee.

The Royal Horticultural Society, though not a learned society in the sense of, say, the Linnean, had nevertheless become a very influential body, and attracted the support and assistance of many of the established figures of the botanical world. Its Head Office and Gardens were at South Kensington where, in 1859, it had leased 21 acres of land, bounded by Cromwell, Exhibition, Kensington and Prince Albert Roads, from the Commissioners for the 1851 Exhibition. The beautiful gardens, which included statues, fountains, a bandstand and a cascade, all surrounded by Italian arcades, were opened to the public shortly before the International Exhibition of 1862 took place on the south end of the site, and were a very popular feature with its visitors (Fig. 5.7). But, elaborate though they were, they were not to survive for long, for by 1877 the site was needed, in addition to that occupied by the Exhibition itself, for the building of the new Natural History Museum. The Society's Head Office, however, remained at South Kensington for another ten years, and it was here that Cooke would have attended meetings of the Scientific Committee.

At that time, the Committee was a large and prestigious body, which included well known botanists from all over the country and even from abroad, so it is improbable that more than a small proportion were ever present at its regular meetings. It was chaired by Sir Joseph Hooker, the Director of Kew Gardens; the staff of the Gardens, and of the Botany Department of the British Museum, would have been regular attenders, as was the Rev. M. J. Berkeley; these were probably joined by notable botanists from Cambridge and other centres within reach of London. Its duties were defined as follows: 'The Committee sits at 4 pm at each of the Fortnightly Meetings, to consider and report on any new or curious plants, diseases, inventions and other objects of horticultural interest or value'. Fellows were recommended to send specimens in, in order to receive the best possible advice.

For Cooke, the invitation to a seat on the Committee must have been a matter of moment, for never before had he sat side by side with establishment botanists as an equal, although he was now 51 years old with a world-wide mycological reputation. He was, as we know, involved with the affairs of a host of natural history societies and other organisations for laymen, but belonged to no learned societies, not even, at this stage, to the Linnean, whose membership at the time included many gentleman amateurs of no particular scientific distinction; in preference to joining the long-established Microscopical Society he had even started a new layman's club, the Quekett. The Royal Horticultural Society, as an important meeting place and forum for the country's botanists, should, if he was prepared to take advantage of it, open up a new world of opportunity for Cooke, and banish for ever the isolation of which he had complained so bitterly to Broome some five years earlier.

He attended meetings regularly and, according to the surviving minutes, took an active part in them. For instance, between March 1878 and May 1879, when the meetings were usually chaired either by Hooker or Berkeley, Cooke exhibited a freak wild primrose, was appointed, with Berkeley and W. G. Smith, to a sub-committee set up to investigate diseased bulbs sent in by members, and read two papers on diseases of vines and one on the secretion from the glandular hairs of *Auricula*.

It may be asked how a man with so slight an experience of the microscopic fungi as Smith had at that time came to be sitting on the Scientific Committee. The reason was that he had been pushed into prominence by the controversy over Potato Blight. Every year, the Irish potato crop was being devastated by Blight. Thanks to the work of Berkeley it was now generally accepted that the fungus *Peronospora* (now *Phytophthora*) *infestans* was the cause of the disease (Fig. 10.12). It was also known how the organism spread from leaf to leaf and from leaf to tuber in the summer months by producing tiny spores which were carried from plant to plant by wind and rain, to give rise eventually to minute, motile zoospores, able to swim in films of surface mois-

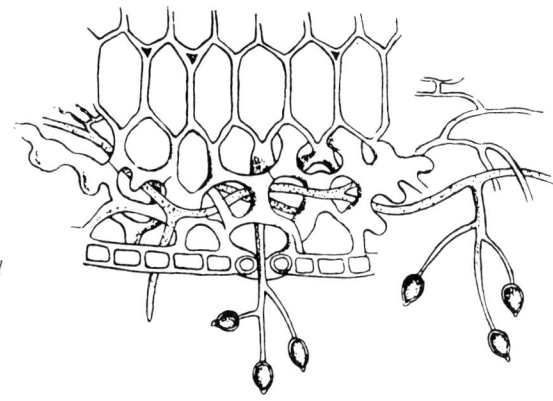

Fig. 10.12 Illustration by M. J. Berkeley of a section of a potato leaf invaded by the blight fungus, *Phytophthora infestans*. (*Journal of the Royal Horticultural Society*, 1846).

Fig. 10.13 Microscopists (*left*, Cooke; *right*, Berkeley) examinining the potato blight fungus. (From a cartoon by W. G. Smith, *Graphic*, 1875).

ture, and germinate to infect a new site (Fig. 10.13). This discovery was a big step forward in elucidating the life of the fungus, but it did not solve the problem of how it overwintered and infected the new crop in the following summer, and hence how the disease might be controlled. Mycologists of the calibre of De Bary in Germany and Berkeley in England were eagerly searching for a hypothetical resting spore which could tide the fungus over the winter when, on July 10th, 1875, Worthington Smith announced in an article in the *Gardener's Chronicle* that he had succeeded. He published his triumph in other prestigious journals, Cooke reprinted it in *Grevillea*, the Royal Horticultural Society bestowed upon Smith its Knightian Gold Medal, and he became the hero of British mycology. He was a guest of honour, with Cooke, at the inaugural dinner of the Cryptogamic Society of Scotland which took place that year, and Cooke made due reference to the discovery in his address: 'by some fortunate experiments he [Smith] was able to add the final link–the great missing link–of the chain of the history of the potato fungus–(applause)'.

But Smith's triumph was short-lived. By December Professor Anton de Bary, in Germany, had shown that the spores and their accompanying male fertilizing organs, which Smith had depicted within the potato leaves, belonged to another fungus which had contaminated his cultures, and that his drawings were even in some degree the product of his imagination. Britain had no experimental mycologists at that date capable of repeating and checking the observations of Smith and de Bary, and the journals which had published Smith's claim with such a flourish dared not risk losing face by throwing doubt on his work, so his reputation in Britain remained high until the true resting spore was discovered nearly 20 years later.

It could only have been Cooke's membership of the Scientific Committee and the consequent widening of the botanical circles in which he moved that led at last, in 1877, to his being proposed as an Associate Member of the Linnean Society. The Linnean, founded in 1788 and named after the great Swedish taxonomist Carl Linnaeus (after whom Cooke had called his eldest son), was then as now, the foremost learned society for both botanists and zoologists with an interest in taxonomy and collecting. But at that time, though most eminent biologists were Fellows, the Society counted among its members many amateur naturalists who had done little for their subject except perhaps make a collection of the local flora or fauna at some time in their lives. In view of this it is astonishing that Cooke, who was now accounted an authority throughout the mycological world, had not become a Fellow, let alone an Associate, earlier. The omission was almost certainly the result of the social gulf between himself and his botanical peers. Membership of the Scientific Committee of the R.H.S., however, brought him into regular and frequent contact with the botanists at Kew, and it was four of them, George Bentham, Daniel Oliver, W. Thiselton Dyer (lately of the Society of Amateur Botanists) and J. G. Baker, who headed the list of nine Fellows who sponsored him for Associateship on February 1st, 1877. The remaining sponsors were zoologists who had probably been asked for their support by the four botanists. Two months later the ballot was held, and 'M. C. Cooke, MA, LLD' was elected. It is difficult to estimate to what extent he took part in the Society's activities or how much he used its fine rooms in Piccadilly, though his daughter told me that he would sometimes meet his friends there. He seems to have been on familiar terms with the Society's Librarian, for in a postscript to an otherwise very brusque letter to him in 1885 (see p. 181) he wrote: 'You shall have a noggin of the whiskey–when I get you on the sly–to damp down the New Year'. He read at least one paper, on the fungi of Texas, at a meeting in April 1878, a summary being published in the Society's *Journal*; in the year of his election he and Berkeley published a joint, 25-page paper on Brazilian fungi in the same journal; the following year there was a six-page article on the coffee diseases of South America, and in 1883 a brief taxonomic note. After that date he made no further use of the Society's publishing facilities.

Cooke's next venture would display yet again his tremendous versatility. In 1875 the Polish mycologist, Dr. Joseph Rostafinski, a pupil of the great German mycologist Anton de Bary, had published in his own language a monograph on the Myxomycetes (slime moulds), using microscopic criteria in their classification for the first time. In August of that year he had arrived in England to visit Berkeley, and the latter had invited Cooke to meet him. Cooke replied: 'I should be glad to meet Dr. Rostafinski–but I fear I shall not do so–I go to Wales on the 23rd for a week, as I have not had an "outing" this year'. We do not know whether the two men ever met, but Cooke was greatly interested in an up-to-date classification of so neglected a group as the Myxomycetes and, having heard about the monograph (Fig. 10.14)

> determined to procure [it] with a dictionary and grammar, and proceed to learn enough of the Polish language to enable me to translate as much of the monograph as referred to the British species, and with the assistance of a commercial Pole whom I discovered in the City to help me over obscure passages, I completed and printed the reprint [as *Myxomycetes of Great Britain*] in 1877, with 24 plates copied from the monograph,–a very few copies of which were sold, and I had the priviledge of paying the printer's bill as my reward (Aut).

Such an expense must have been particularly unwelcome just then, as it would have come at about the time that Annie returned to Junction Rd. with her two young children, Mabel and Frank, and household expenses would have soared. Cooke's remarkable linguistic tour-de-force remained the only treatise on British Myxomycetes for nearly 20 years.

The year 1877 also saw the beginning of Cooke's personal contacts with French mycologists, as in disclosed in a letter he wrote to Saccardo in Sep-

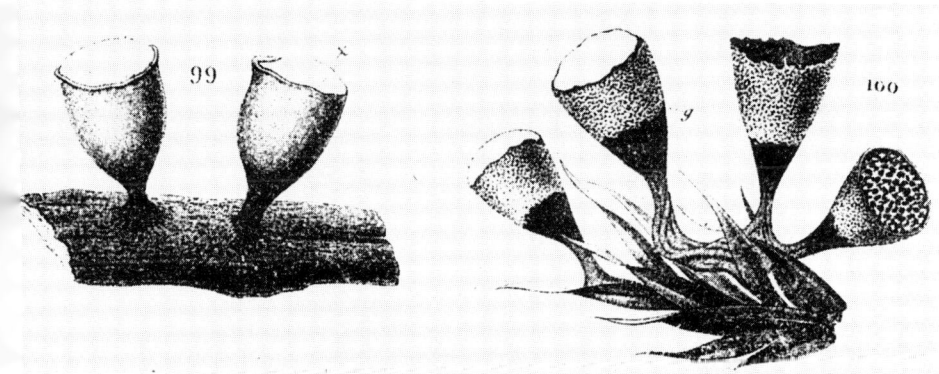

Fig. 10.14   Illustration of *Craterium* spp. by J. Rostafinski from *Myxomycetes of Great Britain*, Cooke's partial translation of Rostafinski's book on myxomycetes.

tember–'M. Jules de Seynes and M. Max Cornu are to be with me for a week making mycological excursions in this country'–and in the same month the following year he attended the Paris field meeting, devoted to fungi, of the Société Botanique de France. He has left a full account of this meeting, the excursions, the exhibitions, the dinner and the toasts, in *Grevillea* [13]. French mycologists were indeed dedicated:

> 'The excursion to the Forest of Fontainbleau raised some of us from our beds at 5 o'clock, in order to catch the 7 o'clock train at the other extremity of Paris. After a ride of 37 miles breakfast was welcome, but before it was partaken of an outlying portion of the forest had to be explored, so that the hotel was not reached until 12 o'clock. Justice to the viands was fully rendered.

It must have been through contacts such as these that Cooke met Dr. Lucien Quélet, shortly to become co-founder of the world's first mycological society. Quélet, a medical man, was an ardent and knowledgeable collector of fungi who persuaded Cooke to join him in the publication of the *Clavis Synoptica Hymenomycetum Europaeorum*,

> which was an enumeration of the species of the larger fungi indigenous to Europe, with descriptions of the species, in Latin so as to be available over Europe, but that was also to a large extent a failure,

although it was published in Paris, Vienna, Berlin, Milan and New York as well as London. The dedication of the book: 'To the Illustrious and Venerable the Reverend M. J. Berkeley, M.A., F.L.S., 'The Prince of Mycologists' ... with the respect and esteem of the Authors', is evidence, if such were needed, of the veneration in which Berkeley was held in the mycological world.

Cooke's own international standing was further recognised when, after a ten-page article of his had been accepted by the *Nuovo Giornale Botanico italiano*, doubtless through the good offices of Saccardo, that gentleman suggested that he should put his name forward for election as a Corresponding Member of the Societa Crittogamologica Italiana. He was duly admitted in 1878, together with Berkeley, Braithwaite (the bryologist and member of the Quekett Club) and a number of European cryptogamists.

All through 1878 and 1879 Cooke and Saccardo were exchanging fungi and opinions, for Saccardo was now deeply engrossed in his *Sylloge*, the first volume of which was to appear in a few years' time. Although the two men disagreed profoundly on certain aspects of the classification of fungi, Saccardo must have valued Cooke's criticism, for as early as June 1878 the latter was acknowledging the receipt of Vol. 2 of the *Sylloge*, though whether of a manuscript or a pre-printing he does not say. Earlier that month the collaboration seems to have been in some danger of coming to an abrupt end owing, as usual, to Cooke's outspokenness and lack of tact. Saccardo had sent him an enumeration of certain Ascomycete genera, and Cooke replied that it omitted all his (Cooke's) species:

> If you had informed me of your preparation of such a monograph I think I could have assisted you, as I have examined a great number of Berkeley's species. Your work is consequently much more imperfect than it might have been, and some errors could have been avoided.'
>
> I enclose a curious species which I suppose you will consider a *new genus*. What a *great mistake* these new genera are. The next generation will certainly sweep most of them into oblivion (PAS)

Saccardo must, understandably, have been both angry and upset by this, for Cooke's next letter, 12 days later, runs:

> ...I do not think you understand me accurately. I fully credit you with the good service you have rendered Mycology, and in my remarks on your genera of Nectriacei I only stated my opinion generally, which I shall adhere to, that the principles of classification now being adopted by many mycologists for *Ascomycetes* is wrong, as it is purely *artificial*–and not a *natural* system. The artificial has been discarded for the Phanerogamia [flowering plants] and I consider it a backward step to return to an *artificial* system for Ascomycetes.
>
> As I cannot write you in Italian I cannot hope to convince you that I have good reasons for my opinion, so, I must leave it, but I think it would be folly to permit this wide difference of opinion to cause any ill feeling between us. I hold I am justified in my opinion as you may be in yours (PAS)

The natural classification which Cooke espoused was an attempt to classify organisms on the basis of their real or supposed relationships to one another, rather than on the purely artificial basis of the number or appearance of certain parts–stamens, spores, septa, etc.- which had been used in the original Linnean system. As Cooke notes, the natural system, which was a by-product of Darwin's evolutionary theory, had largely taken over in the classification of flowering plants, but it was almost impossible to apply it in practice to fungi with the limited knowledge of them then prevailing. Indeed, a few years later Saccardo demands of Cooke why, with his views, he continues to use the highly artificial feature of spore colour as a primary character in the classification of agarics. By October the two men seem to have made their peace, for Cooke is thanking his colleague for paying him the compliment of naming a new genus, Cookea, after him; but he declines the offer on the grounds that there is already a genus *Cookia* in the flowering plants, which would be confusing. The genus was eventually named *Cookella*, the type-species being *C. microscopica*, a fungus growing on the underside of oak leaves.

In addition to the huge amount of other work on which Cooke was engaged he had continued to publish exsiccati intermittently ever since 1865, despite the time-consuming nature of the work. In 1875 he had begun a new series* on which the *Gardener's Chronicle* commented that it was a great improvement on the former series, the specimens being much more neatly disposed: 'Dr. Cooke, who is certainly indefatigable, has, in a large proportion of specimens, figured the spores under the camera lucida†,' Now, Henry Ravenel, the first American with whom he had been in correspondence, wrote to him once more. Ravenel had recovered from his losses in the Civil War and was employed as Government Botanist, thus having excellent opportunities for taking up his mycological activities again. He proposed to Cooke that the latter might like to undertake to edit and publish a series of exsiccati of American fungi made up from material which Ravenel would send him. Cooke promptly acceded to the idea and eight centuries were published between 1878 and 1882. In December, 1877, Cooke publicised the forthcoming event in *Grevillea*:

> H. W. Ravenel, who is well known to mycologists ... is now collecting specimens in Florida and Georgia with a view to their publication. The new series will probably contain about four fascicles of 100 specimens each, and will be issued to subscribers at 21 shillings each fascicle. The series will be edited by M.C. Cooke'.

About 30 sets were prepared.

> Ravenel took no part in the actual preparation and distribution of the series other than to wrap up his collections in more than 40 packages and dispatch them to Cooke. The latter planned the format, named the fungi used, made up the volumes, advertised and distributed them, and collected the proceeds, an unknown portion of which he sent to Ravenel [14].

There are two methods of preparing exsiccati, concerning the relative merits of which there was considerable discussion at the time. In print Cooke held very strongly to the view that, where possible, the dried specimens should be glued directly to the sheets rather than being placed inside envelopes fixed to the sheets, as he considered that the first method prevented any possibility of mixing specimens, allowed for easier fumigation against infestation by mites, and gave better protection against damage by handling. However, as speci-

---

*Seven intact folders of this series, complete with specimens, are lodged at the Perth Museum and Art Gallery, Scotland. They are titled 'Fungi Britannici Exsiccati, Editio Secunda, Cura M. C. Cooke, Fasciculus ...', and are dated between 1875 and 1879. Unlike the First Edition, the folders were prepared by Williams and Norgate, not Hardwicke, probably as a result of Cooke's quarrel with Hardwicke in 1870 rather than on account of the latter's death in 1875. I am very grateful to Mr. M. A. Taylor, Keeper of Natural Sciences at Perth Museum, for this information.

†An instrument allowing the microscopist to trace on to drawing paper the image he sees.

mens must be freed from the sheets for detailed examination, and this will inevitably result in more damage than removing them from an envelope, as well as giving the same opportunity for mixing them up, the use of envelopes would seem preferable and is the preferred method to-day. In practice Cooke used both methods, the reasons for his choice in individual cases seldom being obvious.

Chapter Eleven

Royal Botanic Gardens, Kew. 1880-1884

Cooke was a lifelong missionary, that overriding facet of his nature never altered. But over the years his interests changed, and so, necessarily, did his mission field. At Trinity School it had seemed that his ambition was to ensure for the biological sciences the place that was their due in an elementary school curriculum fitted to the modern world. He would do this by breaking down the antagonism of the Church, which wielded absolute power over the subjects taught, and by educating the teachers through lectures at the meetings of their societies, articles in their periodicals, and special museums which he worked hard to set up. Towards the end of his scholastic career he began to spread his gospel further, to the working man, for whom he organised evening classes in botany, always accompanied by the field-work which he regarded as of such prime importance.

Not surprisingly, in view of the circumstances of his departure from Trinity School, he had broken completely with the world of pedagogy by the time he reached the India Museum, but he was still actively engaged in adult education, with special emphasis on field work. In addition, however, his newly-found interest in fungi was growing rapidly, as was his expertise, in a field in which there was very little competition in this country at the time. Very soon he was corresponding regularly with Britain's foremost mycologists, and gradually he would discover the whole world of professional science. These two concerns, natural history for everyman, and the scientific world of mycology, were to dominate the rest of his life, the first, as already related, taking precedence until the late 1860s, and thereafter increasingly giving place to mycology. By 1880 he was recognised nationally and internationally as second only in Britain to the ageing Berkeley. By then his pioneer work for amateur nature lovers had given place to irregular attendance at field meetings of established societies, often as a specialist in his own subject, and to a continuing flow of popular books.

He was in his mid-forties, and his work on fungi was still being carried out entirely in his spare time, before he achieved full recognition from his peers as a mycologist. In 1880, when the opportunity came at last of a new job which would allow him to devote his time uninterruptedly to the subject he had made so much his own, he had reached the comparatively advanced age of 55. In addition to his taxonomic studies, which he would be able to continue as a professional in his new job, he still gave high priority to the field work he had loved since his youth, but which from henceforward would be almost entirely devoted to fungi. This latter aspect of his activities will be examined in Chapters 13 and 14, the present chapter and its successor being devoted to his job and to certain events in his family circle.

His responsibilities to his large household may have been less disrupting to his work than they would have been to many other men of slender means, for by 1879 he had three women to run it; Sophia, his wife, now aged 57; Annie, the mother of his children, aged 36; and the customary maid. These three would be able to manage the house and Mordecai's five surviving children (Ada (16), Willie (14), Ernest (12), Frank (5), and Herbert (1), together with Annie's daughter, Mabel (9)) with a minimum of interruption to the work of the master.

Such was Cooke's personal situation when, late in 1879, negotiations began between his employer, the India Office, and the Royal Botanic Gardens, Kew, for his transfer to the latter. Dr. Watson had been trying for years to obtain Government agreement to the establishment of a great new India Museum and Library under the same roof, not, as they were then, in separate and highly inconvenient buildings. But he had been unable to achieve this, it had finally been decided that the Museum's collection of 'curiosities' must be dispersed, and they were now to be distributed between the Food Museum at South Kensington, the Animal Products Museum at Bethnal Green, the Royal School of Mines, and Kew, all the vegetable products going to the last named. It took 20 large vans to move Kew's acquisitions to their new home, and the India Office made the Gardens a grant of £2000 to extend its Museum to house them. It was proposed that Cooke should be seconded to Kew as curator of the collection, a job that would have been no sinecure. Looking back now at the correspondence which passed between Kew and the India Office on the matter of the secondment, which continued for seven months from November, 1879, broken into occasionally by an agitated letter from Cooke, it is remarkable that the transfer ever took place at all.

On behalf of Kew the exchange is sometimes conducted by the Director, Sir Joseph Hooker, himself (Fig. 11.1); sometimes by his son-in-law the Assistant Director, who was none other than William Thiselton Dyer who, nearly 20 years earlier, had been a youthful member of Cooke's Society of Amateur Botanists, and a few years ago had been one of his sponsors for associateship of the Linnean Society. The Directorship of the Gardens had passed to Joseph Hooker (1817–1911) in 1865 from his father, the great botanist Sir William Hooker, and a worthy successor the son proved to be. His botanical expeditions to Antarctica on the 'Erebus', and to the Himalayas, are renowned; he and his friend Thomas Huxley were the leading defenders and champions of Charles Darwin and his contentious new theory of evolution; and under his reign Kew grew and flourished exceedingly, especially its Herbarium and the taxonomic aspects of its work. He was a skilled and sympathetic Director too, keeping his finger on the pulse of his empire, choosing his staff himself, and tactfully guiding their work along the lines of his overall plans. Though he insisted on an honest day's work, and brooked no interference with his management, he was very solicitous for the welfare of his staff and made great efforts to improve their pay and working conditions; there are few records of staff discontent under his regime.*

Dyer (1843–1928) (Fig. 11.2) was quite a different character. Although a noted botanist in his own right, he was primarily an experimentalist. He belonged to the new generation of biologists, taking a degree in natural history at Oxford instead of qualifying in medicine as most of his predecessors had done, and his career since then had been brilliant. After several moves he became one of the first Demonstrators at Huxley's classes at the Natural Science Department at South Kensington, after they had been moved there from the School of Mines in Jermyn Street. In 1873 he took over from Huxley as head of the Department and revolutionised the teaching of Botany along the lines that the latter had developed for zoology. He had held this post for only two years when Hooker appointed him as his Assistant Director at Kew. There Dyer concentrated more on experimental botany than on systematics, though his gifts lay in the direction of promoting research rather than in undertaking it himself. In 1877 he married Hooker's daughter, Harriet, thus becoming a member of the Hooker dynasty which ruled at Kew for so long. But he was an autocratic and unapproachable man, not generally liked by his staff and with

---

*It was Hooker who was responsible for sending the first seedlings of the rubber tree, *Hevea brasiliensis*, from its native Brazil to Malaya, via Kew and Ceylon, so starting Malaya's rubber growing industry. From the turn of the century the acreage under rubber grew, and year by year more native jungle fell before the planter's axe. The Colonial Office kept a paternal eye on this process, and the Conservator of Forests in charge of the Malayan Forest Service in 1915 was one George Eaton Stannard Cubitt (p. 5), who collected Indian and Malayan plants for Kew. Soon after Eaton Cubitt arrived in Malaya his younger sister, Gladys Nellie Cubitt, was sent out by the family to join him. There she met and married a young rubber planter, Marcus Claude English, and in due course gave birth to the author of this book. Thus was Sir Joseph Hooker responsible for the multiplication of Cooke's collateral relatives, and hence for providing him with a biographer!

Fig. 11.1 J. D. Hooker, Director of the Royal Botanic Gardens, Kew, 1865–1885.

Fig. 11.2 W. Thiselton-Dyer, Director of the Royal Botanic Gardens, Kew, 1885–1905. Dyer hyphenated his surname in 1891

little concern for its welfare, so when he replaced Hooker as Director on the latter's retirement in 1885, Kew became a less happy place.

The earliest of the letters concerning Cooke's transfer to Kew to have survived are from the subject of the correspondence himself, and are dated November 5 and 6, 1879: they press for a speedy decision on the matter as he has a large family to keep and has been offered a job as a sub-editor. Dyer replied on Hooker's behalf that after December 31st his services would be transferred to Kew, where he would be employed for three days a week at a salary of £200 paid by the Indian Government. His duties would be: firstly to curate the Indian collections, secondly to report on plant diseases due to fungi, and thirdly to arrange the fungal collections at Kew, working under the Keeper, Dr. Oliver. But then the complications began.

On January 12th Hooker wrote to Sir Louis Mallet of the India Office that he had received no confirmation of the arrangements for Cooke from that Office, and that Cooke had not appeared at Kew. The next day Mallet replied that Cooke's services had been retained by the India Office, as had the sum of £200 which would have been paid to Kew for his salary, 'it being understood that Cooke's services are no longer needed at Kew'. A prompt and strongly worded answer from Dyer castigated Mallet's unannounced change in the arrangements, quoting chapter and verse of previous correspondence, and demanded compensatory arrangements. At this point Mallet, who was seemingly recognised at the India Office as a difficult man to deal with, went away for a few months, and subsequent correspondence is signed by various officials, not all with legible signatures. The next letter, to Hooker himself and marked 'Private', explains that Cooke had asked for a higher salary and had been told by both Dyer and Mallet that this was out of the question and that he must take the offer as it stood or leave it. Apparently Cooke was objecting to being asked both to continue to curate the Indian collections, and to take on the duties of a mycologist, all for a salary of £200; he considered that if he was to undertake both jobs, his salary should be augmented by £100. The letter continues:

> In the cause of peace and quietness it is worth considering whether we may be a little mean and patch up a peace with Cooke [in the absence of Mallet] . . . consider how far an accommodation can be reached with Cooke on the old basis . . . there will be no difficulty about a little more money.

For three weeks after this there is silence until on February 7th the matter is referred to no less a person than Lord Cranbrook, Secretary of State for India.

However, three days earlier Cooke had again rocked the boat, and this time he almost sank it. On February 4th he had attended a meeting of the Pharmaceutical Society to which, though he was not a member, he would have been well known through his contributions to its *Journal*, in order to speak about a collection of drugs which had been presented to the Society by the India Museum. In introducing Cooke, the Chairman had remarked that he:

> had done good service to the Society in helping to obtain for it a portion of the contents of the India Museum. Some members of the Society might very likely deplore the resolution which the Government had arrived at as to the distribution of the Museum but . . . it was important that that portion of the specimens which touched upon pharmaceutical science should find its way within the Society's walls.

When he got up to speak Cooke prefaced his remarks by saying that:

> he had attempted to get into [the India Museum] as complete a series as was possible of the drugs which were employed in all parts of India . . . When it was resolved that the India Museum should be dissolved he took measures to put the Society into communication with the Secretary of State, in order that the whole of the materia medica collection might be transferred to the museum of this Society. . . . Subsequently, official arrangements, of the nature of which he was even now totally unaware, were made,

whereby the whole of the collections were transferred bodily to Kew. ... But he was under the impression that the whole of the special materia medica collection would ultimately be transferred to the Pharmaceutical Society. He must confess that he was disappointed to find that what had been given to this Society was but a small proportion of what he had hoped it would receive. He could only regret that the whole collection had not been disposed of in the best possible way. The samples of opium and opium apparatus were parts of a collection, and he had no doubt that when this was understood at Kew the remainder would be sent [1].

Cooke, who was obviously regarded by his superiors as a mere technician, unfit to be consulted about the uses to which his years of labour might best be put, had made a serious blunder, for on February 12th Dyer wrote to the Curator of the Pharmaceutical Society concerning the Society's recent meeting as follows:

Sir J. Hooker considers that the action and language of Dr. Cooke has been such as to place this establishment in a somewhat false position. Owing to negotiations which appear to have been carried on between representatives of the Pharmaceutical Society and officials of the India Museum, but of which Sir J. Hooker is only indirectly informed and has no official knowledge, the Pharmaceutical Society seems to have been led to expect the transference to them of the collections of Indian materia medica intact.

Every specimen that could be spared was handed over, but Sir J. Hooker could not hand the whole thing over as he felt bound to keep a complete collection of vegetable products for his public.

Under these circumstances Sir J. Hooker is surprised to find that Dr. Cooke appears to have both claimed and received the credit for having obtained for the Pharmaceutical Society the donation which they received from Kew. He has gone further than this and has expressed 'his regret that the whole collection had not been disposed of in the best possible way.' Sir Joseph Hooker thinks that it is highly undecorous that a gentleman who is understood to be in some way still officially attached to the India Office should permit himself to pass censure of this kind upon an arrangement which is part of a policy sanctioned by the Secretary of State in Council.

Sir J. Hooker trusts that he may be favoured with some explanation of Dr. Cooke's intervention in the matter...

There followed a flurry of letters between Kew, the Society and the India Office, attempting to put the record straight, including one from Hooker in which he claims that Cooke's conduct was 'insubordinate':

'The serious quesion was whether an official so little habituated as Dr. Cooke had proved himself to be to official discipline, was likely to work well with so large and complicated an establishment as Kew, the business of whch can only be carried on by special subordination and obedience to orders.'

But shortly before the above episode there must have been another development which worried Hooker, for on the same day that Dyer wrote to the Pharmaceutical Society a letter went to the Secretary of State for India pointing out that now that the plan to second Cooke to Kew had been abandoned in favour of giving him 'an entirely unfettered pension' (a fact not previously mentioned in the correspondence), this did not relieve the India Office of its obligation to pay Kew £200 in lieu of Cooke's services. The India Office returned a soothing reply to the effect that it was sure it would be possible to resolve the affairs of the Museum before the return of Mallet. On March 5th a top level meeting duly took place between Hooker and Lord Cranbrook at which Cooke's indiscretions must have been among the matters discussed and Hooker must have supported the proposed cancellation of his secondment to Kew, for an India Officee official wrote Hooker a 'Private' letter the next day to say that he had 'urged upon' a senior officer:

> ... the advisability on public grounds of not appointing Cooke to Kew. ... Should [the Council] insist on holding us to our bargain I stipulated that Dr. Cooke must be instructed officially to obey your orders on all points, and that he should be told semi-officially that he must apologise to you for his action in regard to the Pharmaceutical Society.'

Promptly on March 9th an obedient but obviously quite uncomprehending Cooke wrote to Hooker apologising for any disrespect shown to him at the meeting, but emphasising his ignorance of having made any such remarks. Which, of course, he had not done. As Hooker pointed out to the India Office, it was not any personal afront of which Cooke was guilty, but misconduct of an official nature, which appears to have been quite beyond Cooke's understanding: he was a political innocent. Nevertheless, it is difficult to avoid the conclusion that he was being used as a scapegoat by the two bodies, to smooth over their obvious failure of communication. Certainly Cooke lacked any sort of tact or finesse; on the other hand, with his seniority–he was Assistant Curator of the whole Museum–he might have expected a certain amount of say in its affairs. Instead he was being slapped down like a naughty schoolboy and warned to keep his fingers out of matters that were the prerogative of his masters. One wonders if this would have happened had he had the same educational and social advantages as they.

Hooker must already have known Cooke fairly well from meeting him on the Scientific Committee of the Royal Horticultural Society, and he may therefore already have formed some opinion of the difficulties which could ensue should he take the mycologist on to his staff. There had also been an episode back in 1875 which both Hooker and Dyer may have remembered. A colleague at the India Museum had, one day, shown Cooke an unsigned paragraph in a current periodical, which must have been either offensive or controversial in nature, and had attributed it to Dyer. Cooke disagreed and, as he knew Dyer, had volunteered to write and seek his denial. As usual, Cooke's words ran away with him and Dyer, taking the letter as a personal attack, showed it to Hooker. Cooke's letter of apology to Hooker, in which he says he has also written to Dyer, survives.

Returning to Cooke's secondment, the India Office must have insisted that the arrangements proceed despite Hooker's objections, for by March 23rd it had written to the Office of Works (the Department responsible for Kew) that all previous proposals, including the pension were withdrawn, and Cooke was to work at Kew for three days a week at a salary of £200, plus an additional £100 to be paid to him personally by the India Office until December 1st, 1884. On these terms Cooke would undertake any work required of him by Hooker. It seemed that the months of negotiation had at last reached a happy conclusion and that Cooke could start at Kew. He wrote to Hooker, apparently having understood as much from the India Office, only to receive a further snub from Hooker:

> Dear Dr. Cooke, The arrangements to which you allude are not, in so far as I am informed, concluded. When they are, and I am informed of it, I shall at once have pleasure in communicating with you.

Hooker was a shrewd administrator. He wanted not only a suitably subservient addition to his staff, but an assured grip on the money which came with him; he had no intention of losing that money when Cooke left his service. It took two more months of intensive negotiation before financial matters were concluded to his satisfaction, and it was not until May 31st that the India Office was able to write to him that: 'Dr. M. C. Cooke has now been directed to report himself to Sir J. Hooker for duty at the Royal Gardens, Kew'. But still he did not start work. June was occupied with more correspondence between the India Office and Kew concerning methods of accounting for the separate sums of £200 and £100 (the former sum only would be transferable to Kew on Cooke's departure), interrupted by a bemused Cooke who, having been told off once by Hooker for presuming that he could start work before receiving a formal notification from Kew, was now being accused by the India Office of 'unnecessary delay' in reporting for his new job. To the India Office he wrote quoting in his own defence Hooker's letter of March 31st; from Hooker he wanted to know before reporting again i) when to report in person ii) which days of the week he is to work, iii) how he will draw his pay. The saga ends on June 29th with the following note, counter-signed by Cooke:

> To attend at Kew Herbarium on Mondays, Tuesdays and Wednesdays, by train reaching Kew Bridge at 10.15 am and leaving at 4.55 pm, making attendance at Herbarium from 10.25 to 4.45 with 20 minutes midday for luncheon when required, assuring six clear working hours per day on duty.

His starting date for salary purposes was taken as June 21st, 1880, but the payment of the salary continued to cause confusion for some months, especially when it was discovered that the Treasury had forgotten to deduct income tax at 6d in the pound.

Kew had to make special preparations to house the 4,000 specimens it would be receiving from the India Museum. In its grounds the gracious, stone-fronted Museum No. 1 (Fig. 11.3) stood, as it stands to-day, on the opposite side of the lake from the huge and magnificent Palm House which had been built 10 years earlier. But the Museum was already full, so a new three-storey wing, with a fine teak staircase and wrought iron banisters, was built behind it, nearly doubling its floorspace. Into this the Indian collection was moved, eventually to be integrated with Kew's existing collection of economic plants, and it was here that Cooke would preside when information was required.

The space occupied in 1880 by the Library, Herbarium, and administrative offices of Kew Gardens was, in comparison with to-day, minute, consisting essentially of the original Hunter House (Fig. 11.4) on Kew Green to which the Herbarium extension at the rear (C-Wing) had been added in 1877. Hunter House is a large and elegant Georgian building which had once been the home of the Duke of Cumberland. On the Duke's death in 1851 William Hooker had been allowed to take the building over for a Herbarium and Library, and from then on Hunter House had been the hub and centre of Kew Gardens. With space so very limited in Cooke's time the entire mycological collection was housed in Hunter House itself, in the room to the left of the front door with windows looking on to the entrance drive. The intensity with which Cooke

Fig. 11.3   Museum No 1, Royal Botanic Gardens, Kew.

Fig. 11.4 Hunter House, Royal Botanic Gardens, Kew.

worked would have allowed him little time to watch the comings and goings of the world's botanists through those windows.

Cooke was the first of a line of taxonomic mycologists to be appointed to Kew that is unbroken to the present day, and even though a large part of his duties were to the Indian collection, he must have spent every available moment of his three-day week on his first love–mycology. His coming was announced in the *Kew Report* thus:

> Dr. Cooke, besides being available to give information in respect to any matter connected with the Indian collections lately under his charge, has undertaken the rearrangement of the collection of Thallophytes [Fungi, Algae] in the Herbarium, as well as the duty of reporting on questions connected with plant-diseases produced by fungoid organisms which are submitted to Kew. Dr. Cooke has at once attacked the arrears of the section of the Herbarium devoted to fungi, which ... has been somewhat neglected.

He had obviously thrown himself into his new job with his accustomed vigour, something which he himself confirms in a letter to Broome written at about this time:

> I am just now so fully pressed by demands upon my time that I have not a spare hour––nor shall I until my work at Kew is in a more forward state.

An event which may have precipitated his appointment to Kew, and which certainly increased his workload when he arrived, was the presentation to that institution in 1879, by the 76 year-old M. J. Berkeley, of his entire fungal herbarium of over 20,000 specimens. Cooke had the huge task of incorportaing this into the existing Kew collections. Sir Joseph, despite his misgivings as to whether his new employee would be a suitably subservient member of his staff, could not have found anyone to carry out his duties with more zeal and enthusiasm, as he admitted in a letter to Berkeley a year later.

The accusation of the wildly prejudiced American mycological commentator, C. G. Lloyd (2), that Cooke's herbarium consisted 'in the main of what he had taken from Berkeley's herbarium' must here be refuted. Of course Cooke's herbarium contained a number of Berkeley's collections, as Berkeley's contained a number of Cooke's, for the two men and been exchanging specimens for years. But to suggest that the huge herbarium of a born collector such as Cooke consisted 'in the main' of another man's specimens is nonsense.

It seems that Cooke even now was unable, or saw no need, to moderate his customary bluntness, and he soon put Dyer's back up. Early the following year the Deputy Director apparently referred to Berkeley, rather than to Cooke, a fungus that had been sent in for identification, and in thanking Berkeley for his trouble he wrote:

> What a contrast between your mode of handling a little matter of this kind and Dr. Cooke's. The newer generation of naturalists may have a deeper insight into important principles but I do not see that they can ever come near their predecessors in ... a knowledge of forms.

As Dyer himself had no interest in forms, let alone in fungal forms, and as he was at least ten years younger than Cooke, this gratuitous attack on a member of his staff would seem to be wholly unwarranted.

Now, Cooke had four whole days a week to devote to those other activities which he had previously had to fit in around a full-time job, and he wasted no time in filling up those precious extra hours. In his own words:

> In 1797 Sowerby published his 'Coloured Figures of British Fungi', since when no effort has been made to produce figures of all the known species of the larger Fungi, belonging to the order *Hymenomycetes* ... To meet the requirements of the British student of Fungi a work was long felt desirable which could include coloured figures of all the known British Species, and I was often urged to undertake such a work, but the magnitude and responsibility of the undertaking deterred me for a long period from making the attempt.
>
> At the meeting of the Woolhope Club, in the autumn of 1880, the subject was again considered, and I was induced to commence these *Illustrations* provided a sufficient number of subscribers could be secured to guarantee the continuance of the work, and limit the pecuniary responsibility within reasonable bounds [3].

In fact it was 'the excellent series of drawings of Fungi which have been made by Dr. Bull', the doyen of the Woolhope Club and a man whom Cooke came to revere almost as much as he did the Rev. M. J. Berkeley, that originally put the idea for the *Illustrations of British Fungi* into Cooke's head, and over the years he had been collecting drawings of the rarer species of agarics from colleagues in the hope of publishing a complete series. The move to Kew provided the ideal opportunity and the first of the 76 parts, which were eventually to make up eight volumes, appeared in 1881. It took the whole of the ten years he was employed at Kew to complete the 1200 plates that made up the most exhaustive work of its kind ever published. As with so many of his other works, the *Illustrations* was financed entirely by private subscription. (Figs. 4.2, 11.5, 11.6 & 14.3).

In the Preface, written in 1883, he thanks ten friends and colleagues, including Berkeley, Bull, Broome and Smith, for allowing him to use their drawings, and the great majority of figures bear either his own initials or those of one or other of these contributors (131 figures out of 149 in Volume 1). It is therefore strange to find Smith making a bitter posthumous attack on Cooke for tracing and then altering one of his (Smith's) drawings without acknowledgement:

> Plate 585 ... Ag. campestris–This is an unacknowledged tracing of a dg. of mine in B. Mus. colln. My plant is correctly named A. amiensis, but C. altered the name to *campestris*–My locality Nottinghamshire, he alters to Northamptonshire. My orig. is act. size but C. is stated to be 2/3rds real size [in fact Cooke's is about 2/3 the size of Smith's] There is *no annulus* on my dg. but Cooke added an ample annulus on his plate ... My pileus is smooth, C. put on a lot of scales. My stem is smooth. C. added scales, etc.,

[4] (Fig. 11.6).

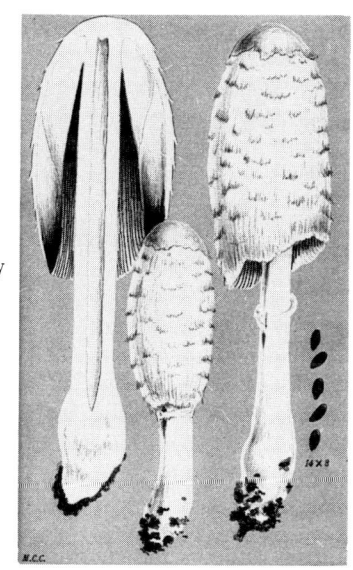

Fig. 11.5 *Coprinus comatus*, the lawyer's wig, or shaggy cap fungus, from *Illustrations of British Fungi*. Original in colour.

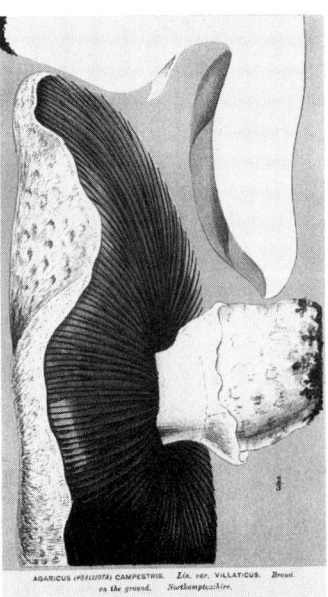

Fig. 11.6 Plate 585 of *Illustrations of British Fungi*, considered by W. G. Smith to be an unacknowledged tracing of his own work.

It is true that the two figures bear a strong resemblance to each other in the shape of the fungi they depict and the way in which they and their sections are disposed on the page; but Cooke's is not a tracing. If he had wanted to fabricate a fungus it would have been much easier, with his long experience, to have drawn it out of his head than to have bothered to visit the Natural History Museum to copy the outline of Smith's, only to go to the trouble subsequently of making all the alterations the latter complains of. As with other outbursts of Smith's, it seems likely that he was really giving vent to a deep-seated jealousy of his friend.

The plates nearly all have backgrounds of various shades of grey or fawn, ensuring that any pale or white parts of the fungus stand out. Usually several fruit-bodies are shown, and a vertical section is nearly always included, but unfortunately the spores are usually omitted and, if shown, are not very satisfactorily depicted; for as Cooke himself notes, the method of reproduction used was not suitable for small objects.

Leila Cooke was nine years old when the last part of the *Illustrations* came out, and she had vivid memories, even in her old age, of her father's technique for making the drawings, and of the trouble both he and his printer went to to ensure the best possible reproduction.

> Nobody was allowed in [to her father's study] except me as a small child who must sit still and not talk, while Father sat at his table and worked ceaselessly. At the time he was receiving parcels of fungi from all over England, and the world. I think he must have

been working on the plates for his big work 'Illustrations of British Fungi'. He cut the paper to the size required. He arranged the fungi as he wished them and drew them in outline. Then he stood against the glass doors of the greenhouse [the lean-to conservatory attached to his study], and holding another sheet of paper against the outline, traced them on to the sheet of paper he was going to colour. If he had not got the colour he required in his box, he mixed up a sample and took it to Windsor and Newton and had a few pans made specially to match. Sometimes a Mr. Griggs came, who was printing the plates, and the colours were compared and discussed.

Cooke estimated that had he not been able to prepare the patterns himself he would have had to have paid an artist at least £600 to do so: in the event, each of the 76 parts sold for 5/-, or £19 for the completed series.

Mycologists of the time welcomed the *Illustrations*; it was much used then, and it remains a valuable work of reference to-day. Inevitably there have been many criticisms of his identifications, some justified, others not. Some of those made in his life-time he himself acknowledged as valid, others he fought vigorously. In many cases critics disagree among theselves, reflecting differing views on nomenclature rather than any inaccuracy on Cooke's part. Ironically, in view of his strongly held opinions on unnecessary 'splitting', a number of his more recent critics regard some of his species as mere environmental variations, and 'lump' them with other species they resemble. Only rarely is any adverse comment made on the figures themselves.

In the Archives of the British Museum (Natural History) to-day are 14 large, hard-backed scrap-books entitled 'British Fungi arranged on the basis of Fries' and numbered 1–13 and 15. All are in Cooke's hand-writing. They contain original drawings of fungi by Cooke himself, by Berkeley and other mycologists, cut from the sheets on which they were drawn, together with illustrations of fungi cut out from the works of a number of authors, again including Cooke himself, all pasted in, labelled and annotated. A book-plate in each volume states:

Free Public Library
Tullie House, Carlisle
Reference Department.
Presented by Dr. Carlyle. (Bequest)

The last volume is stamped with the date '1 Oct. 92'. The Natural History Museum obtained the set from Cumbria County Library, Tullie House, Carlisle, at the end of the 1960s. The first three volumes are devoted to the 'Agaricini', the next three to the other fleshy Basidiomycetes, and later volumes contain the Ascomycetes and microfungi, many being very imcomplete: Volume 15 is the 'Synopsis'. Because a number of illustrations from different sources, sometimes as many as seven or eight, are included for many fungi, especially the agarics, it has been conjectured that the first three volumes are those in which Cooke collected over the years the figures from which he would eventually pick the most suitable for his *Illustrations*. However, the figures collected for a given species do not by any means always include that used in the published work, and in a number of cases a figure actually cut from the *Illustrations* (presumably a proof) has been pasted into the scrap-book. It is also difficult to explain the remaining volumes of the series unless one supposes that Cooke hoped one day to undertake the task of illustrating the entire fungal kingdom. But there is no hint that such a scheme was ever in his mind, and even he must have realised that it was a practical impossibility. In addition to these objections to the volumes forming the preparatory work for the *Illustrations* there is also a question of timing. One must suppose that Cooke had presented the volumes to Dr. Carlyle before they passed to the Tullie Library in October, 1892, following the Doctor's death. But the last volume of the *Illustrations* did not appear until that year, and Cooke would hardly have parted with the drawings on which the great work was based before it was completed.

We know that the preparation of scrap-books, whether simply for his own pleasure or as a basis for a future publication, was a favourite hobby of Cooke's, and I believe that 'British Fungi' was just another of these, probably started for himself, but given later to a keen amateur to whom Cooke had taken a liking, and who he expected would make good use of it. He may well have included in the 'Agaricini' section some of the original figures he had collected for the *Illustrations* once he was sure he had finished with them, but I believe that it is unlikely that this was his main working collection.

During the 1880s and 1890s the pursuit of toadstools became an ever more popular occupation of natural history societies all over the country, and Cooke's help and advice was sought by many of them (Chaps. 12 and 13): it would seem that it was by some such means that he came to know Dr. David Carlyle, of 2, The Crescent, Carlisle, a Scot and a bachelor, who had been in practice in the city since 1885, when he was 28. It was in 1877 that the Carlisle Literary and Scientific Society and Field Naturalists' Club was inaugurated, and Dr. Carlyle was first mentioned in its minutes in 1880, when he gave a lecture 'On Fungi' which was to be the first of many on various aspects of mycology. He attended Club meetings and excursions assiduously, was elected to the committee in 1881, and became Vice-President four years later, thus being in an excellent position to promote his hobby among his fellow members. We do not know whether he and Cooke ever met, or whether their's was a pen friendship, but Carlyle would have been just the type of committed amateur whom Cooke would have liked to encourage with a gift such as his scrap-book.

Not content with embarking on such a major new project as the *Illustrations*, Cooke chose 1880 to launch another book of popular science. The idea of bringing some of Darwin's biological discoveries to the notice of the lay public had been at the back of his mind for some time, and he had been collecting notes and memoranda for 15 years. In March, 1880, following the success of *Ponds and Ditches* and *Woodlands*, he had written to the publishing department of the S.P.C.K. offering it a book to be called *Curiosities of Vegetation*, which title the publishers accepted at the time, but which later became *Freaks and Marvels of Plant Life, or Curiosities of Vegetation*. It is another of Cooke's 'scissors and paste' volumes, consisting chiefly of verbatim quotations from the published works of others, usually Darwin, laid out, explained, and fully illustrated so as to be attractive to the layman. It covers such subjects as insectivorous plants, plant movements, seed dispersal mechanisms, mimicry etc., and in the Introduction Cooke defends himself against possible criticisms that he has:

> made very free use of the many researches of Dr. Charles Darwin ... without adding to them, in number or in illustration. To this we plead guilty, with the excuse that by so doing we should contribute something towards the diffusion of a knowledge, and, as we hope, of a more general appreciation of the important additions he has made to our knowledge of vegetable life. Some there are who have been content to associate his name only with a theory which they may not comprehend, but do not fail to condemn. With that theory we are not now concerned; but there is another aspect in which we desire that this accurate and indefatigable observer should be known and remembered, outside an exclusively scientific circle; and that is, as a collector of facts, the results of patient observations, illustrative of the history of plants and animals.

The book was due for publication early in 1881, and half of it was already in type when a dreadful event took place; he was proceeding by omnibus with the remainder of the manuscript when:

> ... I had to alight at Oxford Street, and unfortunately left my parcel on the seat, and did not discover it for some time after, when I followed the omnibus, and made all possible enquiries of the time keepers, and at Scotland Yard, but never obtained a trace of it. In this dilemma I hastened home and soon commenced rewriting the missing pages–which caused me the more trouble as all my notes, when used had been cast into the waste paper basket. By dint of close application until midnight for several days I completed the missing pages ... (Aut.)

There is a note at the end of Chapter 7 marking the point at which the disaster happened and apologising to readers for any errors or omissions which may have resulted from it.

A year after the book was published Charles Darwin died, and Professor Thomas Huxley and Sir Joseph Hooker were pall-bearers at the funeral at Westminster Abbey of their friend and fellow scientist. *Freaks and Marvels* remains Cooke's humble tribute to the great man.

The strain of the traumatic episode of the loss of the manuscript, on top of his other heavy commitments, proved too much for Cooke's health. On

July 18th, 1881, he wrote to Professor Oliver, Keeper of the Herbarium and Head of the Museum at Kew, telling him that he was suffering from severe diarrhoea and 'an affliction of the head', and had been advised by the doctor at the India Office to spend a week by the sea. The next day, having reached Neatishead, he wrote again requesting a month's leave. His autobiographical notes continue the story:

> I was soon compelled to telegraph back to town for a nurse to accompany me and was soon on my way to Cromer, to be in close communication with my friend Dr. Knaggs who was then taking his holiday at Sheringham and I could be sure of his advice. It was soon evident that I had fallen a victim of paralysis.

The 'nurse', it turns out, was none other than Annie, now pregnant with their last child; Sophia, presumably, was once more left in charge of the children. On August 8th Annie wrote to Kew from Church St., Cromer, apparently with some difficulty:

> Dr. Cooke desires me to inform you, that his illness appears to have reached its maximum and to have settled in paralysis he is thankful that it is a mild form of one arm, at intervals daily. There is a relief from headache, and hopes there is slight improvement generally, with kind regards
>
>   I am   Yours obediently   Annie Cooke

Mordecai's friend and medical practitioner, Dr. Henry Guard Knaggs, as well as being a doctor, was a keen lepidopterist, the author of the *Lepidopterist's Guide*, and one-time editor of the *Entomologist's Monthly Magazine*, and it was no doubt their mutual interest in natural history that brought the two men together. On Dr. Knaggs' return from Cromer at the end of August he wrote to Kew:

> Mr. M. C. Cooke of Junction Road N.W. has been under my care for some weeks. Whilst in the country at Cromer he must have had something very like a slight apoplectic fit which has left him in an excessively weak and desponding condition which renders him quite incapable of concentrating his thought and energies on work–indeed I consider that an absolute rest. . . .

In to-day's terms it seems likely that the 'apoplectic fit' was, in fact, a series of slight strokes leading to temporary paralysis of his arm. His case clearly caused concern, for early in September Sir Louis Mallet of the India Office wrote to Sir Joseph Hooker:

> I am afraid that Dr. Cooke's case is a serious one, but we have thought it fair to give him a chance of his leave before we raise the question of his resignation.

Baron von Mueller, the energetic botanical explorer based in Melbourne,

Australia, who had a special interest in fungi (see below), wrote to Hooker that 'poor Dr. Cooke should be sent on a sea voyage to a healthy part of Central America'. More practically, he offered to contribute £10 towards a collection to help Cooke financially.

But they had all reckoned without Cooke's determination and resilience. Altogether he was 'six months on the sick list', but he soon ceased to take the absolute rest recommended by Dr. Knaggs, for by the end of the year he was writing to Berkeley:

> As for myself I do very well, so long as I restrict myself to examination etc. of the larger fungi, but as soon as I sit down to the microscope or spend an hour or two in continuous writing, my head compels me to desist, and I fancy it must be some considerable time before I shall be able to undertake any continuous microscopic work.

In fact, as he wrote to Saccardo, he was still weak from his illness a year later.

On October 1st, 1882, only a few weeks after her return from Cromer with her sick 'husband', Annie gave birth to their last child. Leila Annie, called after one of Dr. Bull's daughters, was to become the apple of her father's eye, to adore him in return, and to live to pass on to posterity the few personal details about him that we have.

Cooke's foreign correspondence in no way abated when he moved to Kew; in fact, it may well have increased in volume. His brisk exchange with Saccardo continued, as did the long-standing correspondence with his various American contacts, and to these was added a constant stream of letters and specimens from naturalists in Australia, large tracts of which country were still geographically, let alone mycologically, unexplored. Specimens poured in too from many other parts of the world–from India, especially the mountainous north, from Malaya, South America, South and East Africa; from Java, Japan and Tahiti–all to be examined and described in various journals, usually *Grevillea*.

In 1882 Saccardo completed the first volume of his monumental *Sylloge Fungorum Omnium Hucusque Cognitorum* and sent Cooke a copy in gratitude for the assistance he had received from him over the years. Cooke, in his postcard acknowledging the gift, comments disarmingly: 'It is excellent. I shall write to you with a few minor *errata*, if you will permit . . .' His brief review in *Grevillea* in September of that year is very favourable, though the differing views of the two men on classification are inevitably mentioned. But this review turns out to be a preliminary comment only, for it is immediately followed by a detailed, four-page critique entitled 'The Perisporaceae of Saccardo's Sylloge Fungorum' in which Cooke [5] pulls no punches. He comments on what he considers to be Saccardo's 'archaic' artificial system of classification, based primarily on the ascospores (which he himself had advocated for another group of Ascomycetes only a few years earlier (p. 190)) and relegating to comparative unimportance the various structures in which those spores are borne. Of Saccardo's system he writes:

> Of course it will be expected that a new system will obtain adherents, because love of novelty will ensure that, but to suppose that it will be *permanent*, will argue all lack of faith in the human mind. Inasmuch as children and uncultured or deficient intellects prefer counting stamens to any more elaborate method for the determination of flowering plants, so will they accept the counting of septa as the perfection of simplicity.'

The 'few minor errata' of the postcard occupy the next 2 pages of the critique.

A few months later Cooke prints, at Saccardo's request, a translation of the latter's reply, in which Saccardo gives examples to support his classification and points out that the great Fries himself, the father of the natural classification of fungi, has found it impossible to improve on spore colour, a thoroughly 'artificial' character, as the basis of classification in the Agarics.* Moving on to the attack in his turn, he considers that the classification of Cooke and Fries is:

> equivalent to that of Tournefort [about 1680!], in which plants are divided into trees, shrubs and plants, without taking any account of the characteristics, which are much more important, but much more difficult to preserve [misprint for *ob*serve?]. In fact both Fries and you distinguish groups and genera (what genera!) by the naked eye, as Tournefort did; by carpologists, on the other hand [those using Saccardo's method], characteristics which are more intimate (and more constant), are studied and unveiled ...

The argument was, of course, being conducted within the limits of the scientific knowledge of the age; neither man could be 'correct' according to modern views on the basis of the facts then available.

Cooke, as it was his own journal in which this cut and thrust was taking place, could not resist having the last word, and printed a final 'Rejoinder', in which he notes that 'the above letter in no way alters my opinion, and I have nothing to retract', and continues the argument at some length, ending with a handsome tribute to the 'indefatigable author of the *Sylloge*'.

In 1885 he reviewed Volume 3 for the *Journal of Botany*, beginning:

> We are glad to announce the issue of the third volume of this work, for although confessedly only a compilation, it is exceedingly useful for collecting together into one focus all the stray scintillations of mycological genius in the direction of species-making, which have hitherto been too much diffused to enlighten anyone ...'

He continued with some outspoken criticism of Saccardo's nomenclature but did not, on this occasion, take issue with him over the principles of classification.

*Even to-day, spore colour remains an important character in the classification of the Agarics.

The two men continued to correspond until well into the 1890s, albeit less and less frequently, and Saccardo sent Cooke the later volumes of the *Sylloge* as they appeared, each one with a dedication in his own hand-writing to 'his friend' M. C. Cooke. This set was acquired by the Library of the Biological and Chemical Research Institute, Rydalmere, N.S.W., Australia, in the late 1920s or early 1930s (their provenance is unknown, for Cooke had sold his library in 1898). The blank pages of the interleaved edition are heavily annotated in Cooke's hand-writing.*

All this time Cooke had continued to work on American fungi with his customary intensity. As well as Peck, Ellis and Ravenel, he was in contact with other transatlantic mycologists notably W. G. Farlow (1844–1914) of Harvard, with whom he had started to correspond in 1876, only a year after Farlow had published his first mycological paper. After obtaining his degree in medicine, the young Farlow (Fig. 11.7) had worked in Europe for two years, joining Rostafinski (the Pole whose monograph on Myxomycetes Cooke had recently translated) as a student of De Bary. On his return to the States in 1874 Farlow was appointed Assistant Professor at the Bussey Institute of Harvard University, where he worked on the fungal diseases of plants, Cooke's letters to him up to the end of the decade being concerned exclusively with pathogenic fungi. In 1879 Farlow became Professor of Cryptogamic Botany at Harvard, and immediately the subject matter of Cooke's letters changed to the larger Ascomycetes and the Agarics. Now Farlow began to build up a superb cryptogamic library and herbarium which he eventually bequeathed to the University, and to make his name as the foremost American cryptogamic botanist of his time. Through his teaching he had a profound influence on succeeding generations.

From the start of their correspondence Cooke's letters to Farlow show a respect and restraint that is evident only towards a few other mycologists, even though Farlow was twenty years his junior and was barely established in his career at the time. For instance, in 1876, after asking Farlow to amplify some points, he continues:

> I have no other object than that of truth in making these enquiries . . . but you will pardon me if I indicate any weak point which requires to be strengthened. As I am as much interested in American as in British Fungi I hope you will not consider me impertinent. (WGF, 3)

And in 1878:

> As to your last paper permit me to say that I have read it with very great pleasure. I am pleased indeed to see any one giving attention to U.S. Fungi, in the manner and with the efficiency that you are doing. I could not but look with regret at the small reputation which a few very inefficient men were achieving on your side of the Atlantic in species

---

*I am most grateful to Mr. J. Walker of Rydalmere N.S.W., Australia, for this information.

Fig. 11.7 W. G. Farlow of Harvard University, U.S.A.

making ... There are so many interesting and useful points in your paper to which I would gladly refer but I must content myself with expressing my thanks for such a useful aid in clearing up many obscure points in American mycology. (WGF, 9)

Cooke also took Farlow into his confidence to a greater extent than he did many of his corresondents, for instance, over controversies with Ellis, with whom he collaborated for many years on a series of papers on New Jersey fungi. In 1879 Ellis published in the *Proceedings of the Acadamy of Natural Sciences of Philadelphia*, apparently without first notifying Cooke, a list of 20 species of several genera of the Sphaeriales, many of which the two men had described jointly, which Ellis now considered to be conspecific because they all had large, elliptical ascospores. Cooke reprinted the list in *Grevillea*, adding his own scathing comments to the effect that the said spores were, in fact, variable in size, shape and colour, that no account had been taken of the widely differing fruit-bodies, and that:

besides, on what grounds were 20 other species excluded which have similar sporidia [ascospores]? surely they have been known to the author. We can only regret that Mr. Ellis was so impolitic as to commit his conclusions to print. That an aspirant for scientific honour should have done so was simply an act of premediated suicide [6].

In the next issue Cooke printed Ellis's courteous and reasoned reply but, as in his exchange with Saccardo, added the last word himself–a note to the effect that the correspondence was now closed, that it would rest with posterity to give a final judgement, and that he was confident it would be in his favour. He wrote to Farlow at this time

> ... Ellis is unfortunately unable to discriminate between closely allied species–at least, if he sends me the same supposed species two or three times, it is invariably different each time. I am afraid he will never become a critical botanist. Hence specimens from him supposed to have been described by me will need verification. (WGF, 19)

The last of the 'Cooke and Peck' papers appeared in *Grevillea* in 1880, and in 1881 Cooke again wrote to Farlow:

> I believe that I have given him [Ellis] offence as he has not written to me for some time. As a rule the mycologists of the States cannot brook criticism. So long as one agrees with them all proceeds well, but as soon as one dares to differ all hopes of harmony come to an end. Thus it was with Peck–and so I suppose with Ellis. Nevertheless I can afford to wait for the confirmation of my opinions. I criticise freely, discuss fairly–and enjoy the friendship of all the best European mycologists of the day and yet with these all proceeds most harmoniously, whilst it is most difficult to retain relationship with some of our thin-skinned Transatlantic friends. Perhaps criticism in scientific matters is not acknowledged in the States–of this I am ignorant. (WGF, 21)

To-day no scientist would brook the scathing, sometimes personal, attacks rained on his contemporaries by Cooke, but even compared with the normally outspoken polemics of Victorian times, his are often extreme. Usually he seems quite oblivious that his onslaughts may go beyond scientific criticism and cause personal offence, for he pleads time and time again that they are not meant to be in any way malicious. His English colleagues and Saccardo had probably become inured to him, but it is understandable that some foreigners found his diatribes unacceptable. However, it seems that fear of reprisals did sometimes restrain his pen, for some years after his first onslaught on Saccardo's *Sylloge* he wrote to Farlow:

> I should have written in much stronger terms about it, but there are some who are ever ready to cry out about jealousy, malice etc. etc. if a candid opinion is ventured upon, and, no longer since than last month, one of our British Mycologists, a young one, it is true, who never fails to go out of his way to have a fling at one and my antiquainted notions–has himself written a fulsome eulogy of Saccardo in which he says that he 'is head and shoulders above all the mycologists of the day'. From which you may judge what his opinion is worth. (WGF 47)

Despite being himself impervious to any point of view other than his own, there is no doubt that he believed in the benefits of free discussion. In his own younger days he had suffered the miseries of isolation (p. 181) and knew how such deprivation affected a scientist. After he had retired he wrote of the Woolhope Club field meetings that they:

> were most calculated to break down the egotism and individuality which is most associated with isolated study. Free interchange of ideas, of doubts, of fears, is antagonistic to exclusiveness.

As with so many of us, there is a certain inconsistency in the way in which he applies to himself the precepts he propounds for others!

A problem which was excercising the minds of many biologists at the time, and had been worrying Cooke in particular, was the question of the priority of scientific names. Not infrequently, an organism is 'discovered' by a number of workers at different times and in different places, each discoverer describing and naming it in ignorance of the fact that others have done so before him: later workers then have to decide which name is to be the valid and universally accepted one. In 1905 a set of internationally agreed rules was drawn up to deal with this problem, but in the 1880s no such system existed. That the first name given to an organism should become its valid name was an easy concept to agree upon: it was much more difficult to decide which name should be accepted as the first. For new names cropped up in all sorts of places, in papers in scientific journals, in exsiccati, in unpublished reports and in jottings in private note-books. Could scientists be expected to search all such sources for the first name given? In *Grevillea* (1877–1878) five English cryptogamists–M. J. Berkeley, R. Braithwaite (bryologist), M. C. Cooke, J. M. Crombie (lichenologist) and F. Kitton (phycologist)–set out their considered opinions on the matter in a statement entitled 'Priority of Name'. Very reasonably they state that, to be recognised, a name must have been *published*, either by the circulation of accessible specimens, or in a journal or book. A private or exclusively printed report which is not circulated is 'no security for priority of name'.

Acceptance by British mycologists of this rule led to difficulties in Cooke's relationship with the American, C. H. Peck, for in 1882 Cooke confided to Farlow:

> We on this side–conclude that–as Peck's Reports [to his State government] are private documents not published or sold or freely distributed–they do not establish any priority for his names except by courtesy–and that his species–except such as have been published in other works–are not binding upon us. I have not seen them for some years. Am told by my publishers that they are *not* sold and *not* published. There is not one of them in any public library in London, either of the Linnean Society–the British Museum or the Herbarium at Kew. (WFG, 27)

Farlow evidently disagreed, considering Peck's names to be valid, but Berkeley backed Cooke, who wrote another lengthy letter to Farlow justifying the British point of view and ending:

> It is *unreasonable* and unjust to bind us to respect names printed in an obscure private report, which certainly never *has* been on *public* sale although copies may perhaps be obtained by private influence. (WFG, 30)

This was only one of Cooke's complaints against Peck. In other letters to Farlow he grumbles repeatedly about the latter's refusal to distribute specimens of the new fungi he describes:

> I know practically nothing of Peck's species ... as he takes most studious care not to let his species get over here although the truest course would be to let them be known as widely as possible. (WFG, 39)

and:

> No-one that I hear of can ever say that they have seen the authentic specimens of any of Peck's species. He is remarkably disinclined to permit any one to *see* them. (WFG, 34)

Peck's output was so enormous that he must indeed have been a sore trial to mycologists.

Nothing daunted by his disaster with *Freaks and Marvels* and the illness which it precipitated, Cooke now turned his attention to yet another aspect of botany:

> It was proposed to me to publish an illustrated work which should give figures of all the known species of British Fresh Water Algae, as the only work by Hassell was old and out of date. As usual I obtained suscribers ... (Aut.)

The two-volume work was published in parts by Williams and Norgate between 1882 and 1884. The undertaking would have been made possible by the author's position at Kew, where he was in charge of the algal, as well as the fungal, collections, and would have had free access to a superb library of phycological literature. According to the Introduction the book was not written as a scholarly scientific text, but for:

> ... the Microscopists who desire some acquaintance with the organisms met with on their excursions to ponds and ditches. For the absolutely scientific algologist it will be only fragmentary.

In other words it was written for those whose interests he had especially at heart, the members of the Quekett Club and similar bodies, for which purpose he made little pretence at originality, following compact descriptions of each organism with lively quotations from other authors describing life cycles and habitats. It is surprising, in a book aimed at laymen, to find for each species an exhaustive list of synonyms: such lists would, if original, have demanded so much research that he must in the circumstances have been quoting from previous authors to which he would, of course, have had easy access at Kew. Nearly every species is well illustrated in colour (Fig. 11.8).

Although it was priced comparatively highly at £4.10.0, the book sold out quickly, showing that, as usual, its author had filled a need. But it was never reprinted for, as Cooke had observed, it was not complete–a phycologist writing in 1927 doubted if it covered 25% of British species–and more scholarly works by specialists began to come forward.

The last page of a seven-page letter to Farlow written at this time gives a vivid idea of the way Cooke organised his days to cope with the extreme pressure under which he worked:

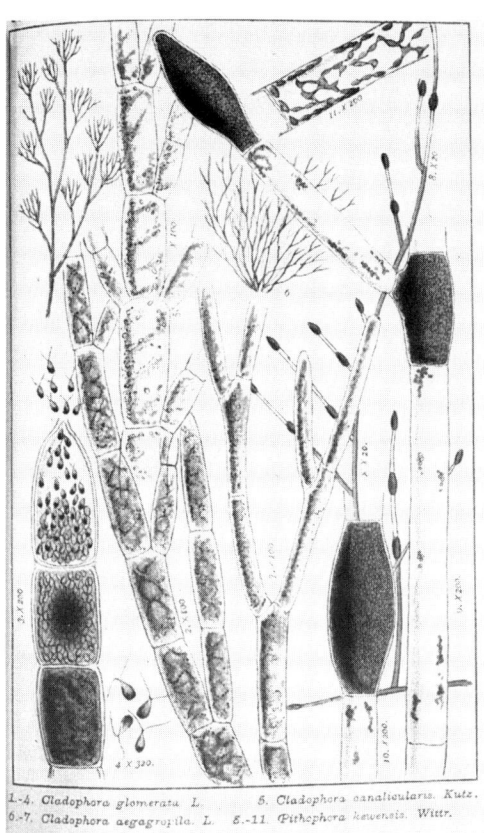

Fig. 11.8   Illustration from *British Algae*. Original in Colour.

I am sorry they [the answers to Farlow's questions] have been so long but the truth is–that the drawings for my works 'Illustrations' and 'Fresh-water Algae'–which have all [to] be made carefully for the photographer with my own hands keep me constantly employed as I am 3 days weekly at Kew–and one day is fully occupied in business matters with printers, paper makers etc. as all the business details of these books are in my own hands, and I have to buy paper etc. So that I have only 2 days a week free. The average this year will be about 4 plates a week to draw fairly and colour the patterns–you can understand how little time I have at my disposal, and how much correspondence must remain undone . . . This letter has been a month practically finished–which I much regret. (WFG, 29) (Fig. 11.9).

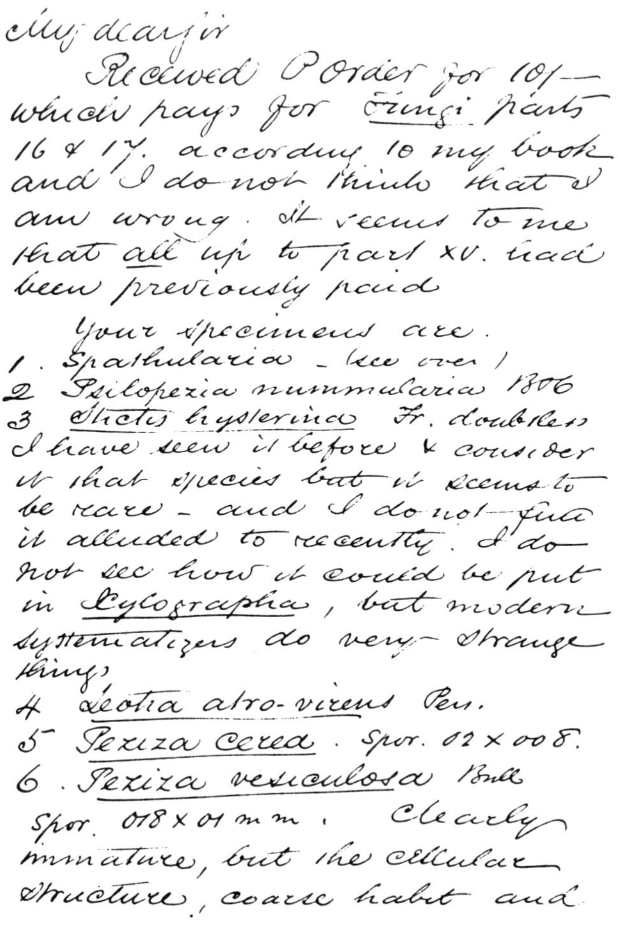

Fig. 11.9 Part of a letter from M. C. Cooke to W. G. Farlow, in which both business matters and mycology are discussed.

To this must be added his continuing role as contributor to, and editor and manager of, *Grevillea*, and in the autumn, the seasonal priority of collecting and painting agarics for the *Illustrations*. A great deal of Cooke's correspondence with Farlow, and doubtless with other mycologists all over the world had it survived, consisted of postcards concerning the despatch and sale of his various publications, one of which is worth quoting:

> I cannot remember at all amongst the number of orders I get from some 600 different people–any single one. I make it a rule to acknowledge *immediately* every order before it is removed from the table. I am compelled to do this systematically.' (WFG, 36).

One wonders what his mycological output might have been had he been relieved of clerical duties by an efficient secretary (Fig. 11.10).

At the height of this furious activity disaster struck again, and the indomitable Cooke was forced to send to all his subscribers a printed postcard running thus:

> I regret to inform you that the premises of my plate-printers were entirely destroyed by fire on Monday last, and with them all the work in progress, including Part xviii of the 'Illustrations', which was just completed. This will cause uncertain and unavoidable delay, but as soon as possible some arrangement will be made to continue the 'Fungi' and 'Fresh Water Algae.'

The first intimation we have that Cooke had added Australian fungi to his workload is a paper in *Grevillea* in 1880 written jointly with a Hungarian mycologist, the Rev. C. Kalchbrenner. But long before this Baron Ferdinand Von Mueller (Fig. 11.11) had been sending collections of Australian fungi to

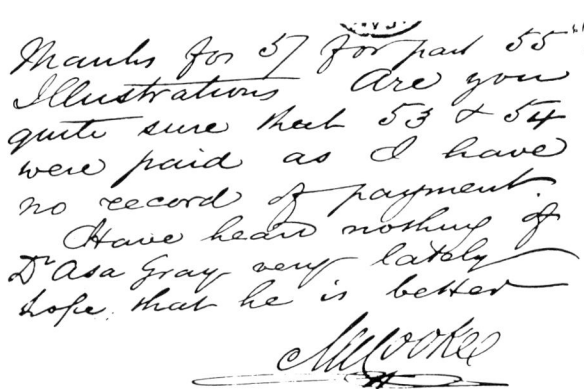

Fig. 11.10   Undated postcard from M. C. Cooke to W. G. Farlow.

Fig. 11.11 Baron F. Von Mueller of Melbourne Botanic Gardens, S. Australia.

Berkeley, and of flowering plants to Hooker at Kew. Von Mueller was an Austrian baron who had made Australia his home, and was well-known as an intrepid explorer and plant collector in that then little known country. His headquarters were at Melbourne Botanic Garden and National Museum. When Cooke moved to Kew, Von Mueller began to send him parcels of the larger fungi which duplicated those he sent to Berkeley, thus casuing the two mycologists much unnecessary work, made no more palatable by the fact that the specimens were 'very wretchedly preserved'. Cooke suggested to Berkeley that in these circumstances it would be best if they were both to send all their specimens to Kalchbrenner first for sorting, as he was already dealing with Von Mueller's collections and finding their condition and labelling such that only 10% could be named. Eventually Cooke gave Von Mueller instructions as to the minimum information required for the identification and description of the larger fungi, and on how to label and wrap them, but looking at some of his collections in the Kew Herbarium to-day one cannot but feel that a little imagination must have been necessary in the preparation of the finished paintings Cooke and Kalchbrenner produced.

His first papers on the larger fungi of Australia appeared in *Grevillea* between 1881 and 1883, and consisted of check-lists of those identified up to that date with the localities in which they were found and Latin diagnoses of those being newly described. When the series was finished it was reprinted, together with four pages of Kalchbrenner's illustrations, as a monograph entitled *Fungi Australiani*.

All through 1884 the unremitting grind continued, until at the end of the year his appointment came up for renewal. One might have hoped that this time the administrative machines would have run smoothly, but that would have been too easy, though on this occasion it only took one month to sort the problems out. Cooke had asked for a rise of £100, and on September 5th the India Office duly wrote to Hooker re-engaging the mycologist for a further five years at a salary of £400 but for a six-day week. Cooke wrote indignantly to Sir Joseph refusing these terms and suggesting that there might have been a misunderstanding over his request for an increment. This indeed proved to be the case, for on October 3rd the India Office admitted that it had all been a mistake and re-engaged Cooke at the increased salary but for the original three-day week.

Chapter Twelve

Years of Disaster. 1885-1892

When Mordecai Cooke's appointment at Kew was renewed he was already 60, yet his eldest child, Ada, was only 21 and his youngest, Leila, three. Three of his own children and Mabel were still dependent on him. Ada was now teaching at St. Anne's School on the Holly Lodge Estate, but she still lived at home, as did her brother, Willie. Willie, who had inherited the Cubitts' artistic talent, had left the Cowper St. Schools in the City Road, and taken up an apprenticeship with a chromolithographer, but he also attended painting and drawing classes and was fast becoming a competent artist.

Mordecai's third child, Ernest (Fig. 12.1), had only recently left school, had immediately joined the Merchant Navy as an apprentice seaman, and was now on his first voyage. The terms of apprenticeship at that time make extraordinary reading for us to-day. Boys were usually indentured for four years, their pay for the whole period being from £24 to £30, sometimes paid out to them in instalments but often only handed over as a lump sum at the end of the voyage, together, usually, with an additional 10/- per year 'in lieu of washing'. The apprentice was expected to supply all his own bedding and clothes–if the owner supplied them the cost was deducted from the boy's wages. The boy was also bound over to good behaviour and total obedience, in return for which the ship's master was bound, on a surety of from £10 to £30, to 'the proper means to teach the apprentice'.

Ernest's apprenticeship was served on the full-rigged sailing-ship, 'Argo', Master, W. Thompson. She was a two-decked, iron-hulled vessel of 1561 tons gross, built in Glasgow four years before at a time when many owners were already turning over to steam. According to Leila Cooke, the 'Argo' was bound for Australia, a tremendous first voyage for a boy of 15; but all that is known for certain is that early in 1885 she was on her way home and had set

Fig. 12.1 Ernest Frederic Cooke, 1868–1885.

sail from San Francisco, bound for Liverpool with a cargo of wheat. On May 27th, as she was nearing Queenstown (now Cobh) in Cork Harbour, Ernest was in the rigging–against regulations, according to Leila–when, in the stark words of the *Register of Deaths of Masters and Seamen in British Merchant Vessels*, he 'fell overboard from upper-fore-top-gallant yard, striking the top-gallant sail in his descent' and was killed. It is a horrifying reflection on the times to find that Ernest's death was in no way unusual; in the four years 1885–1888 the official Register of apprentices lists a total of about 800 boys of whom no fewer than 52 died at sea, almost all of them by drowning or other accident. The similarity between Ernest's death and that of his elder brother Harry is, however, so marked that one cannot but wonder whether both tragedies were due to some unfortunate hereditary reaction to height.

Leila Cooke was three years old when her brother was killed, so could only have been two when he left for his first and only voyage, yet she remembers him:

> bringing me downstairs and giving me a 'flying angel' ... and he brought me downstairs in his uniform and danced, and there was a big mirror over the fireplace–and he danced with me, with my legs over his arms.

She can remember, too, Captain Thompson's visit to her father after the tragedy, but she must have been repeating an oft-told family memory when she wrote the following:

> When his [Ernest's] chest arrived there were in it many objects that he had collected; seaweeds and shells and corals, two Venus Flower Baskets, a sea horse and the wings of flying fish. My father mounted these, and under a glass case they stood on the dining-room mantlepiece. These objects must have inspired him to write his book *'Toilers in the Sea'*, for on the cover there was a Venus Flower Basket.

Whether this supposition of Leila's is correct we cannot know. *Toilers in the Sea* was published in 1889, four years after Ernest's death, and it is true that it is Cooke's only book about marine life and that he could have been collecting material for it in the intervening period. However, nowhere in its pages is there any hint of the family tragedy. The reason for the undertaking is given as the need to fill a gap in the literature for amateur microscopists and naturalists for whom no suitable book existed on the Foramenifera ('Chalk-makers'), corals, sponges and similar organisms. As to the Venus Flower Basket (*Euplectella aspergillum*), Cooke himself records that it was a favourite collector's piece of the period, being imported in ever increasing numbers for display in drawing-rooms and naturalists' cabinets, indeed the price asked had dropped from £30 for one of the first specimens sold to the current rate of 10/- each. He was certainly greatly attracted by their beauty: telling his readers that in shape the Flower Basket resembles a narrow cornucopia about 10 inches long, he continues:

> The entire fabric is cylindrical, and hollow, looking like a delicate fairy-like basket-work, formed of elongated fibres ... The top is covered with a similar network lid ... Throughout its entire length this fairy basket-work is ornamented still further with oblique concentric ridges, or furbelows of still more delicate network, and a collar of like material forms a fringe round the lid. As seen in museums, and in the windows of naturalists, the whole structure is of a virginal whiteness ...

As usual with Cooke's non-mycological books this one consists of judicious selections from the literature, woven together by his own comments and explanations. He did not approach the S.P.C.K. about its publication until May, 1888, when the Society accepted it subject to the cost of the illustrations of which, as they would not be original, the rights would have to be bought from their respective owners. Some difficulty must have arisen over them, for in October the Secretary of the General Literature Committee 'reported that Dr. Cooke, having been put in a perplexity about the illustrations for this book', it had made no further progress. The 'perplexity' must have been satisfactorily resolved, for the volume appeared the following year.

1885 was a tragic year for Cooke. Not only was his son killed in May, but in August his mother, to whom he was deeply attached, died at Neatishead.

Mordecai, who had visited her earlier in the year, was with her when she died, and in one of his very rare surviving references to his family he wrote to Broome:

> Painful as the occasion of my absence from Kew must be, it is the source of very great comfort to me that my dear mother quietly passed away without any decided pain of any kind. She did not keep her bed more than 10 days up to which her appetite was good and all her faculties clear . . . It was such an end as anyone might envy at such an age as 83 [81, in fact]. Quiet, peaceful, happy, without the least suffering, almost a continuous slumber and terminated without struggle.

Mary Cooke was buried in the graveyard of Neatishead Chapel, with her husband. Mordecai seldom visited Norfolk after that. In September he wrote to an unknown mycologist:

> I have had on the whole a most eventful six months of trouble and anxiety and it has been under most exceptional circumstances that anything [mycological] has been done at all'.

At about this time the Ebenezer Cookes moved from Falkland Rd., Kentish Town to be once again near neighbours of the Mordecais, settling at 15, St.

Fig. 12.2 Ebenezer Cooke, 1837–1913.

John's Road,* a northward extension of Junction Road. The two large families (Ebenezer and Ellen had eight surviving children) saw a good deal of each other, little Leila, in particular, becoming very friendly with her cousin Winnie. Ebenezer, who had retained his interest in natural history, often accompanied Mordecai on his fungus-hunting expeditions on Hampstead Heath, but the brothers did not always see eye to eye, and there are family reports of considerable friction between the two. To Mordecai's family Ebenezer was known as 'Uncle Teazer', but his children's nick-name for Mordecai (and they must have had one) is not recorded (Fig. 12.2).

For some years Ebenezer had been devoting all his time to developing his theories on art teaching and to trying them out in practice. To this end he first made detailed studies of the development of drawing ability in young children, and there is considerable evidence that his own numerous offspring were the source of his observations. Then in 1878 he gave his first lecture on the subject, the ideas that he put forward being truly revolutionary for that date and eventually becoming the foundation of modern methods of teaching art to young children.

> ... in drawing the lecturer should not have the child troubled with the grammar of art, but encouraged to make his own natural efforts of expression which would give a chance to the development of any artistic feeling or talent that might be within him, and save him from becoming at best a poor imitator of the great originators of art [1].

He used children of different ages to demonstrate his method while he was speaking, and thus became the first ever to consider in a practical way the relationship of the teaching method to the natural development of children's drawings.

That same year he was appointed art teacher at Camden House School, a combined kindergarten and teachers' training college which used Froebel methods.

> He came to Camden House for many years, giving lessons on nature study and on drawing and painting both to children and students. He was a wonderful teacher ... He taught them to see, to look for and try to find truth for themselves, and he opened up to them a world of loveliness of form and colour ... He gave unique lessons in nature study, making pupils discover facts and deduce reasons for themselves, by bringing carefully chosen specimens that helped to bring out some special conclusion [2].

How very reminiscent this is of what we know of Mordecai's teaching methods; the elder brother must have had a great and lasting influence on the younger while they were both living under the same roof, and both, of course, owed much to the Pestalozzian teaching methods they encountered in their

---

*Now St. John's Way.

childhood and youth. Mordecai must also have been the inspiration for Ebenezer's continuing and profound interest in natural history, particularly botany. Throughout his career the younger brother combined the teaching of the subject with that of art, using one to illuminate the other, and the success of his methods was so great that in 1882 he was elected to the Executive Committee of the Froebel Society. Art was Ebenezer's recreation as well as his livelihood, the countryside and plant life being sources of undying pleasure to him. For many years he spent part of every summer painting in Norfolk, a custom which only ceased with his mother's death.

Ebenezer's change of occupation from lithographer to pioneer in art teaching did not, as far as we know, alter his attitude to the world and the society in which he lived, but a similar improvement in Mordecai's circumstances had a profound effect on his social and political convictions. Even though Mordecai was never accepted as a social equal by the scientific establishment amongst whom he now worked, and could never afford to live in the salubrious districts in which they lived, he was rubbing shoulders with them daily, corresponding with them freely, and acknowledged by them for his contributions to science. Long past were the days when he kept the accounts for the Working Tailors' Association and ran evening classes for operatives. By 1884 he was saying:

> Are we not in great danger of exaggeration in the direction of popular education? Not the less so because it has its sentimental side . . .[3]

And advocacy of societies with 5/- subscriptions had given place to membership of the gentlemanly Woolhope Naturalists' Club. As to what he regarded as the general deterioration of modern times, he had written as far back as 1871 that:

> there is a lacking of respect, almost a defiance of authority, an impatience of restraint, a desire for change, a restlessness and dissatisfaction at old barriers and landmarks, an assertion of individual right to think and act independantly of everyone else, and everything else, all tendencies which manifest themselves in times of political activity, and which expend themselves in various directions when political activity subsides [4].

In 1887 he quoted this earlier opinion and saw no reason to alter it.

All this is reflected in his political affiliations. Though no written record survives of his youthful ideas, it seems safe to assume that, at least until he left Holy Trinity Schools, he subscribed to his parents' radical views. But from then on, as his station in life changed and he began to mix with, and aspire to become the social equal of, the gentlefolk of the world of mycology, his sympathies must gradually have drifted to the right, for in 1885 he was already a member of the Primrose League and a holder of its medal.

The League had only been founded two years earlier to commemorate Benjamin Disraeli, the great leader of the Tories who had died in 1881. Its objects were to perpetuate Disraeli's tradition and to infuse new life into the Party, and it was a far more influential body in those early days than it is to-day. Disraeli's favourite flower, the primrose, was adopted as its emblem, something on which Cooke could not forbear to comment in his nature study book for children, *Through the Copse* (p. 286). He makes the little girl, Cissy, ask, 'Why do we have a Primrose Day, when nearly everybody carries primroses?' To which Uncle Matt replies:

> Because on that day a great man and a great statesman died, and we are sad for the past. The flower tells us of sunny days to come, and then we hope for the future.

Cooke must have been an active member of the League for quite a long period, for not only had he four bars to his medal, dated from 1885 to 1888, but he also held the Grand Star, Second Grade, a magnificent, ten-pointed, enamelled star with a crown above it and a primrose at its centre. The medal, with its bars, was designed to be pinned to the chest, but the Star hung round the neck on a broad, purple and white ribbon (Cooke's having been lengthened with a piece of black ribbon sewn on with remarkably coarse stitches).* Leila Cooke recalls that when the local branch of the League held meetings adressed by famous speakers her father would always attend, sitting on the platform proudly wearing his Star; he was 'altogether in favour of Disraeli and thought that Gladstone was just a spouter–couldn't bear Gladstone.' By the time Leila was old enough to understand such matters her father was no longer an active member of the League, neither did he 'bother very much about politics except at election time'. Indeed, the only evidence that Cooke must at one time have been politically active is his possession of the Grand Star (Fig. 12.3), for it was awarded specifically to branch officers who had given devoted service over a period of years. What he did is unknown, for despite an exhaustive search through the archives the present Secretary of the League has been unable to find any surviving record of Cooke or of the granting of his medal or Grand Star. It is astonishing that with all his other commitments he could have found time for political activity, but for those four years at least he must have done so.

Over the years of ceaseless collection and exchange of fungi from all over the world, Cooke's private herbarium had grown to enormous proportions and now contained 43,500 specimens disposed in '14 cabinets and sundry smaller packages', all of which would have been kept in his study at 146, Junction Road where they could be locked safely away from the prying hands of the children. There also, he seems to have kept most of his books and to have carried on all his microscopy, writing, drawing and business affairs, for with a household of ten there could hardly have been another room to spare for him. As well as taking up space, the herbarium would have needed constant atten-

---

*Both Medal and Star are now in the possession of one of Cooke's grand-daughters.

Fig. 12.3 M. C. Cooke's Primrose League Medal with bars.

tion, especially when new accessions were added and names or classifications changed. By 1885 Cooke finally realised that a limit had been reached and that his employment at Kew offered the perfect solution for what must have been a very hard decision to take. So on October 24th he wrote to Dr. Oliver:

> It has occurred to me that the best method for me to adopt, in the face of my precarious health of late, would be to dispose of the herbarium to some public institution where I could consult it when desirable. It is contained in . . . deal cabinets. . . . The price fixed is £250 provided it is disposed of to Kew or the British Museum . . . Having a large family to provide for I cannot afford to be generous and *Present* my Herbarium to the country . . . I could myself transfer the specimens to the general Herbarium as part of my work at Kew . . . The space available in a private house is not sufficient to work a herbarium of this size.

A fortnight later Dyer wrote to the Office of Works recommending the purchase at the price suggested, which he considered 'modest and reasonable'. The acquisition, with Berkeley's herbarium, would put Kew in advance of all other botanical establishments. The transaction was completed with remarkable speed, for the herbarium was delivered to Kew exactly one month after Cooke's initial approach, although he had to wait three months for payment. Two years later Kew also bought, for £75, Cooke's collection of figures of fungi. At the time it numbered 22,000 drawings, 6,000 of which were by his own hand, but as he had not yet completed the *Illustrations* he accumulated many more later.

The year 1885 marked the end of an epoch at Kew, for it was then that Sir Joseph Hooker retired from the Directorship and was replaced by his son-in-law, the pompous William Thiselton Dyer, who was disliked by the staff in general and loathed by Cooke in particular. According to Leila Cooke:

> Thiselton Dyer represented the new ideas of organisation–the old days of freedom to work at what you liked and think independently were gone. The Herbarium became a Government Office. All work done by members of the staff had to be submitted to the new Curator [W.T.D.?] and published under the name of the institution. This seemed, as far as I remember, a cause of friction. And I don't think my father liked being called over the coals for wasting time (this is my impression of the chief cause of friction.) New men with science degrees from the Universities were in charge. Honorary Degrees counted for nothing. It was the beginning of the new age of industrial organisation applied to free scientific endeavour.

W. G. Smith records that Cooke's hatred of Dyer was such that at one stage he could not even bring himself to speak his name, but this could well be one of Smith's more extreme exaggerations.

Cooke's next book, *British Desmids, a Supplement to British Fresh-Water Algae*, appeared in 1887. It covered that large group of unicellular green algae, of immensely varied and beautiful shapes (Fig. 12.4), which are usually symmetrical about an axial constriction. Most of the book had, in fact, been completed some years earlier, as the author explains:

> This is the third time which the present work has been prepared for the press during the past 4 or 5 years. In the first and second instance it was made ready at the instigation of a publisher, who intended immediately to issue it on his own responsibility, but in both cases a collapse occurred, which put an end to all business, before publication commenced [5].

Only after these two disappointments did he issue it in parts at his own risk, and by then, to his embarrassment, specialised phycologists were about to fill the gap since the last study had been published in 1848. As he does not specify the type of reader for whom the book is intended it must be assumed that, like *Fresh-Water Algae*, it was meant for general microscopists, consequently the criticisms of superficiality which were levelled at him would seem to be unjust. Even though the book had little value in the long term, Cooke was performing a considerable service at the time simply by collecting together and making readily available the widely scattered literature.

The illustrations were, however, a different matter. The great Swedish phycologist, Dr. O. Nordstedt of Lund, wrote a scathing article in the *Journal of Botany* giving a list six columns long of drawings from the book which he accused Cooke of copying from other authors, mistakes and all, and claiming

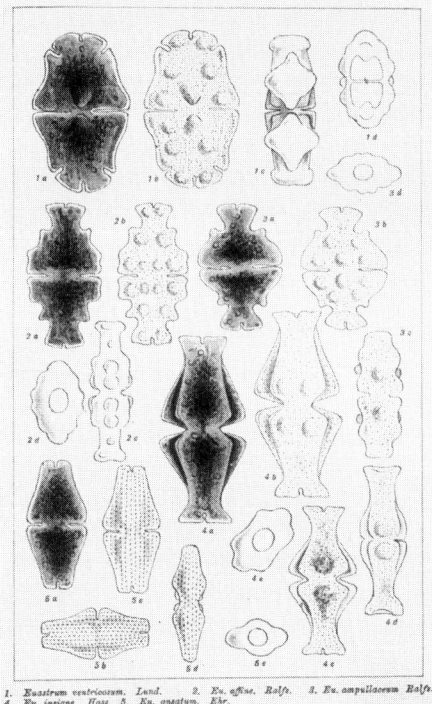

Fig. 12.4 Illustration from *British Desmids*. Original in colour.

that they were original. The concensus of opinion among modern phycologists whom I have consulted* seems to be that Nordstedt was correct, at least in part, and that the omission of acknowledgements was very reprehensible. However, the origins of the figures can, in fact, be deduced from the very comprehensive list of authorities which Cooke does include, and the utility of the book at the time probably compensates for the crime.

His trilogy of books on the Algae was completed in 1890 with the publication in the International Series of Kegan Paul and Trench of the *Introduction to Fresh-Water Algae, with an Enumeration of All the British Species*. He had himself suggested the book to the publishers 'in consequence of the work on *British Fresh-Water Algae* being so rapidly sold out', but his main objective was to produce a cheap and concise handbook which might attract new interest to a subject covered until then only by large and expensive works. To this end nearly half the book consists of an account of the life cycles and biology of the organisms, the remainder being given over to brief descriptions of the species; a black-and-white illustration is supplied for one member of each genus, and there is a glossary. Unfortunately he could not resist devoting one chapter

---

*Professor F. E. Round, Professor A. J. Brook and Mrs. J. A. Moore.

to his views on the 'dual lichen hypothesis', but apart from that the book served a useful purpose. It was reprinted seven years later, even though the series, which had been so prestigious in its early days, when he and Berkeley published *Fungi, their Nature, Influence and Uses*, was now in decline, its managers having failed to move with the changing demands of the market.

It brought forth one damning review, based on inuendo rather than on fact, from a sworn enemy of Cooke's, George Murray, the cryptogamist at the new Natural History Museum at South Kensington. Murray had been appointed in 1876, two years after Cooke's break with Carruthers, as a junior assistant in charge of algae and fungi, his own interests lying primarily with the algae. Nothing is known of the grounds for the hatred between the two men, but it was certainly mutual. At his best Murray was a kind-hearted and well-liked man, but a few years after his outburst over the *Introduction* he suffered a mental breakdown which eventually forced his retirement from the Museum; perhaps in the early stages of the illness he was not sufficiently stable to be able to tolerate Cooke's forthrightness. Murray's 'review' was so obviously a thinly disguised personal attack on Cooke that it really hardly merited a reply, but its victim chose to reprint it in full in *Grevillea*, following it with a paragraph pointing out its total lack of scientific foundation.

Cooke's second five-year term at Kew came to an end in December, 1889, and the following year he would reach the normal retirement age of 65. But he himself had no wish to go, and he was still fulfilling a very useful function at Kew, so steps were taken to renew his appointment yet again. Despite the fact that Hooker had retired it seems that he was still actively involved in the affairs of the Gardens and willing to use his huge influence on its behalf, for in October, 1889, he, not Dyer as would have been expected, wrote a long letter to the India Office setting out the reasons for re-appointing Cooke for a further term. The request was granted, but this time for three years only.

The year 1890 was another unhappy one for Mordecai. On April 22nd he wrote to Kew as follows:

> Pray excuse me to-day. By some mishap the upper part of my house was on fire last night. It was fortunately discovered early and no very serious damage done except one room burnt out–but it was such a shock and we have had the firemen on the premises the best part of the night–that I am unfit for anything, and I have to see the Inspectors from the Insurance offices. It was a most marvellous escape. . . . My books were all safe from damage by fire or water but elsewhere there is desolation.
>
>   Yours sincerely, M. C. Cooke.
> N.B. violent diarrhoea resulting from fright is so very inconvenient to travel with.

In October he wrote again, offering a formal apology for once more letting his tongue run away with him in an apparent criticism of the administration at Kew (the circumstances are not known), though he denies that it was in any way deliberate, and considers that he must have been misinterpreted. In mitigation of his gaffe he continues:

> I have been under pressure of the most severe troubles and anxieties during the last few months known only to my relatives and most intimate friends.

Considering the extraordinarily tangled relationships which already existed in this family, and which had been kept hidden for so many years, it is difficult to imagine that there could be any new troubles and anxieties sufficiently terrible to warrant such secrecy, and it is unlikely that the whole truth behind this letter will ever by known. However, it is certain–and no secret was made about this either by Leila Cooke or Mordecai's grand-daughters–that while Leila was quite young, Annie left Mordecai for the second time, and that this time she never returned. Leila thought that she herself might have been about four years old when this occurred, which would date the separation at about 1886; but other evidence, particularly that of one of the grand-daughters, suggests a slightly later date, about 1890, and if this were so the letter quoted above could well refer to Annie's departure, and more particularly to the events precipitating it. Leila herself offered two different explanations. To one of her nieces she wrote that she had been told by her elder sister, Ada, that 'it was because she refused to have any more children'–she had borne eight, and was 38 at the time of Leila's birth. To me Leila said:

> I think that the real cause of the difference was *money*, as far as I can see. Any spare money, instead of going on the maintenance of the house, went in books. I can remember her saying once–it was always books, books, books, when what they wanted was boots, boots, boots! She was rather a jolly person.

Probably some specific and deeply traumatic event had taken place which was never divulged to the younger children or to the next generation.

Ada and Willie, both now grown up and self-supporting, did their best to persuade their mother to stay, and when they failed to do so they decided to continue living with their father. Frank would have liked to have stayed too, but was given no choice in the matter and forced to go with his mother, as was his younger brother, Bertie; but Bertie was particularly fond of his half-sister, Mabel, and she, of course, went with Annie. Mordecai refused to part with Leila, to whom he was devoted, so she remained behind, presumably in the care of Sophia, Ada and the maid.

Annie moved no great distance, and despite constant changes of address in the years which followed, she never strayed far from Upper Holloway. Indeed, it is likely that her first move was no further than Poynings Road, a turning off Junction Road a few hundred yards from No. 146, for Mabel was certainly living there in 1891. Mordecai paid Annie an allowance, which she came to collect every week, and in Leila's memory the two always seemed very friendly when they met, though she was once told that after they had been separated for seven years Annie refused a request from Mordecai to return home.

Now that Mordecai had two households to support his financial situation must have been more precarious than ever. Mabel had become a dressmaker, so was probably self-supporting, and Annie may have taken in lodgers, but the boys were, at first, too young to help. Then, in June, 1891, Mabel, now aged 20, married George Henry Stone, a compositor who, as he lived at the same address, is likely to have been her mother's lodger. The wedding took place at St. Peter's Church, Highgate Hill, and was witnessed by 'M. C. Cooke' and 'Annie Cubitt', surely a remarkable gesture of goodwill on Mordecai's part. It would seem that John Cubitt was not present at his daughter's wedding or he would, presumably, have acted as witness instead of Mordecai; but he must still have been in touch with his family, for on the marriage certificate his occupation is given as 'Stud groom', a change from his previous job of ticket collector.*

By whichever of her surnames Annie was known while she was still living with Mordecai, it seems likely that after she left him she became Mrs. Cubitt, for not only did she sign herself thus on three official documents up to the year 1897, but one of her granddaughters, when taken to see Mordecai as a small child, remembers being told by the maid that an old lady who was also visiting was her grandmother, Mrs. Cubitt. Both Frank and Bertie used the name Cooke when they grew up, although Frank, as we have seen, had been registered as a Cubitt at his birth; whether they took their mother's name as children and changed it to Cooke later it is impossible to discover. Frank did not remain dependent on his parents for long, for he was 16 at the time of Mabel's marriage, and would already have begun his apprenticeship for his future trade of letter-engraver.

All through these family troubles Cooke carried on with his mycological work with his usual indefatigable energy. The *Illustrations* continued to appear, the last part in 1891, as did *Grevillea*, which still carried a large proportion of contributions by the Editor himself. In particular he had, since 1884, published a series of 17 articles under the title 'Synopsis Pyrenomycetum', 'being a revision of the classification and enumeration of pyrenomycetous fungi included in Saccardo's "Sylloge",' together with seven papers on the Polypores, 'Monographia Polyporum'. In 1891 the Pyrenomycete papers were collected and issued as a single volume.

Ever since his monograph *Fungi Australiani* had appeared in 1883 Cooke had continued to publish in *Grevillea* descriptions of new fungi sent to him by von Mueller, and to identify those species which were already known from other parts of the world. Eventually von Mueller suggested that Cooke might

---

*I have been unable to find any further information about John Quincey Cubitt. He is not mentioned in his mother's will, made in 1912, so he may have been dead by that date; on the other hand, I can find no death certificate for him before then, so perhaps she disinherited him.

like to prepare a *Handbook of Australian Fungi*, which the various colonies (now states) could possibly be persuaded to sponsor. Cooke jumped at the idea, the colonial governments agreed, and the book was published in 1891, all but a few copies being shipped straight to Australia. It gave descriptions of 2079 species of the larger fungi which the author had received over the years from the Baron himself, from Mrs. Martin of New Brighton, and from Mr. Bailey, the Government Botanist at Brisbane.

Even though Cooke discarded 90% of the specimens sent as useless, the remaining 10% were seldom of very high quality, and he had to rely entirely on the drawings and descriptions sent with them to gauge the appearance of the fungi when they were fresh. It is hardly surprising, therefore, that the illustrations which accompany some of the descriptions are idealistic to say the least. Those species which were not new to science had already been included in Saccardo's *Sylloge*, and Mr. Hilton, of the University of Western Australia, points out that the descriptions Cooke gives of these are in fact translations of Saccardo's Latin diagnoses, duly acknowledged–the only occasion on which such translations have been published. The limitations imposed by time and distance on the accuracy of a work on such transient organisms as fleshy fungi inevitably detracts from its accuracy, and Cooke was fully aware of this. Yet, though it has been roundly condemned by generations of mycologists [6], no-one in Australia was equipped to carry out a similar task at that time, and without Cooke's attempt it would have been many years before a fungus flora of Australia was produced. Mr. Hilton considers that though the book is of purely historical interest to-day, it was a valuable contribution for its time.

At the end of the year Cooke offered to the S.P.C.K. another book of popular natural history under a title described by him as 'eccentric', namely *Vegetable Wasps and Plant Worms*. The General Literature Committee agreed to examine the manuscript but was not impressed, because in February, 1892, its members decided that they 'could not see their way' to accepting his offer to revise and rewrite it; nevertheless, the book appeared later that year. It is an excellent account of the fungal parasites of insects (Fig. 12.5), in which Cooke combines his own mycological outlook with quotations from an entomologist who had published on the same subject many years earlier. It is interesting to find that at the time he was writing Cooke was exchanging publications with Roland Thaxter (1858–1932) of Harvard, who was then engaged on his own monumental studies of the fungal pathogens of insects. Cooke wrote to him that he was 'especially glad of *Laboulbeniaceae* [one of the fungal families concerned] as I have a popular volume of 'Entomogenous fungi' just ready for press, and can now add your new species'. (WGF 110).

Despite their quality, neither it nor *Toilers in the Sea* seem to have been commercial successes, for when, later that year, Cooke suggested a subject for yet another volume, namely the lower plants or Cryptogamia:

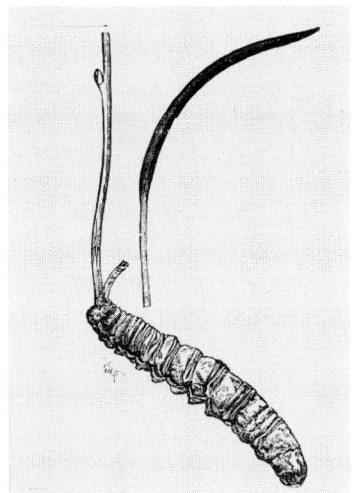

Fig. 12.5 Illustration from *Vegetable Wasps and Plant Worms*, showing *Cordyceps robertsii* parasitic on a caterpillar in New Zealand.

the Secretary pointed out to Dr. Cooke that the Society could not give him the same terms . . . as for the previous books on the grounds of the commercial history of his last two works. Dr. Cooke now wrote to say that he was willing to take £50 for the first edition of 2000, and a small royalty.

The Committee agreed and the book duly appeared in 1893 in the S.P.C.K.'s 'Romance of Science', series, which was directed to young people; its terrible title, *The Romance of Low Life*, cannot therefore be blamed on its author.

It covers all the Cryptogamia from the ferns to the fungi, though inevitably a disproportionate amount of space is devoted to the last named. Cooke states his intenion thus:

> It is not so much to the results of new or recent investigations that I have desired to give predominant interest, as to the general influence which increased knowledge of the structure and history of these minute and obscure plants has had upon the romantic beliefs and unsubstantial theories of a less enlightened age.

To this end he has marshalled much material scattered through the scientific literature that would otherwise have been inaccessible to the layman, and written it up in popular terms. Inevitably, the book attracted a vicious review from George Murray, who disapproved of all such pandering to the lay taste. He does concede, however, that:

> He [Cooke] has much experience in writing popular books and knows just the kind of thing that tickles. He also knows when to do some 'fine writing', and to drop into poetry etc. Judged by the standard of such literature it is not badly done [7].

Murray also draws attention to an odd aberration in Cooke's indexing.

> 'What is Nostoc?' indexed under 'What' is amusing enough; but 'The Comma Bacillus', 'The Destroyer', both indexed under 'The', without anything under 'Comma', 'bacillus' or 'destroyer', is surely a record.

Early in 1892 the following 'Notice' appeared in *Grevillea*:

> For 20 years we have continued periodically to issue the numbers of this Journal as a labour of love; with the next part the twentieth volume will come to a close. . . . It is not without feelings of regret that this resolution has been taken, but fickle health, increasing years, and diminished vigour have been the excuses which have presented themselves . . . Our only concern now is to make known our determination to stand open to any proposal whereby our pecuniary interest shall cease, so that we may rest from the periodical anxieties of the last two decades. Whether we have earned this repose is not for us to determine, but we are never blind to our own failings, any more than to those of others, and can only wish that what we have done had been better done, even although we have endeavoured to do our best.

The journal did not have to wait long for a new proprietor, for later the same year it was bought by 42 year-old George Massee and a colleague of his, Massee becoming the new Editor. (Fig. 15.7)

From the time that Cooke first went to Kew he had been in touch with Massee, who was then living with his mother in Scarborough where he was teaching botany and making a special study of the fungi of the district. Cooke had been much impressed by the younger man's figures of fungi, and eventually used many of them in his *Illustrations*, so when Massee and his mother came to live in Kew in the early 1880s it was natural that the former should gravitate to the Herbarium where he worked as a freelance, supporting himself the while by lecturing on botany at a number of institutes. Cooke helped him greatly with his mycological studies, took him to forays all over the country, and gave him every opportunity to become more widely known. By the time Massee took over *Grevillea* he had published numerous papers himself and had started work at the Natural History Museum. Though undoubtedly a brilliant mycologist with a vask knowledge of his subject, Massee had grave weaknesses; he was careless, he had no capacity for taking pains, and he often totally disregarded the work of others and so made some extraordinary mistakes of his own [8].

Cooke continued to write papers for *Grevillea* after selling it, but although the journal had not only survived for 20 years under it founder, but paid its way as well, a fact of which Cooke was extremely proud, Massee was only able to keep it alive for a further two years; volume 22, 1893–1894, would mark its demise.

As Cooke must have been very well aware, his appointment at Kew would terminate on December 30th, 1892, unless he was granted a further extension by the India Office. But he was now 67 years old, two years over the normal Civil Service retirement age and, by his own account, in failing health, so he would have realised that such a renewal was highly unlikely. Dyer, however, had no intention of letting him go until he was sure of a permanent replacement for him, and deliberately kept Cooke's hopes of reappointment raised until the last possible moment. As late as December 19th the India Office wrote: 'My dear Thiselton Dyer, What about Cooke? . . . ' And in a letter to a friend the following year Cooke said:

> I have not written to you since I had to submit to banishment from Kew, . . . Of course I knew of the restriction [on age], but, up to the last day, I was assured that as a specialist on Scientific duties I could be retained for another two years [9].

In his innocence, he blamed the government of the hated Gladstone for refusing to take the risk of bending the Civil Service rules. In fact, unknown to Cooke, Dyer had been ignoring the India Office and dealing with the Office of Works, to which department he had written as early as November 16th requesting the establishment of a new, permanent post for a cryptogamic botanist at Kew (to be occupied by George Massee) so that the subject could be put on a less precarious footing than it had been in the past. Laudable though this was as far as Kew was concerned, his callous use of Cooke to safeguard his position during the negotiations was quite unprincipled.

Inevitably the axe fell and Cooke was 'banished'. The India Office granted him a fairly generous pension of £240 a year but, his domestic situation being what it was, he retired very fearful for his financial future. In due course Massee was appointed in his place.

Chapter Thirteen

'Forays Amongst the Funguses'. 1880-1889

When Cooke was seconded to Kew to take up his post as professional mycologist, his enthusiasm as a field naturalist in no way diminished. In theory, at least, he now had more time to devote to an activity which had given him life-long pleasure and satisfaction, as well as providing the living specimens that were essential to his studies. In this chapter and the next, therefore, we shall return to the year 1880 to trace his continuing contribution, both mycological and general, to the field club movement which, in London rather later than in the rest of the country, was now reaching its peak.

There was, at this time, a change in the emphasis of his field work. While he never entirely forgot his goal of helping the amateur to an enjoyment of nature as great as his own, as witness his continuing ties with the Quekett Microscopical Club, he became ever more closely associated with the specialist fungus forays of the Woolhope Club. Also, as a result of the increasing recognition he was gaining as one of Britain's most eminent mycologists, more and more natural history societies would from now on be asking him to lead their members on an annual foray in their own locality, giving him honorary membership in return, though not expecting him to take part in their non-mycological activities.

One of these clubs, with which he was to retain close contact for the rest of his life, was the Epping Forest and County of Essex Naturalists' Society, a title soon to be shortened to the Essex Field Club. The Society was launched on January 10th, 1880, with the chemist and naturalist Professor Raphael Meldola, F.R.S., of Finsbury Technical College, as its first President. He was an excellent choice, for not only had he played a very active part in the formation of the Society, but he had for long been associated with the adult education movement, and was currently a member of the Technical Committee of Essex County Council. He was a prolific writer on natural history and was for some time the Correspondent on this subject for Frank Buckland's magazine, *Land and Water*. He wasted no time in enlisting the infant Society in the conservation lobby.

Enclosures had been proceeding for many years and at an alarming pace in the Essex forests of Epping, Hainault and Waltham. The commoners (i.e. the local people) had begun to petition the Crown against these as far back as 1845, but to no effect; indeed, in 1851 an Act of Parliament had been passed allowing the complete disafforestation of Hainault, and this had been completed by 1863. Now only 12,000 acres of Waltham Forest and 7,000 acres of Epping Forest remained, and even there enclosures were continuing fast with no compensation being paid to the commoners. By 1870 it had at last been realised by those holding the reins of power that if the dismemberment of what remained of Epping Forest were to be allowed to continue, a vital lung for the seething and disease-ridden population of a grossly overcrowded London would have been lost for ever. Proceedings were started in parliament, the Corporation of London took a hand, and it was eventually decreed, not only that enclosures must cease, but that all those made during the last twenty years were illegal. Further Acts were passed, and in 1878 the Corporation of London bought 3,000 acres and was appointed Conservator. What remained of Epping Forest was thrown open to the public by Queen Victoria on May 6th, 1882.

But even after the Corporation took charge the Forest was not safe, for one of the first actions of the new Epping and Essex Naturalist's Society after its formation was to help to fight off a proposal to run tramways through the area. In 1881 the Society again joined an active campaign, this time to protect the Forest against an attempt by the Great Eastern railway to bisect it with one of its lines. Among other organisations whose backing was sought was the Quekett Club, which carried unanimously a motion of support. Other attempts to desecrate the Forest were fought off over the years, often again with the support of the Quekett Club, including one in 1888 during which Meldola and Cooke collaborated to draw up a statement to the Conservators, copies of which it was planned to distribute to interested organisations. Cooke's editorials in *Science Gossip* and his remarks in his book *Woodlands*, so many years ago, show that such a cause would be very close to his heart.

He may first have heard of the Epping and Essex Society from one of its founder members, James Lake English (1820–1888), who had become well known in mycological circles some years earlier through his invention of a method for preserving toadstools in wax so cleverly that they were almost indistinguishable from fresh specimens. He sent displays of his preparations to meetings in many parts of the country, and in 1875 even to Perth for a foray led by Cooke, and he often carried off prizes at Royal Horticultural Society exhibitions. English had been born and bred in the Forest, and had studied its flora and fauna all his life. He was a self-taught and brilliant mechanic and taxidermist, using his skills to earn his living as well as to further his interests as a naturalist. He even made his own compound microscope. At that period there were considerable openings for taxidermists, for not only were animal displays required by the rapidly expanding museum movement, but no country house was complete without its glass-fronted showcase of stuffed animals and birds; English was much in demand for this purpose by the Essex gentry. While he was an excellent all-round naturalist, his chief interest in his younger days had been as a lepidopterist; only later did he begin to notice the larger fungi, but he soon became extremely knowledgeable about them. Cooke regarded him as a 'shrewd observer', and in 1881 collaborated with him to publish a list of the hymenomycetous fungi of Epping Forest in the Society's *Transactions*.

Worthington Smith was a founder member of the Essex Society and had a seat on its first committee, so it was doubtless he who was responsible for organising a fungus foray in that foundation year, and establishing forays as regular annual events thereafter. He saw to it that this inaugural outing was a distinguished affair, inviting Cooke to join English and himself as a joint leader, while Plowright, two other Woolhopeans, and the Frenchman Le Cornu, came as guests. Smith had for many years been the licenced chronicler of the Woolhope Club's forays, and now he turned his practiced hand to describing the Essex Society's outing for the *Gardener's Chronical* [1], from which account the following is taken:

> Dr. Cooke, furnished with a large leather travelling trunk (in place of a hand basket or tin collecting case) was one of the first to arrive in the Forest. By 4 o'clock the Doctor's phenomenal portmanteau was full of funguses. Where one generally looks for a toothbrush might be found a Phallus, in place of a sponge was a bloated Boletus, in lieu of writing paper sheets of dry-rot. Shirts were shirked, and fungi both fresh and frouzy were in all the compartments of the valise. No one but an advanced fungologist could so treat a portmanteau.

It should be noted that Cooke did not usually collect in a travelling trunk. His daughter Leila told me that:

> 'he had a big black bag, about 12" or 14" long, stiff and high; and the sides were firm and they didn't give. And everything that he found he used to put in this black bag.

She remembered that when, as a small child, she accompanied her father to forays of the Essex Field Club, she too put the fungi she collected into the black bag. As to his apparel, he of course wore the compulsory black suit, and on his head he always wore a hard hat–a a sort of chimney-pot, but not really a chimney-pot, a hard black felt [Fig. 15.7]. Until quite late in life. Later on he wore a felt with a dent in the middle.

To continue with Smith's description of the foray, during the afternoon:

> came down such a terrific and unexpected storm of rain that half the party got wet through to the skin in the first two minutes. The ladies (poor things) were taken under a tree and covered over with newspapers.

But the day ended happpily enough with a meat tea at the Forester's Arms at which some of the day's finds were cooked and eaten.

> After tea Dr. Cooke delivered an admirable extemporaneous lecture on the discrimination of fungi, especially adverting to the edible and poisonous species; the leather portmanteau was brought into requisition, and the rich, odorous and dripping fungi were taken from the valise and commented upon.'

It was in April, 1881, shortly before his paralytic illness, that Cooke delivered his Inaugural Presidential Address to the Hackney Microscopical Club and Natural History Society which, with the Quekett, was to be the only society in the affairs of which he would retain a general, rather than a purely mycological, interest. The Address was a very solemn one entitled 'The Student of Nature'[2]. First he took the opportunity to outline to the young Society what the attitude of its members to their hobby should be; the joys and rewards of a serious study of nature as distinct from the more frivolous approach of the mere rambler and collector. And he did not omit to mention the part which the President should play:

> ... what diverse creatures we [the members] are. We follow different occupations by day; we are interested in different recreations at night. Herein lies the advantage of societies with a catholic programme. The individual man retains his individuality [but] it seems to me that your President at least should forget that he is a specialist when called upon to preside over you. His sympathies should be extended to all; a general–in fact as well as in name.

Then he went on to emphasize the need of the human mind for variety of occupation:

> Nothing exhausts and shatters the mind to fragments equally with one eternal round of the same dull theme. The same brick walls, the same squaring of the circle, the same click-clack of bobbins and spindles, the everlasting £.s.d. Let this dull round go on from year to year and the product is a mummy. Intellect there is none–soul there is none. Nothing but a perambulating ready-reckoner.

Contrariwise, if he changes his own shop for nature's workshop, and shuts one but to open the other, the change steals like a tonic over his debilitated mind, and rehabilitates him for the work of the morrow.

Then, what was the supreme advantage of natural history over all other recreations?

Happy the choice that selects for recreation activities that become a *discipline*. . . . Of all studies the classificatory sciences are the best disciplinarians. The drill sergeant is invisible; there is no sense of constraint; there is no feeling of rebellion against authority; the impulse is gentle as the touch of a feather, potent as a lance of steel. The habit of grouping or classing objects according to affinities or resemblances, arranging in genera, families and orders, commends itself to the mind with experience, that, all unconsciously, the love of order spreads from the hours of recreation to hours of work . . . Let no one dream that time thus spent is lost. Atom by atom the little worms may build unseen beneath the surface of the sea; but by-and-bye the coral island becomes a hard, inflexible rock . . . A little experience may be of no use to you to-day, but twenty years hence it may be worth gold. It matters not *now* that the imprisoned steam blows off the lid of the kettle; in twenty years it drives a railway train from Darlington to York.

Cooke's views on the part that amateurs can play in the genuine furthering of scientific knowledge, while they inevitably apply to a much narrower range of activities to-day than they did in his time, are still largely true for the systematic botanist, and certainly for the mycologist.

Amateur scientists, if you will, lovers of science as a recreation. . . . may do much, and might do more, for universal knowledge . . . We who profess to follow some branch of natural science as the business of our lives, cannot afford to laugh at or despise those who use it as a pastime; we ourselves began it, perhaps, in like manner. We were not always so earnest; we dallied with the tempter, played and danced about it, but at last were caught in the toils. Many a man has played himself into science in earnest.

The address ended with a lengthy description of the ideal naturalist taken from Charles Kingsley's *Glaucus*, followed, as in his speech to the Amateur Botanists so many years ago, by his favourite verses from Wordsworth's 'Tintern Abbey', beginning 'Nature never did betray the heart that loved her'.

The naturalists of Hackney do not seem to have responded quite wholeheartedly to their President's exhortations, for when he addressed them again in 1888 (he did not retire from the Presidency until the following year) he found it necessary to chide them gently for their backslidings. First he held up before them as an example to be emulated, the Society of Amateur Botanists, pointing out the astonishingly high proportion of its earnest students who, later in life, had become acknowledged authorities in their own fields. Then:

What I want above all to do is to infuse all I can of a spirit of earnestness into your association with us, as members of this little Society. I want to urge upon you with all the force that I can command, that you do not, as a rule, make the most of your privileges, that you do not set a sufficiently high value upon the acquisition and mental arrangement of scientific knowledge . . .

Gentlemen! if you are content to play with science, I do not object for a moment, but I like to see a boy thoroughly earnest even in play. Play with it if you please, and do not make a toil of it, but let your play have some soul in it, for you won't get half the pleasure out of your play that you might get, if you do not put a little earnestness into it.

It seems likely that it was after this address that Cooke presented to the Hackney Society as a memento, the handwritten records of the Society of Amateur Botanists, for it was from it that Dr. J. Ramsbottom procured this important archive for the Natural History Museum.

Meanwhile the Quekett Club, of which Cooke was now Vice-President, was flourishing; as well as Ordinary meetings, the Conversational meetings or Gossip Nights, continued to be a regular feature of its life, and though soirées had had to cease for financial reasons, they had been replaced by an annual dinner in town in addition to the Excursionists' Dinner at Leatherhead. Cooke's last action before his illness had been to move successfully in the Committee that the existing President, Mr. T. C. White, should retain his office for a further year, but on May 26th, 1882, Cooke's was the only name to be put forward as his successor. Though a co-founder of the Club, he had had to wait 17 years for this recognition, but having attained the post he held it for two years, never once missing an Ordinary or a Committee meeting; and taking part in nearly all the excursions as well. Indeed, he initiated a new scheme whereby each excursionist listed and, if possible, sketched, his day's finds, so that when all the papers were laid on the table at the next ordinary meeting, members would have before them a complete picture of the flora and fauna of the selected site. In addition, he contributed several papers on both fungi and algae, two of them extempore, to fill gaps left by cancellations.

The first of his Presidential Addresses was an extremely lengthy and contrived discourse on 'Biological Analogies', a comparison of the ways in which plants and animals solved similar problems in their life cycles and in their reactions to the outside world. The moral drawn was the absolute necessity for the biologist to have some knowledge of both kingdoms even though he specialised in one (a precept which he had certainly followed himself) for a difficult problem in one kingdom could often be solved by noting analogies in the other.

The second address [3] was a plea to young members, in an age of restlessness, craving after excitement, and growing habit of exaggeration, not to be seduced by modern scientific hypotheses which were as yet unproven. Subserviance to a hypothesis tempts the investigator to untruthfulness and exaggeration, to weighting his observations in favour of what he wishes to prove.

This tendency is but a less extreme aspect of the current social phenomena of the:

> ... exaggerations and sensationalism which seems to pervade everything in these latter days, politics, religion, science, art, business, and even common conversation. ... It matters not what the special subject, there is a decided and marked identity in the restlessness, fanaticism, dogmatism, energy, excitement, recklessness, and consequent suspension, or rather distortion, of healthy mental action.

It is this tendency which has led to the current extraordinary explosion of short-lived 'crazes'–the 'aesthetic craze', the 'Darwinian craze', etc.–which bear little resemblance to the carefully propounded ideas of great men which were their nominal source.

> How much of this unhealthy development is to be traced to the restlessness of the age ... I am content to leave to individual opinion. For my own part, I think such causes at least contributory to the class of phenomena alluded to.

Whether or not Cooke's theory was correct, or the grounds for his concern were justified, he was certainly within his rights to warn his audience against the biassed interpretation of scientific observations. Unfortunately, however, his examples included the 'dual lichen hypothesis', 'which is so tenacious of life that it still retains a semblance of vitality', as well as the bacterial causation of disease, which he accepted in specific cases but does not seem to be prepared to extend beyond these:

> Can we really say that there is no danger [of exaggeration] with the Bacilli? It is not possible to take too much for granted, and exaggerate the relations of these minutest of organisms with zymotic disease? It it not well to be particularly guarded in such cases lest zeal should outrun discretion? The danger is all the greater since the subject would accommodate itself so readily to sensationalism.

'Zymotic' was a term used in the nineteenth century for fevers and contageous diseases, which were thought to be caused by an undefined morbid principle in the system acting in a manner analogous to, but not identical with, fermentation. By the turn of the century, with the discovery of bacteria and the rise of the science of bacteriology, the term was rapidly falling into disuse. Cooke's words may imply that he had foreseen the discovery of infectious agents other than bacteria, but it seems more likely that he still favoured the zymotic theory as the explanation for certain diseases.

Despite a lifetime spent in improving the scientific education of the masses he goes on to resist strongly 'incessant appeals' to 'the intelligent public',

> The translation of scientific terminology into the vulgar dialect loses in accuracy, in

proportion as it is diluted, ... exaggeration supplies the deficiency ... Once carry a disputed subject out of the circle in which it could be intelligibly discussed and investigated, into a new circle in which prejudice takes the place of knowledge, and it is condemned at once ... as a failure in search of compensation for disappointed vanity.

Comparison of this statement with his outlook twenty years earlier when he had written the *Manual of Botanic Terms* specifically to give the unschooled working man a chance to understand a terminology based on the classical languages, shows clearly the increasingly reactionary nature of his thought. It must have been at about the time that he gave this address to the Quekett Club that he first embarked on political activity with the Primrose League, and it has been noted, too, how very conservative his taxonomic ideas had become by then. Though he still had many years of useful mycological work before him, it would be in no way innovative, and 1884 does seem to mark the final closing of his mind to new or progressive ideas in any walk of life.

During the business meeting that immediately preceeded the Presidential Address a member had suggested that, following the example of many other scientific bodies in recent times, the Club should change its rules to admit ladies. Replying from the Chair, Cooke pointed out that 'it was not everyone who had been a Queketter long enough to remember the very strong battle which they had upon this very question years ago'. The memory of that event would have been deeply imprinted on his mind, for shortly afterward he had written a satirical account of an imaginary Club meeting held after the admission of ladies, telling how, among the donations to the Club that night had been a book, 'Etiquette for Scientific Gentlemen', presented by the Authoress. Next, a proposition was put forward by Mr. Scroggs 'That no perambulators be allowed in future to accompany the field excursions of the Club', for Mr. Scroggs had recently been knocked into the ditch from which he was collecting by such a vehicle. There followed a very animated discussion during which Mrs. Grundy 'denounced the proposition as being only another attempt on the part of those horrid batchelors to curtail the rights of women', and the motion was, of course, heavily defeated by the female majority. And so on—until finally a paper which should have been given by a gentleman was taken as read because a lady had just discovered the meanings of the Latin words and considered them not quite 'nice' [4].

The original proposal referred to by the President had been made on March 27th, 1868, the event being concisely recorded in the Club's *Journal* thus:

> The meeting was then made special, pursuant to notice, in order to consider a resolution for admitting ladies to the Club. The resolution, having been put from the chair [Dr. Tilbury Fox], was negatived.

Cooke now gave a rather fuller description:

> On that occasion all the energies of the members were called out against it [the proposal] in a way that few who took part in the matter were likely to forget, with the result that when the motion was put ... only one vote–that of the mover–was recorded in favour of it. He thought it was as well to remind the gentleman who had just sat down that the same feeling was still in existence.

The motion that ladies be not admitted was then carried unanimously; Cooke had successfully fought off change.

However, he was less successful when, at a meeting 2 yrs later, the Club decided to alter the date of its Annual General Meeting from July to February. Cooke was one of the few who spoke against the idea without, as fellow members pointed out, giving any good reason for his objection save a general dislike of change. After considerable wrangling over the constitutional propriety of the proceedings, the motion was put and was carried overwhelmingly. This was too much for Cooke, who resigned from the Committee (on which he had sat as Vice-President since retiring from the Presidency) on the grounds that the decision was unconstitutional. The Committee expressed the 'unanimous regret' of its members, but from then on Cooke seems to have withdrawn almost entirely from the life of the Club.

The fungus forays established so long ago by the Woolhope Club had now become a central feature of British mycological life, drawing to Hereford each October enthusiasts from all over the country and from beyond its shores. Cooke's friends, Mss. Quelet and Le Cornu (Fig 13.1), were very frequent visitors, and on a number of occasions Cooke returned their visit and took part in mycological field meetings of the Société Botanique de France in Paris. 'I remember' he wrote later,

> with pleasure some excursions with French mycologists in the forests of Compeigne and Saint Germains. They were under the direction of an energetic and esteemed friend who joined one of the Woolhope meetings afterwards, to the great delight of everybody. In addition to keeping the party under direct control, which is imperative in such a place as Compeigne, this was facilitated by calling all the party round him at intervals, by means of a preconcerted signal, and demonstrating for five or ten minutes, on some rare or peculiar species which had just been found. ... I am not aware that this particular method has ever been adopted in Britain, but, at the time, I considered it an admirable plan for introducing the educational element, without seeming to do so, and, at the same time, consolidating the party ...[5]

All British mycologists who could possibly get to Hereford did so, intent on the enjoyment of the heady mixture of days spent in the beautiful countryside in excited pursuit of rare species of toadstool, and evenings of friendly social intercourse over good food and wine, followed by learned discourses and earnest exchange of views. Cooke usually managed to attend but was occasionally prevented by pressure of work.

'FORAYS AMONGST THE FUNGUSES'. 1880–1889

Fig. 13.1 The French mycologist, M. Le Cornu, at a Woolhope foray. (From a cartoon by W. G. Smith, *Pictorial World*, 1877.)

Fig. 13.2 Setting out on a foray. (From a cartoon by W. G. Smith, *Pictorial World*, 1877.)

One remarkable man who was frequently present, and who was made an Honorary Member in the 1880s, was a personal friend of Cooke's and was probably introduced by him; this was Dr. Henry Wharton (1846–1895) of Kilburn. Wharton was not only a qualified medical man, but a Greek scholar of distinction and an ornithologist, as well as a mycologist. Leila Cooke can remember, as a child, being taken by her father to small dinner parties at the Whartons', her special entertainment being provided by the family cat, which jumped on to the table at the end of the meal and walked round from place to place eating the crumbs left by the diners.

Wharton was present at the foray of 1879 when, owing to the absence of W. G. Smith, the regular chronicler of the proceedings, Cooke undertook this responsibility. The meeting may be taken as typical of the Woolhope forays, and quotations from his description of it follow, augmented by a few highlights from Smith's records of other years. The first day of the 1879 foray, a Tuesday, was exceedingly wet [6] (Fig. 13.3) but:

> The generous host [the owner of the land] compensated for the weather, and the patriarch of the party [probably Berkeley, who was present] confessed that 'grouse pie' and 'fifty-two-year-old port' were charities which covered a multitude of sins. It cannot be reported that the fungi collected were many of them either new or rare.

Fig. 13.3 Foray sketches by M. C. Cooke.

Fig. 13.4 Collecting bracket fungi. (From a cartoon by W. G. Smith, *Graphic*, 1873).

Wednesday was fine, the party travelled to the collecting site by train, had a very successful day (Fig. 13.4), and:

> the last hour before the arrival of the returning train was spent in a lively discussion, which took place in a commodious refreshment room near the station, on the structural relationships of Tremella, Guepinia, Calocera and allied genera. The French visitors entered very freely into the discussion.

On Thursday the generality of Club members joined the specialist mycologists for a morning foray, then all returned to Hereford for the Annual Meeting followed by the now traditional dinner at the Green Dragon at 4 p.m., of which 80 persons partook. Cooke passes over this event briefly, but Smith's description of the 1874 occasion will give the atmosphere:

> Comatus soup (Coprinus comatus–the 'Agaric of civilisation'* being the principal ingredient) was a piquant and tasty novelty, for which there was a great demand; one member (we regret to record) most irreverently called it 'fungus soup'–a sad exhibition of levity which all thought it best to let past unnoticed. A very large tureen full of 'vegetable lamb's kidneys' was passed round, and judging from the halo of gastronomic and epicurean delight which flickered upon the features of the guests during its consumption, we may safely say it was hailed with general approval. We have a dim, dreamy consciousness of certain other delectable fungi being passed round at the very end of the dinner, but for the life of us we cannot remember the names, Latin or English; we can, however, call to mind the divine aroma, the luscious flavour, and the appetising appearance of the tureen or tureens as they went steaming past (Dr. Bull will tell you the names) [Figs. 13.5, 136].

Returning to Cooke and the 1879 event, after the usual toasts, 'some facetious remarks by Dr. M. C. Cooke, under the title of *Croûtes au champignons*, set the table in a roar.' But the evening was not yet over: at 8 o'clock the party repaired to the home of one of the members for a conversazione, where no less than six papers were read, including two by Cooke, before the party broke up.

Nothing daunted by their exertions on Thursday, the mycologists were out again on Friday, this time to Holme Lacey, where the first foray had been held 12 yrs before, finishing in style with lavish hospitality from the landowner.

That evening they repaired to the residence of Dr. Bull where papers were again read and there was a display of members' drawings of fungi.

> With the advent of Saturday, the visitors were all dispersed, and the ancient city of Hereford resumed its accustomed serenity. Doctors again breathed freely, all fears of fungus poisoning vanished, the eccentric individuals in thick boots and gaiters, carrying suspicious baskets and candle-boxes, had departed; the locusts had spared the city, and in a few hours Sunday would arrive, and with it the opportunity for thanksgiving for providential deliverance.

The Woolhope foray of 1884 was a very special one centering around one

---

*A reference of the association of the fungus with disturbed ground such as building sites, and hence with the spread of suburbia, a phenomenon concerning which Cooke left an unpublished essay (*The Builder's Foot* (Tol)).

Fig. 13.5 Menu for the Woolhope Club dinner, 1877, by W. G. Smith.

man, Dr. Henry Bull. Mycology was not Dr. Bull's only interest; according to Cooke he:

> could find exercise for his natural activity ... in magisterial duties, endless committees, fungus hunting, listening to the notes of Hereford birds, gathering wild flowers, entertaining friends, keeping up a large and punctual correspondence [8].

And as well as all this, he had undertaken in 1876, on behalf of a special Pomological Committee set up by the Club, to:

Fig. 13.6 The Rev. M. J. Berkeley presiding at a Woolhope Club dinner. (From a cartoon by W. G. Smith, *Graphic*, 1875).

investigate the varieties of apples and pears grown in the district [Herefordshire was then, as now, one of the main fruit-growing centres of England], to enquire into their origin and history, and ascertain their value and uses, and to name such varieties as are not known elsewhere, and have a really distinct character, with a view to the publication of a 'Herefordshire Pomona'[9].

This monumental work, with each variety beautifully illustrated by the Belgian artist, M. Severeyns, had been issued, unbound, in 7 parts over the years, and the last, double-sized, part had just appeared. The momentous event was being celebrated at that year's Green Dragon dinner, and Cooke had composed an epic poem in honour of Bull, the manuscript of which, embellished with delicately drawn portraits of Bull, each peering out of the cut face of an apple (Fig. 13.7), he later presented to the Royal Horticultural Society [10].

Entitled 'Flamen Pomonalis, or Pomona Unbound', the work covers 75 pages of an exercise book, and is arranged so that the right-hand side of the double spread carries the text of the poem while opposite are given sources, explanations and references as necessary. It is entirely characteristic of its author, displaying his sense of fun, his verbosity, and his familiarity with classical myth and legend. Pomona, he explains, was the Roman goddess of fruit trees and gardens, whose High Priest, Flamen Pomonalis, representing Bull, offered sacrifices to her divinity.

'Canto 1, Temptation', takes place in 1874 at the Club banquet:

The guests were assembled and seated around;
Meats and drinks disappeared with a guttural sound,

> Dish followed dish, with pace measured and slow,
> 'Tis not *here* the custom too rashly to go;
> .....
> Free and more freely, flowed cider and wine,
> And sparkles of wit made an effort to shine;
> .....
>   When–through the dim gloom
>   Of that cheerless long room
> Came the odour of apples–in long drawn perfume;
> 'The lights they burnt blue;
> .....
> Mid silence of horror, a 'lady in white'
>   Passed in at one door and out at the other
>   Before there was time consternation to smother;
> And then she was gone out of sight.

The lady was, of course, the Goddess Pomona, who set about tempting Bull to become her High Priest, i.e. to write *The Pomona*.

'Canto 2, Trial', describes the experimental aspects of the High Priest's duties:

> All over the table, all over the floor,
> Blocking the window, stopping the door,
> Piled in the fender, heaped on the stove,
> In lines on the mantelpiece groaning above
> .....
> Swung from the ceiling, loading each chair,
> Apples, apples! everywhere
> Their dense heavy vapour in clouds arose
> Condensing in teardrops down his nose;
> .....
> There were Codlins hot and Codlins cold
> Codlins in the pot, nine days old
> .....
> Seedlings many, and nameless more,
> Rolled and tumbled over the floor
>   In search of a reputation
> .....
> There's nought like experience sages declare
> for teaching the merits of apple or pear,
>   And he came to a like conclusion.
> .....
> He had apples roasted and apples fried,
> Apples baked and apples dried
> And apples in all conditions tried.
> .....
> There were apples in tarts and apples in pies
> And apples in every form of disguise
> Apples in puddings, and apples in cakes

Apple fritters and apples in flakes
Charlotte of apples, *a la Française*
Apple butter and apple cheese
There was apple jelly and apple jam
. . . . .

Eventually Flamen Pomonalis–

. . . thus soliloquising said –
For seven long years, on board or bed,
I've had apples swimming in my head;
. . . . .
And, spite of struggles made in vain
Apple sauce is creeping over my brain
    But of apples cooked, and apples raw
        I'll eat no more
    From apples garnished and apples plain
        I'll now refrain
    For of apples tender and apples tough
        I've had enough.

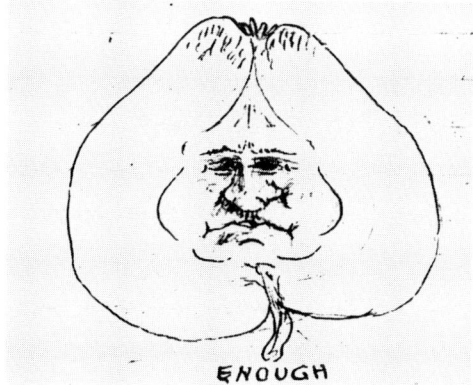

Fig. 13.7   Dr. Henry Bull as seen by Cooke in his illustrated ode, 'Flamen Pomonalis'.

In 'Canto 3, Triumph', the book is completed and Pomona lays her hand upon the head of her High Priest. With much punning on the seven unbound parts of which the *Pomona* was made up, the poem ends:

Ten busy years, that seem to fleet so fast
That every one runs shorter than the last
. . . . .
Ten years oracular to end this story
And call Pomona back in all her glory.
. . . . .

Called back, unbound, without a rag to cover her
Called back, in parts, the mystic number seven
Although in size remarkably uneven
The latter part the largest–not uncommon
In nymphs and goddesses–and mortal woman.

The Green Dragon dinner of 1886 was as sorrowful an occasion as it had been joyful in 1884, for in the interval Dr. Bull had passed away. It fell to Cooke's lot to give the after-dinner address on that sad evening, and a heartfelt and dramatic, if lengthy, obituary speech it was [8]

Gentlemen,–There are times and seasons when the most ready and apt amongst us experience a most profound difficulty in facing the situation. When courage seems to shrink dismayed from the prospect of an unequal contest. When the fulness of the heart overpowers all the efforts of the head. To-day, more than ever I remember to have felt, I am conscious of rising to a task and a duty which, were it not a duty as well as a task, I could not have attempted.... This Woolhope dinner, of which we have so often partaken together, and yet so different from all its predecessors. This Green Dragon room, in which we have met so often with smiling faces, if not with uproarious mirth, and yet to-day our hearts are heavy and we cannot smile. We have eaten and we have drunken, almost in silence, as if a spirit hovered over us, and overpowered us by its presence....

It is that 'vacant chair' which reveals the secret of the change in the dinner of to-day. It is the consciousness of that unseen presence which is gazing upon us with those deep and sympathetic eyes, that fills us with such strange emotions.

The speaker goes on to eulogise the author of the *Pomona* and the inspired founder of the Woolhope forays. He tells his audience that only last year Bull had sought his co-operation in the preparation of a catalogue of the fungi of Herefordshire, which was well advanced and which he was anxious should be his next important work.* The address ends fittingly with a stirring call to the local members of the Club to carry on its work with all energy and enthusiasm as Dr. Bull would have wished:

Shall not the voice of the dead have some influence with you, and stir you up to some firm resolution for the future? Will you disregard now the mandate you have so often listened to and obeyed? But, if you fail, how can you ever enter a wood or forest again in these dull October days, watch the leaves falling ...

If you would maintain a kindly and perpetual remembrance of one who was ever a staunch friend and earnest worker for the Woolhope Club, you could not do this more effectually than by keeping up in its integrity the annual reunions which he established, and did so much to maintain, in the interest of a pursuit which to him was almost a passion, through all changes, all mutations, like the babbling brook––

*Cooke completed and edited the catalogue, and it was published in Purchas and Ley's *Flora of Herefordshire* in 1889.

> 'For men may come and men may go,
>   But we go on for ever.'

Sadly, though, this was not to be, for death was to take a further, heavy toll of the older generation of mycologists. Bull's decease was followed in 1886 by that of C. E. Broome. Cooke left a delightful sketch of him (though unnamed, the forayer could only have been Broome) at a Woolhope foray:

> Amongst old foragers was a tall quiet man with a little stoop, and a genial manner, who was seldom absent and always welcome. Perhaps, of all men with an equal amount of knowledge, he was the most modest and retiring. It was the greatest difficulty to extract an opinion from him, but he might suggest a possibility, and the possibility was usually right. Of all branches of fungus lore, he was the strongest man of the day upon Truffles, and all underground fungi. It was amusing to see him unroll a small square of carpet, and spread it upon the ground to kneel upon, and then extracting a small rake from his capacious pocket, he would scratch around the mossy base of trees, hour by hour, with all the patience of a true disciple of Izaac Walton, to be rewarded perhaps by two or three fleshy little balls between the size of a pea and that of a marble [11]. [Fig. 13.8]

The Rev. M. J. Berkeley, too, was over 80 and was not at all well. Cooke's letters to him from 1888 show increasing concern for his health, and though he continues to write about mycological matters, he does his best to minimise any effort which the old man may need to make. He is still sending him the *Illustrations* as each part comes out, and in July 1888, by which time Berkeley is too infirm to collect fungi for himself, he writes:

> ... it will be a source of satisfaction to me to know that you sometimes derive pleasure from looking at some of your old friends [in the *Illustrations*].

The following year Berkeley died, at the age of 86, mourned by mycologists all over the world. Cooke, in a warm, but not entirely uncritical obituary for

Fig. 13.8   Truffle-hunting at a Woolhope foray. (From a cartoon by W. G. Smith, *Pictorial World*, 1877).

his teacher and friend, wrote that 'as the "Prince of Mycologists" his name will go down to posterity.'[12] Some years later he reminisced:

> It was my good fortune, on a few occasions, to join in a foray with the prince of mycologists, or, as I prefer to remember him, as the 'father of British Mycology'. The genial paternal veteran could never undertake a pedestrian excursion, over a few hundred yards, ... but he would constitute himself a centre, as on one memorable occasion in the mosshouse at Downton [probably Herefordshire], and the fielders would scour the surrounding nooks, and bring him whatever was doubtful or of interest. Such opportunities were gala days for all who were privileged to his anecdotal reminiscences, and hear him discourse, almost like the sapient Jewish King, for 'he spake also of beasts, and of fowl, and of creeping things and of fishes. He spake of trees from the cedar tree that is in Lebanon, even unto the hyssop that springeth out of the wall;' With a truly marvellous memory, and the most extensive knowledge he was at home on every subject, from the pedigree of race-horses, to the pedigree of cultivated flowers, and the most abstruse problems of vegatable pathology [11].

In the space of a few years Britain had lost three of her leading mycologists and, as well, W. G. Smith was now more interested in archaeology than he was in fungi. The Woolhope Club continued to hold forays which Cooke and others attended, indeed Cooke became their leading spirit. But he was in his mid-sixties, there was no local member sufficiently devoted to replace Bull, and inevitably Hereford gradually lost its mycological pre-eminence, attendance slowly declining until the last foray in 1892 when Cooke[13] tried to rally members with an address on 'Fungi, Past, Present and Future', a call to them not to allow the meetings to lapse. But to no avail. The Club had, however, started a movement which was to grow and flourish and to become the foundation stone of a national mycological society.

Cooke would never forget the Woolhope forays; many years later he wrote:

> Old folks are proverbially given to glorifying the past, hence I shall be quite in order in recalling those days when the Hereford District was to me an Eden, a fairy land, teeming with fungi, so that the difficulty was that of determining which kinds were best left behind, and the latter half of the day was occupied partly in throwing away sufficient of the spoils of the morning to make room in the basket for the rarer trophies of the afternoon. There was no difficulty in filling big baskets then, for they always seemed too small [5].

Chapter Fourteen

Missionary for Mycology. 1890-1892

Although by now Cooke had considerable reservations about opening up the world of science to the lay public he was still keenly interested in the activities, especially the mycological activities, of the field clubs that were so popular a means of recreation in the Victorian era, and he was very willing to make use of them to spread the mycological gospel. He had never forgotten his own conversion when, as a schoolmaster who had never looked at fungi before, their fascination was pointed out to him by an experienced mycophagist, and he knew that instruction in the field was an excellent way to make converts to the cause. In recent years many local clubs had begun to devote one of their autumn meetings to a fungus hunt, and Cooke was much in demand to lead these excursions, a function which indubitably gave him great pleasure. He not only led the forays of the Essex, Hackney and other clubs already mentioned, but in 1882 he had been made an Honorary Member of the Hertfordshire Natural History Society, which still possesses records of the forays attended by him, and in 1890 of the Glasgow Natural History Society. During the 1880s he was visiting the Birmingham district and contributing to the mycological section of J. E. Bagnall's *Flora of Warwickshire*, though whether he actually attended Club meetings is not known. Again, in 1887 he was the specially invited referee at the first foray ever held by the Hampshire field Club, when that body instituted the 'first organised inquiry into the Fungus-Flora of the New Forest' [1]. It is more than likely that he also attended the forays of other field clubs as their expert and honoured guest, and altogether he would have a huge experience of such excursions and of ways of making the best use of them.

Over the years 1890 to 1898 he wrote the manuscript of a book, *Confessions of a Mycophagist* (Tol), which was clearly aimed at the members of field clubs but, except for one chapter which appeared as an article in *Grevillea* [2], it was never published: it is far too rambling and sometimes contrived. However, though written in a lighthearted way, the first few chapters give us some idea of his more serious thoughts on the organisation and uses of field club forays.

> I have always been disposed to estimate them as recreative. ... There is a general banishment of all petty cares, and personal anxieties, at least for the day, and an abandonment to the main object of the excursion, a pleasant day in the woods, a healthy stroll in the fresh air, and agreeable companionship. The man who does not possess the faculty of making himself agreeable is out of place at a Naturalists' Field Meeting. Criticism, disputation, good natured banter, sly satire, and attempts–occasionally weak ones–at sallies of wit, are to be indulged in. ...
>
> The official mind contemplates [forays] from a utilitarian eminence. Are they not the means by which a long list of species can be economically compiled, to fill the pages of the Society's transactions, and shed glory upon the county? ... Individual members may regard the Field day as a privileged opportunity to cling to the skirts of some invited guest, and by contact imbibe elementary education.

A few pages on from this tongue-in-cheek, but somewhat disparaging description of field-club forayers, Cooke reverts to the subject with a much more sympathetic, though still slightly jocular, account. First, predictably, he dismisses a class that is:

> made up of the ladies, who have joined the excursion because it is a novelty, or out of curiosity to discover what forays are like, or for some other reason concealed in the feminine breast; and young men, whose chief occupation is to pick off the 'burrs' and disentangle the briars from the dresses of the ladies.

But a smaller group he describes as:

> composed for the most part of steady going old fogies, with books in their pockets and a basket on their arms, directing a keen, restless gaze in all directions, quick and reserved in their demeanour, but evidently meaning business. This is the scientific section, each individual of which is on the hunt for something new or rare ... Little regard is paid to the scores of familiar forms scattered over the ground, familiarity seems to breed contempt ... The only service they seem to be to the general company is that of a court of appeal, a peripatetic store-house of Latin names, to be called upon whenever required, but alas too often incomprehensible and unsatisfactory to the enquiring spirit. [Fig. 14.1]

Then he comes to –

> what an old hand calls the 'pot-hunters', those who look upon all fungi as divided into

the edible and the useless, and whose ulterior object for the day is confined to the prospect of a mushroom breakfast for the following morning... It has been very much the custom for scientific fungus hunters to underrate and depreciate those who disavow all scientific interest, and confine themselves to the utilitarian object of fungus eating. This is manifestly an error of judgement since the ranks of the former are mostly recruited from those of the latter.

There follows a light-hearted account of the organisation of a foray, including a description of the drive to the site on a wet day in hired, open breaks, each passenger cowering under his own umbrella. Next, in the same facetious style but clearly speaking from personal experience, Cooke describes the responsibilities of the 'Conductor' of the foray: (Fig. 14.2)

... there is a periodical aggregation of baskets around the conductor for the day, on whom devolves the responsibility of wounding the susceptibilities of the collectors, by deciding how many* shall be thrown into the nearest ditch, and which shall be reserved and carried over the course... Some consolation must be offered for wholesale confiscation, and this is supplied by a pair of Latin words, which are pronounced over every specimen thrown away. The binomial benediction is never remembered, even though repeated for twenty consecutive times, and this falls as a blessing on the heads of the proprietors of the baskets, like sunshine follows rain, for they count only the number of blessings which are thrown away. There are two paramount ideas which are universal to the party, firstly, that not a single fungus should be left to mark the ground over which

Fig. 14.1 'Steady old fogies' examining an earth star. (From a cartoon by W. G. Smith, *Graphic*, 1875.)

*Specimens, one presumes!

Fig. 14.2 An Essex Field Club foray, 1910. George Massee (second from left) acts as referee.

the foragers have passed, and secondly that all the bright coloured and attractive specimens should be reserved to decorate the table at the end of the day [the table is, of course, intended to be of scientific, not of aesthetic, interest]. If there is a third idea, it resolves itself into a collection of all the binomial Latinities, pronounced by authority, which shall constitute a record of the scientific results of the expedition.

A number of anecdotes are recounted about untoward incidents which have occurred during picnic lunches, and finally:

A fungus foray is never complete without an exhibition of spoils at the end of the day. Theoretically all the specimens are supposed to have their names legibly written on slips of paper, placed beside them, and arranged on the exhibition tables. Such arrangement to follow the sequence adopted in books, and to present a bird's eye view of the whole system of classification. Practically this has to be accomplished by two or three pairs of hands, in about half an hour, under the most unfavourable conditions of a crowded room, and continual interruption. Voluntary assistance is by no means unknown or unappreciated. It usually exhibits greater regard for picturesque effect than for scientific accuracy . . .

There have always been differences of opinion among biologists as to the need to insist that the lay naturalist uses only Latin binomials when naming his finds. Should not the well known and easily remembered common names be allowed, at least for beginners, so as to avoid discouraging them for ever from taking up so fascinating a hobby? Cooke, in a chapter of the *Confessions* called "Words, Words, Words", which he notes is a quotation from *Hamlet*, goes into his views on the subject at length.

> I cannot help sympathising with unscientific people when they protest against Latin names, and especially those applied to fungi. To them, without doubt, they appear to be unnecessary or ridiculous, but much as I sympathise with the objectors, there is no alternative, for scientific purposes, to binomial designations.

He then describes, in layman's terms, the object and usage of Latin names, and continues:

> But to return to vulgar names, for local purposes, retaining the scientific names for scientific purposes, no feasible objection can be urged. Meanwhile the scientific names are not abrogated, or superseded, they are only held in abeyance.... A few agarics are already known, although not extensively, by popular names; and it would be a great advantage if more popular names came into use for common and well known species.... The 'mushroom', the 'horse mushroom', the 'champignon', the 'chanterelle', the 'blewit', the 'liberty cap' and a few others [which] might be named, are widely known, and others will be added in proportion as they come to be recognised as worthy of being distinguished for gastronomic or other purposes....
>
> If all who were really interested in the spread of popular knowledge of fungi would only agree amongst themselves to encourage the application of reasonable popular names to some thirty or forty of the commonest species, they would perform a good work; but there must be unanimity in the selection of names...
>
> In what a perplexing dilemma is one placed when attempting to interest an unsophisticated villager in the values of some toadstool, being compelled to designate it by a triplet of Latin names.* Have I not felt myself that to descant upon the virtues of *Agaricus (Lepiota) procerus* [Fig. 14.3] to ever so sympathetic an auditor, was an anomaly? Of course under the name of the Parasol Mushroom there was no such difficulty, speaker and listener were at once at ease; there was no blushing sensation of pedantry on the one hand, nor of embarrassed ignorance on the other.... Scientific names and designations are entirely out of place in ordinary conversation, except as between scientific people, and to parade them unnecessarily is simply to 'talk shop'.

Many years earlier, in his very scientific *Handbook of British Fungi*, he had carried his ideas on common names to extremes by inventing one, usually a translation from the Latin, for every microfungus. Needless to say, these were never adopted, even in his own lifetime.

From the time of his first introduction to fungi in the 1850s Cooke had taken a great interest not only in the eating qualities of the different species but in cooking methods. The first book he ever wrote on fungi encouraged

---

*Today, only two names would be necessary.

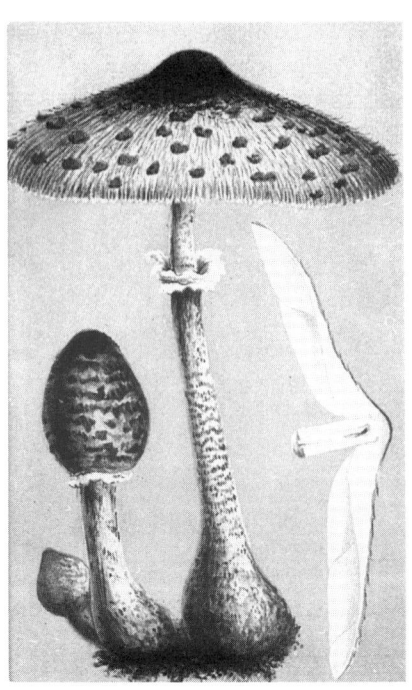

Fig. 14.3 The parasol mushroom, *Lepiota procera*, from *Illustrations of British Fungi*. Original in colour.

readers to try the delights of mycophagy; in fact his enthusiasm was so great that one reviewer felt the book might be a danger to the public. From then on allusions to the pleasures of fungus eating are scattered through his writings until, in 1891, he published a most excellent popular book on the subject, *British Edible Fungi; How to Distinguish Them and how to Cook Them*, beautifully illustrated with his own coloured engravings. Leila Cooke told me that, unusual though it was for a Victorian *pater familias* to enter the kitchen, her father always prepared fungus dishes himself, so the recipes given in this book, probably including those attributed to the Woolhope Club, would have been personally tested by the author.

Cooke's own favourite species seems to have been the giant puffball (*Lycoperdon giganteum*) (Fig. 14.4), for its gastronomic delights are mentioned at least five times in his writings. To quote from the first occasion in 1862:

> A gardener [presumably one of Lady Burdett Coutts' men] brought us a large puff-ball, equal in size to half a quartern loaf, and which was still in its young and pulpy state, of a beautiful creamy whiteness when cut. It had been found developing itself in a garden at Highgate, and to the finder its virtues were unknown. We had this specimen cut in slices of about half an inch in thickness, the outer skin peeled off, and each slice dipped in an egg which had been beaten up, then sprinkled with breadcrumbs, and fried in butter, with salt and pepper. The result was exceedingly satisfactory [3].

Fig. 14.4 A giant puff-ball, *Lycoperdon giganteum*, compared with a species of more common size. (From E. Dobbs, *Fungi for Fun*).

Among the recipes given in *British Edible Fungi*, Procerus Pie, made with the Parasol Mushroom (*Lepiota procera*) (Fig. 14.3), and for which the recipe is taken from the Woolhope Club receipt book, sounds very tasty.

> Cut the fresh agarics in small pieces, and cover the bottom of a pie-dish. Pepper, salt, and place on them small shreds of fresh bacon then put a layer of mashed potatoes, and so fill the dish layer by layer, and a cover of mashed potatoes for the crust. Bake well for half an hour and brown before a quick fire.

In the spring the St. George's Mushroom (*Tricholoma gambosum*) makes good eating.

> The Woolhope Club receipt is to 'place some fresh made toast nicely divided, on a dish and put the agarics upon it, with a small piece of butter on each; then pour on each a teaspoonful of milk or cream, and add a single clove to the whole dish. Place an inverted basin over the whole, bake for twenty minutes, and serve without removing the basin until it comes to the table, so as to preserve the heat and the aroma, which, on lifting the cover, will be diffused through the room.'

Cooke also gives instructions for preserving fungi by drying and pickling them, and a whole chapter is devoted to the popular Victorian delicacy, mushroom ketchup, or catsup, a sauce prepared from fungi. 'In rural districts where Ketchup making is an annual event, the meadow mushroom (*Psalliota arvensis*) is prefered [to the common mushroom *P. campestris*] as more highly flavoured'. The author believes, however, that any edible toadstool can be used satisfactorily, and judging by the highly spiced nature of the final brew, which must have submerged completely any delicate nuance of flavour conferred by the species of mushroom used, this was doubtless true. Of a number of recipes given, the following contains as few additional flavours as any:

> A very simple and effectual method of making this excellent sauce is to wipe the mushrooms and cut off the stems, laying the caps in a pan with the gills upwards, and sprinkling them with salt, taking care to exclude those which are maggoty. They should lie 3

or 4 days, and then squeezed with the hand thoroughly so as to extract all the juice. Take one ounce of whole pepper, one ounce of well bruised ginger, and half an ounce of cloves for each pint of liquor. Boil all together for 15 or 20 minutes, and when cold, decant into clean bottles ... The corking must be good and well sealed to exclude air. If at any time afterwards the ketchup shows any tendency to become ropy, it should be boiled again ... with a little more spice, when the ropiness will disappear, and it will be as good as ever.

The measurement of the liquor in pints assumes huge quantities of wild fungi.

Poisonous fungi are only discussed in a short chapter at the end of this book, and are not illustrated, for Cooke held firmly to the belief that mycophagy is a perfectly safe pastime so long as the practitioner has a thorough knowledge of the species he proposes to eat, and never tries experiments except under expert guidance or with the aid of a reliable book.

The Worcestershire Naturalists' Club had been founded in 1847, even earlier than its next-door-neighbour, the Woolhope. It had no strong mycological tradition, though some of its members took a general interest in fungi, but on October 28th, 1892, only two months before Cooke retired from Kew, a meeting took place which:

> ... will be for ever memorable in the annals of the Club by reason of the fact that two eminent specialists, Dr. M. C. Cooke, M.A., F.L.S.,* and Mr. George Massee, F.R.M.S., F.L.S.,* the highest authorities in Mycology, had, through the instrumentality of our member, Mr. C. Rae, been prevailed upon to accept the invitation of the Club to attend their fungus foray [4].

Mr. C. Rea–Carleton Rea (1861–1946) (Figs. 15.7, 15.8)–had joined the Club at the phenomenally early age of 15 and now, at 31, was a leading member. By profession he was a barrister, but to quote his friend and obituarist, Dr. J. Ramsbottom [5]:

> ... he did not apply himself with the energy necessary for success, and from about 1904 took fewer cases and finally ceased to practice in 1907 ... As he did not scorn delights his days were not laborious: billiards seem to have figured more in his activities than did cases. Truth to tell neither his manner nor his speech were such as to inspire the confidence necessary for legal success.

He was a dapper man and even when out in the field wore a white waistcoat and tie and a high collar with his brown knicker-bocker suit, the whole sur-

---

*Both these are incorrect. Cooke was an Associate of the Linnean Society, never a Fellow; Massee was not a member at this time, but became a Fellow in 1895.

mounted by a brown bowler hat. He was 'of ruddy countenance', sported a monocle, and carried a walking-stick for beating down the brambles as he advanced, never for walking with. Ramsbottom sums him up as 'a period piece of the naughty nineties'. However, he had his difficult side:

> He could be as stubborn as a mule and as cantankerous. When he got started off by something he resented, it was difficult to pull him up and he would sulk and fume in open meetings.

As a naturalist his interests were wide:

> To be with him in the haunts he had known from boyhood was always delightful. He would take one to the spot where some rarity grew, and he knew all the plants, the birds, the insects; also the local traditions, the history, the archaeology.

It seems that he first became interested in fungi in the 1880s, when young Miss Amy Rose, the daughter of a Club member and a keen naturalist and talented artist, asked him to identify some specimens for her. Now ladies were not, at that time, admitted to membership, but Rea struck a bargain with her–he would help to name her specimens if she would paint his fungi. It proved a happy collaboration.

It was most convenient for Rea that the Woolhope forays were held within easy reach of Worcester, and he began to attend them during their declining years, thus making the acquaintance of Cooke and Massee; so it was natural that when the forays ceased in 1892, Rea should invite the two eminent mycologists to Worcester with a view to carrying on the Woolhope tradition there. The foray was a huge success. At Shrawley Wood:

> after a light al fresco luncheon, the work of the day began, and went on with unremitting energy until 4 p.m., when the pretty summer house, about the centre of the vast wood and overlooking a beautiful lake, was reached. Here Lady Georgina Vernon, herself a lover of botany and mycology, had most kindly provided tea.

The mycological partnership between Mr. Rea and Miss Rose must by now have been well established, for there is little doubt that Miss Rose, though still debarred by her sex from membership of the Club, attended this foray, and that either she or her paintings, or both, attracted Cooke's admiring attention. That Christmas she received from him a delightful card, painted by himself, depicting a toad sitting under a group of *Cortinarius* fruit-bodies growing among fallen beech leaves (Fig. 14.5). We do not know whether Miss Rose was singularly honoured by this attention, or whether Cooke was in the habit of sending such cards to special friends; but if he was, only Miss Rose's has survived. She treasured it, had it framed, and when, in 1897, she married Carleton Rea, it hung in his study until his death, when it passed into the hands of

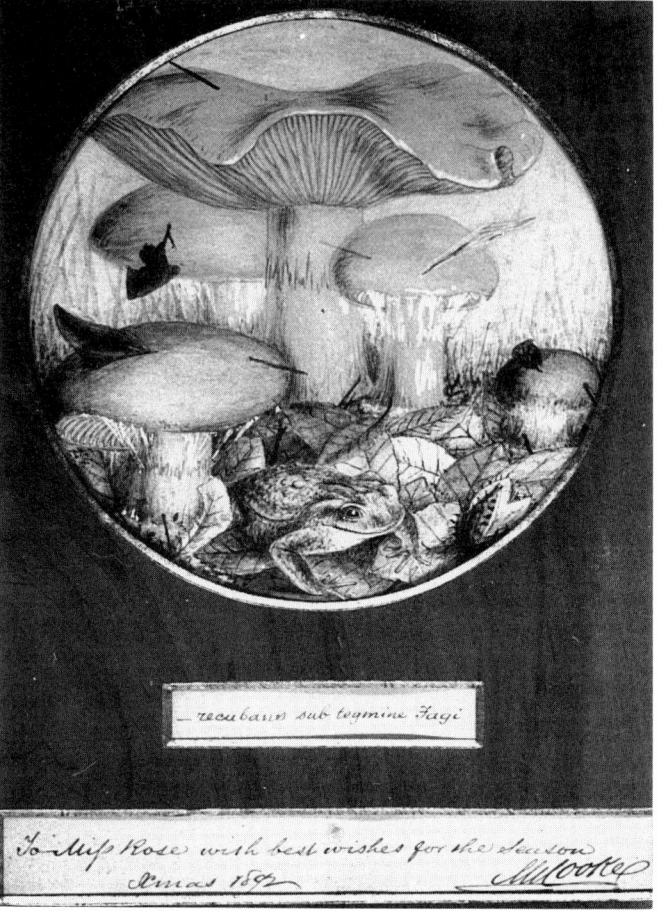

Fig. 14.5 M. C. Cooke's Christmas card to Miss Rose, framed by herself or by her husband-to-be, Carleton Rea.

their daughter Violet, later Mrs. Astley Cooper.*

Cooke does not seem to have attended any further Worcestershire forays, though the Club continued to hold them and Massee was present at some. Nevertheless, both men were made Honorary Members in 1895, in which year also Miss Rose was rewarded for her artistic labours by being elected the first lady member. Rea's interest turned almost exclusively to the larger fungi on which he soon became an acknowledged authority, specialising in the genera *Russula* and *Cortinarius*.

*The picture is now in the possession of the author. In Mrs. Astley Cooper's old age she became acquainted with my brother, Marcus F. H. English, in his business capacity, becoming very fond of him and his family. He introduced us and she showed me her mother's drawings of fungi (bequeathed to the Kew Herbarium) and sold me the framed Christmas Card.

Chapter Fifteen

Troubled Retirement. 1851-1860

The year 1893, the first year of his retirement, was not a happy one for Cooke, as can be discovered from a letter he wrote to James Bagnall (with whom he had collaborated on the *Flora of Warwickshire*) at about that time.

> I have been struggling hard after some employment of a remunerative character, and seem to have failed all round and booksellers seem to be at their wit's ends, so although I have been preparing two or three books I cannot induce publishers to look at them.
>
> Above all I have been anything but well, one way or another, I got so depressed, that I lost all heart. I had good hopes of an appointment for two years at the British Museum, but the 'Cad' who has so persistently attacked me always in the Journal of Botany has had too strong an influence against me, and prevented it. Thus I am at present in an unenviable position [1].

The 'Cad' was, of course, George Murray, whose vicious reviews of Cooke's books have already been noted; but as William Carruthers was still Keeper of the Botany Department (he did not retire until 1895) it may not have been entirely Murray's influence which prevented Cooke's employment there.

By the time he retired the volume of Cooke's taxonomic work had diminished to a minute fraction of what it had been only a few years before, and consequently his contacts with fellow mycologists had dwindled too; the older men had died and the younger must have found his ideas too fixed in a previous age to profit from collaboration with him. His correspondence with Saccardo had virtually ceased by the end of the 1880s, though the Italian continued to send him new volumes of the *Sylloge* as they appeared; the same was true of his mycological correspondence with Farlow, though he continued to write to him on other matters; and the work with Australian fungi was finished. Few papers on American fungi appeared after 1886, and although he had arranged with Sir Joseph Hooker to continue to use the Kew Herbarium freely, he published very few more taxonomic articles after 1892. Only in the autumn was he as occupied as he had ever been, for then he continued his customary attendance at forays all over the country, and must have made the most of the opportunity of renewing friendships which had been in abeyance for the rest of the year. But most important of all to him in his retirement–it must have seemed like a lifeline to a man as depressed as he was–was his continued membership of, and regular attendance at, the Science Committee of the Royal Horticultural Society. It was this Committee that would inspire his last, and by no means inconsiderable, contribution to mycology.

That first year of retirement was probably not helped by the atmosphere at home. What his relations with Sophia were is unknown, but she was now 69 and seemingly not in good health. Ada and Willie, aged 29 and 27 respectively, were still living at home, and Ada at least was married, her husband, Caleb Davis, being a piano tuner by profession. But neither Ada nor Willie were happy people. Ada, of whom her father was very fond, was described to me by her daughter as being a withdrawn and unhappy woman; and Leila Cooke told me that Willie was 'always depressed about something' and 'rather peevish'. These two older children were only seven and five years old when Annie married John Cubitt and left home for the first time, and one cannot help suspecting that her desertion had affected them.

Neither Mordecai nor Leila got on well with Willie, which must have led to considerable tension in the home, for Willie had taken over a room as a studio and so would have been in the house a lot. He seems to have been somewhat jealous of his father, being under the impression that Mordecai could sell his work more easily than he could; but even if this were true later, at that time his jealousy appears groundless, for he had become a very competent artist in both black-and-white and watercolour, and had just completed a commission from the publishers J. M. Dent for illustrations for ten volumes of Jane Austin's novels, a commission which was quickly followed by others. Also, 1893 saw his work hung for the first time at the Royal Academy, where he exhibited book illustrations and watercolours in 1895, 1896, 1897 and 1900. Mordecai was apparently somewhat critical of his work and this may have upset Willie.

Apart from the maid, Leila, then ten years old, was the only other member of this unlikely household, and she became the apple of her father's eye. She remembers that in the evenings, after a day's work in his study:

> sometimes he would take me on his knee and let me look through the microscope. I suppose it rested his eyes . . . sometimes he would read to me for a little while.
>
> About 8 o'clock he would have a glass of ale and two Abernethy biscuits, as he could not eat much because of his perpetual indigestion due, he said, to Norfolk dumplings. He would give me two chocolate creams and tell me to go to bed quietly and don't shake, for the house was old and the floor boards creaked and shook the micrsocope. Then

when the trams and buses stopped running and did not shake the old house, he would go on with his microscope work until about two in the morning.

When I met her Miss Cooke told me that the chocolate creams were always Fry's, and that he kept them specially for her in a drawer at the back of his desk.

Like many fond fathers, Mordecai could not bear the thought of his little daughter growing up, and this irked Leila. On one occasion, perhaps when she was a little older, she asked her father for ten shillings to buy a tennis racquet with, but Mordecai refused, not out of meanness but because 'I should always have been out. And I wasn't supposed to be out. I was supposed to be at home' And again:

He choked me up with *fairy tales*. All *fairy tales*, I remember. And I always had something to read, but it was always the latest fairy tale book. He never talked about his work to *me*. I must have been about 10 or 12 by then. And if I wanted to know about anything, 'Oh, that's not the sort of thing a little girl wants to know'. He *never* seemed to talk as far as I remember. He sat down a wrote a book about it!

Neither were the books in his library very tempting to a ten year-old, although she read some of them later. Leila describes the library thus:

In the study, two sides of the room had books from floor to ceiling. All the poets from Chaucer to his day, all the essayists from Francis Bacon to his day. All the known books on Fungi and Botany, including rare ones like Gerard's Herbal. Couch's Fishes, Morris's Book of Birds, Layard's Nineveh, Champollion's Egypt. Fungi by M. J. Berkeley ... de Bary, Tulasne, Quelet; Lamarck and De Candolle, that I remember.

To this list can be added the journals of many British and foreign natural history socieities, reprints and translations of ancient volumes such as William of Malmesbury's *Chronicles of the Kings of England*, books by the Christian Socialists, Kingsley and Martineau, and *Sappho*, translated by Mordecai's friend Dr. Wharton.*

Leila remembers accompanying her father on frequent trips to 'Bookseller's Row', a street of general shops and second-hand bookstalls that was demolished for the building of Kingsway; he bought many of his books either there or at nearby Wheldon's, the biological bookseller's in Great Queen Street, Drury Lane (Fig. 15.1). She told me, too, that though he disliked the cinema, he was always very fond of the theatre, especially of light comedy and musicals, which he might attend as often as once a fortnight, but always on his own. He particularly admired Charles Hawtrey. Her treat was the pantomine,

---

*Information from Messrs. Wheldon's catalogue No. 15, 1898. The firm (now Wheldon and Wesley of Codicote, Hitchen, Herts.) was entrusted by Cooke with the sale of his library.

to which he took her every year at the Grand Theatre, Islington, and if Dan Leno were in the lead there would be great excitement.

All this time Annie seems to have remained in regular and friendly contact with the occupants of No. 146, Junction Road. She must have been living on her own now, for Mabel and her husband George had moved to St. Pancras and both Frank and Bertie (Fig. 15.3) had left home, Frank working as a letter-engraver, mostly of tombstones, and Bertie being apprenticed to a sign-writer. Bertie became highly skilled at this job for, having Cubitt blood in his veins, he could draw as well as do beautiful copperplate lettering. Outside working hours both young men had other absorbing, and in Frank's case lucrative, occupations. Frank had an excellent voice and was a part-time professional singer, able to charge from three to five guineas for solo performances at dinners and similar functions, and also being a paid member of the chorus of the D'Oyley Carte Opera. Bertie was an amateur boxer. Both still lived in Kentish Town and kept in touch with both parents, but Frank was particularly devoted to Mordecai.

Annie was now a grandmother, for on December 27th, 1892, a son, Herbert George, had been born to Mabel and George Stone; he had doubtless been named after his mother's favourite half-brother. From the addresses given on

Fig. 15.1 Wheldon's bookshop at 58, Great Queen Street, 1857–1891. The then proprietor, John Edwards, stands in the doorway. In 1891 the shop moved along the street to No. 38.

Fig. 15.2 Leila Annie Cooke, 1882–1977.

Fig. 15.3 Herbert Stephen (Bertie) Cooke, 1879–1927.

the birth certificate it would seem that Mabel went to stay with her mother for the birth of her child. But on February 20th, 1894, only a little over a year later, her husband George was dead. There is no death certificate for him at the General Register Office, suggesting that he died abroad, an idea supported by the inscription on a tombstone in Highgate Cemetery which records the fact that he died 'in South . . .', the rest of the inscription being eroded away. It is difficult to imagine what a young, married printer's compositor was doing abroad at this date unless he had enlisted.

After this tragedy Mabel and the baby went to live with Annie at 104, Duncan Road, Hornsey (as far as is known this is the furthest that Annie ever strayed from Kentish Town), but within a few months, on October 1st of the same year, the sad little household suffered a further bitter blow, when Mabel herself died of 'enteric fever and apical pneumonia'. Annie, calling herself on this occasion Mrs. Cooke, bought a plot in Highgate Cemetery in which to bury her daughter, and it is on the headstone of this grave that George Stone's death is also recorded. Now, fifty years old and having raised eight children of her own, Annie set about bringing up her orphaned grandson, still refusing to return to Mordecai but doubtless continuing to receive financial help from him.

All through 1893 Cooke remained listless and depressed. According to

Leila he obtained little relief even from reading, for he was too restless ever to sit down and read a book from cover to cover. 'Probably one evening when he was tired of work and he had a book, he would leave off and read a chapter or two, and one book would last six months'. He enjoyed Ouida and similar authors whose novels lent themselves to this sort of reading, and this is probably why he was so fond of the essayists, especially De Quincey and Carlyle. Leila particularly remembers De Quincey's *The Mail Coach* and Carlyle's *Sartor Resartus*, the second of which he made her read (presumably she had managed to convince him that she had grown out of fairy stories) and which she found incomprehensible. And there was always poetry which was a lifelong pleasure to him. He had a whole shelf of the poetry written between 1860 and 1880, but he did have one pet aversion, Tennyson, whom he described as 'wishy-washy'. Of the poems he chose to read aloud to his daughter she recalls most vividly 'The Ancient Mariner'.

1894 was no happier a year than 1893. On March 8th Cooke wrote to Farlow:

> In consquence of my retiring from much more active work, I am obliged to reduce my library of all such books as I am not likely to require, so that I may be able to subside into a smaller house. I have ventured to send you a list, cheifly of memoirs or tracts on Fungi thinking that perhaps some of them might be required for your College Library.

And he proposed to circulate the list in England too. By November his financial situation was really serious, and he wrote again to Farlow, this time marking his letter 'Confidential'.

> Will you pardon me if I am compelled to justify the latter part of this letter by commencing with a personal explanation. A few months since I was induced with a view to employment and profit to invest all the money I could collect together in a commercial undertaking, which has collapsed and I can obtain neither salary nor capital. This is to me a serious matter so that I am driven to make some determined effort to replace my loss. The only way open to me is to utilize the remaining fifty copies of my Illustrations of Fungi by offering them for sale at such price for ready money as may hope to influence 50 purchasers. It is not a large number but I have no prospect of obtaining so many in England, but with substantial aid from the U.S.A. it may be accomplished.
>
> I would offer the book in 8 vols cloth, published at £30, for *ten* guineas.... This enormous reduction is made to secure the sale of the *whole* of the remainder of the complete work, and cannot be carried out unless the *whole* are subscribed for.

No clue survives as to the nature of the disastrous investment, but the episode is further evidence, if such were needed, that Cooke had little flare for financial matters. After suggesting some possible purchasers for the *Illustrations* the letter continues:

> If this effort *fails*, no alteration will be made in the price, as the small number of 50 copies would be sold in the course of time, but I am stimulated to this effort, not only by my personal losses, but by the additional fear from the disease of my eyes–now compelling me to seek the aid of the Ophthalmic Hospital, in the hope of saving my sight.

This sudden fear for his sight was probably the result of his noticing for the first time the effects of the cataracts from which he was to suffer so severely later on. The apparent recovery noted in his next letter to Farlow only a few weeks later may have been the result of medical reassurance that loss of sight was not imminent.

> I am glad to report that there is an improvement in the condition of my eyes but I am compelled to be careful and not read or write much. It is a great relief to find that there is now no serious danger of loss of sight, as at first was feared.

Farlow succeeded in disposing of a considerable number of sets of the *Illustrations* for Cooke, as well as some of the other books from his library. Cooke also sought the help of some European mycologists, one of whom was Saccardo, but to him he gave as the only reason for the sale his deteriorating eyesight, not confiding to him his financial straits. Whether Saccardo was successful is unknown.

The depression in the book trade about which Cooke had complained to Bagnall soon after his retirement seems to have been brief, for 1894 saw the publication of two more volumes. He had retained an interest in the liverworts ever since he published his *Easy Guide* in 1865, and he must have been slowly accumulating material for a further book since then, for when he sold his library in 1898 it included three 'portfolios' containing drawings and descriptions of nearly 300 species of liverworts cut from the pages of different published works and pasted in. These portfolios, similar to the scrapbooks already described which he kept on other subjects, must have formed the basis of the *Handbook of British Hepaticae* which he now published with W. H. Allen, for in the Preface to that volume he says:

> ... I have, at length, with some reluctance, prepared the following pages for the press. I have felt diffidence, since for many years another branch of the Cryptogamia has absorbed so much of my time and attention, and I have been unable to collect and study the Hepaticae in the field, and therefore have to be content with the chronicle of the labour of others rather than my own. It is acknowledged on all hands that such a Handbook is urgently required ...

The book was not well received by botanists, who criticised the author for omitting to do precisely what he had disclaimed any intention of doing, namely re-examining the group himself in a critical manner and bringing old work up to date. This seems unjust.

1894 also saw the publication by the S.P.C.K. of *Edible and Poisonous Mushrooms: What to Eat and What to Avoid*, a neat little volume illustrated in colour by the author himself (Fig. 15.6). This was the first time he had been persuaded to include both edible and poisonous species in the same book. True to the opinions expressed in his unpublished essay (p. 273), each of the 48 species described is given a common name, some of which he had undoubtedly invented himself. Most are quite acceptable, but there is one serious misnomer—'buff warty cap' for *Amanita phalloides*, universally known to-day as the 'death cap', a name which we must assume had not been invented in Cooke's time. As he had adopted the name 'warty cap' for the whole genus *Amanita*, he presumably felt obliged, for the sake of consistency, to use also it for the one species whose cap is covered with flat patches of volva rather than with warts. But to describe as buff a toadstool that is whitish to dull green in colour is a strange error. As usual, there was some argument with the publishers over payment for the book, and eventually Cooke was forced to settle for £40 and 20 free copies, instead of the 50 copies he had demanded.

Next, in 1895, came a new venture–a series of five little books for children under the general title *Rambles Among the Wild Flowers*. In each 'Uncle Matt', the pseudonym he adopts for the series, takes his little niece, Cissy, for a country walk *Down the Lane and Back in Search of Wild Flowers, Through the Copse, A Stroll on a Marsh in Search of Wild Flowers, Across the Common*

Fig. 15.4 Coloured illustration from *Rambles among the Wild-Flowers. IV. Across the Common*, presumably by M. C. Cooke.

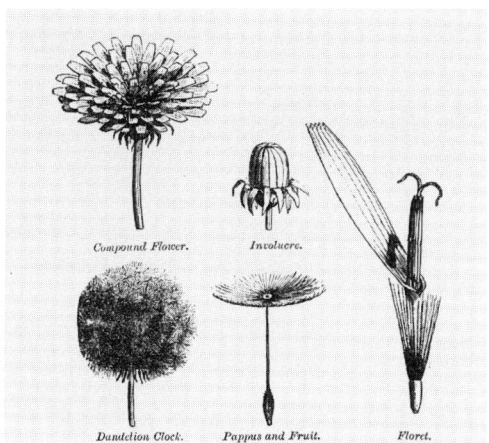

Fig. 15.5 Drawing from *Rambles among the Wild Flowers, I. Down the Lane and Back*, presumably by M. C. Cooke.

after Wild Flowers and *Around a Cornfield in a Ramble after Wild Flowers*, during which the two of them examine and discuss solemnly the flowers they find. It is not just a matter of questions from Cissy and answers from Uncle, as in the *Pinnock's Catechism* of his own childhood; the books are skillfully written so that Cissy is encouraged to observe for herself distinctive features of the plants and their habitats, and then to apply the knowledge she has ferreted out on previous rambles to account for what she finds later. Uncle guides the discussion, and in the process conducts his niece down fascinating byroads of pollination, seed dispersal, folklore, classical myth, ecology, economic botany and culinary lore. The presentation reflects very clearly the influence of Pestalozzian teachings.

The production of the little books by T. Nelson and Sons, Edinburgh, is excellent. The print is large and clear, and each volume has two coloured plates and numerous woodcuts, yet they sold at the realistic price of 1/- each. In the absence of any indication to the contrary we must conclude that the illustrations are the work of the author, and as such are some of the few surviving examples of his competence as an artist of non-mycological subjects. Except where they are deliberately diagrammatic, the plants are delicately and naturalistically drawn in simple style, the result being both pleasing and botanically accurate (Figs. 15.4, 15.5). Several countryside and forest scenes, some including people, are probably the only surviving examples of his considerable competence and skill in picture-making.

After the *Rambles*, towards the end of the year, Cooke published a substantial text-book of mycology, the *Introduction to the Study of Fungi*, which took into account, at last, the new ideas of Brefeld and Saccardo on classification, as well as the many discoveries in other fields made in the twenty years since the publication of his previous text, *Fungi, Their Nature, Influence and Uses*. Among numerous other changes is the inclusion of the yeasts and the fermen-

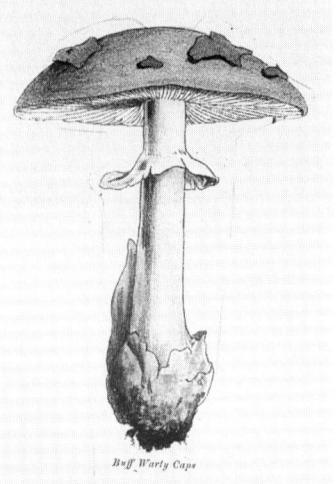

Fig. 15.6 *Amanita phalloides*, the death cap toadstool, from *Edible and Poisonous Mushrooms*. Original in colour.

tative process, which had been entirely omitted from the previous volume, probably because the organisms were looked on by many as being merely a stage in the development of more complex fungi (p. 141). The controversy over the nature of lichens, concerning which Cooke still adhered tenaciously to his now completely outmoded views, is carefully avoided.

In his Preface he says:

> The following pages are the result of an effort to supply an acknowledged want, which I have executed under the impression that it is probably my last contribution of any importance to British Mycology,

perhaps a reasonable assumption for a man of seventy, but one which happily turned out to be unduly pessimistic. Writing to Alfred Clarke, a Yorkshire mycologist (see below), in November he says: 'My forty-third volume [the *Introduction*] good or bad, is just coming out, independently of nearly an equal number of vols of journals which I have edited ... ' and, of course, countless papers, articles and notes. To Clarke he continues:

> ... I hear that the price will be 14/- sold by Discount Booksellers at 10/6. The publishers, A. and M. Black, are Scotch and have 'fiddled' me on account of an omission in the agreement–by which the price was to have been limited to 6/-. So that they pay me a Royalty as 6/- and charge 14/-. So *canny*! [2]

It may well have been the high price that accounted for the slow sales of so useful a book. Strangely, it is the only one of all his works that he fails to mention in his autobiographical notes.

Among the field clubs taking a special interest in mycology was the Yorkshire Naturalists' Union which, since 1881, had devoted its September field meeting to fungi. These forays became increasingly popular with members and, especially as the Woolhope forays began to decline after Dr. Bull's death, mycologists from outside the county started to drift to the Yorkshire meetings, with the result that in 1891 the Union set up a Mycological Committee to develop this aspect of its work. The last Woolhope foray would be held the following year and that year a number of Woolhopeans, among them Cooke, attended both the Hereford and Yorkshire meetings. George Massee was at the 1893 Yorkshire foray, and in the account he and Charles Crossland, a local mycologist, wrote of it, the determination of the Union to take on the mantle of the Woolhope Club was unambiguously, if puritanically, set out:

> The justly celebrated Hereford Foray, which for many years monopolised the first week in October, and was the universally acknowledged meeting-place for exchange of opinion and courteous criticism between British and foreign mycologists, has unfortunately run its course, and it is the hope and ambition of the Yorkshire Union that the annual Yorkshire gathering may–by avoiding the weak points of its predecessor, which were mainly confined to an excess of hospitality–prove at least equally attractive and instructive to mycologists [3].

The writers' hopes were amply fulfilled; the 1895 foray at Huddersfield, attended by Cooke, Massee and Rea, among others, proved to be of crucial importance to the whole future of British mycology, for at it was discussed for the first time the possibility of setting up a national society. The idea bore fruit with remarkable speed; at the next year's foray at Selby the British Mycological Society was inaugurated with the dual purpose of organising forays in different parts of the country and of providing a journal, the *Transactions of the British Mycological Society*, to replace Cooke's now defunct *Grevillea*. Everyone at the meeting, 'about twenty', joined, including Cooke, Plowright, Rea, Massee, and the Yorkshiremen, Clarke and Crossland. Massee was elected President, Rea, who had done more than anyone to press for the Society, became Secretary and Editor (and the following year Treasurer as well) and Crossland, Treasurer for the time being (Fig 15.7). The world's second mycological society had come into existence twelve years after France had inaugurated the first. To-day the British society has about 1,600 members, mostly professional mycologists, and supports a number of committees which organise programmes reflecting the specialised interests of experimentalists of many kinds. But unlike most learned societies, amateurs are still welcomed as members and are held in high esteem by all for their contributions to systematic and field mycology. Every year there are week-long spring and autumn forays; day forays are held locally, and the Foray Committee organises a paper-reading meeting in London each November. The Woolhope tradition still flourishes in this age of professionalism. (Fig. 15.8)

Fig. 15.7 Founders of the British Mycological Society at the Huddersfield meeting of the Yorkshire Naturalists' Union, 1895. *Standing, left to right*, George Massee, Rev. W. W. Fowler, J. Needham. *Sitting, left to right*, Charles Crossland, M. C. Cooke, Carleton Rea. The photograph was taken by Alfred Clarke.

Though a Foundation Member of the new Society, Cooke was too old to take any active part in its affairs; indeed, this was probably the last meeting either of that Society or of the Yorkshire Union that he ever attended. He never, in fact, joined the Union, but he had come to know some of its members well in the years leading up to the founding of the British Mycological Society, and with one of these, Alfred Clarke (1844–1916), he carried on a correspondence for some years and came increasingly to depend (Fig. 15.9). A respected mycologist, Clarke had published on the fungi of the Huddersfield district as early as 1883, but his main interest beyond collecting became recording, not only by means of meticulous labelling, but through his excellent paintings, of which Cooke wrote:

> ... you have no need to hesitate in exhibiting them anywhere. They are far superior to some published plates which I have seen in my time, and equal to the best. Excuse me if you have astonished me by the excellence of your work ... thanks for the treat. (Tol)

Clarke was also famous for his stereophotographs of fungi. In character he was:

Fig. 15.8 The British Mycological Society's autumn foray, 1912. *Left to right*, John Ramsbottom, Carleton Rea, Mrs. Rea (Miss Rose).

haughty, unapproachable and impatient, as though he 'would not suffer fools gladly'. There is evidence that he had a dry, if not cynical, sense of humour ... He was not a sociable creature, but he kept up a steady friendship with a few chosen fellow-mycologists who were congenial to him, and he took keen enjoyment in fungus forays [4].

Perhaps because of his own zeal for collecting, Cooke, not the easiest of men to get on with in his old age, became one of Clarke's chosen few. He seems to have stayed in his house for the 1895 foray and clearly enjoyed his visit, remarking that 'I verily believe that I thought more of enjoying myself than of work', and being specially intrigued by his host's stereoscopic photographs. On his return to London he tried to buy a stereoscope but without success, as he reported to Clarke, but in December he wrote to his friend:

> Thanks for the scope which arrived safe and sound with the excellent photos. I hope that my children and grandchildren will get some amusement and instruction out of them. I shall be pleased to reimburse the cost of outlay if you will let me know. (Tol)

The 'children' to whom he referred could only have been Leila, while the grandchildren must have been Ada's daughters, who would have been small babies for whom a stereoscope would have been very much a toy for the future.

Fig. 15.9 Alfred Clarke of the Yorkshire Naturalists' Union.

Cooke had another spell of poor health in 1896, for on March 4th the following year he wrote to Saccardo after a long silence: 'I have been so unwell for the last twelve months, that I have done nothing, but hope now to be convalescent.' He intended to waste no further time, for he went on to tell Saccardo that he had paintings of 500 foreign agarics mentioned in the *Sylloge* but never yet illustrated, which he intended to publish on a subscription basis, as he had the *Illustrations*. Then once again, the insistence on his imminent departure from this life:

> Such a work would be of immense service, as figures are indispensible for Agarics and with my death all the chance would be lost. It would be the last work that I could hope to undertake.

Cooke could never have seen any of these agarics in their fresh state; like his Australian species they must all have been drawn from dried specimens and the notes of the collectors. Perhaps it was for this reason that the paintings were never published–potential subscribers may have felt that illustrations prepared under such conditions were of little practical use.

Although after 1896 Cooke found it too taxing to travel to forays far from London, he continued most assiduously, despite his bad health at the time, to attend those nearer home, more especially playing a very active part in the mycological programme of the Essex Field Club. In 1895 that Club had opened at Chingford a little Museum illustrative of the natural history of the county, and Cooke, with his unrivalled experience of museums large and small, was asked for his advice on the mycological display. On November 21st, 1896 he gave a lengthy talk on the subject. Starting on a pessimistic note: 'In all our museums, even in our great National Museum, the Botanical portion is relatively very small, very meagre, and very uninteresting', he suggests that money is witheld from botanical displays because they are considered to be inherently dull. However, as far as the flowering plants are concerned the Chingford Museum, tiny though it is, proves beyond doubt that this need not be so with, among other good things, lively displays of plant galls and the insects which cause them, and of dried flowers that do not seem to have 'been put in the family Bible and sat upon for a week'. As for the fungi:

> Your ninety and nine visitors out of a hundred can get no further in a mental definition of a 'mushroom' than a 'toadstool', and these together, represent 'Fungi'–these and nothing more. If 'Arry' brings his 'Arriet' into the Museum on Easter Monday he does it for a 'lark', or out of curiosity. He is not in the 'pursuit of knowledge under difficulties' . . . Supposing that for a brief space of time, you . . . admit him to the museum with a laugh, and let him out to think. No harm will be done, and he may be educated a little against his will. But–how is it to be done? Not by Latin names in couples fixed upon strange objects, but here and there by something he can understand, when it is explained in plain English. One thing he might see would be a group called 'Fungi', either the real thing or a picture, but at any rate *an ocular demonstration*, that under the general name of Fungi, there are Toadstools, Puffballs . . . Blue mould . . . Candle-snuffs . . . and a number of parasites which infest and destroy plants and animals; so that if he only gets an idea of the meaning of one word, he will have been educated ever so little' [5]

However, the education of the complete novice is not the only function of a museum–it must also cater for a more knowledgeable section of the public, in this case for the local naturalist. In short, the Chingford Museum must have a herbarium. Its fungal section should be built up slowly, group by group: 'all that would be wanting would be the willing hands, ready to *do* something more practical than talking'. Rounding off this down-to-earth lecture with a practical warning about the menace of damp, he concluded with the lament of the museum curator through the ages–the building was already too small.

The next autumn, in the year of Queen Victoria's Sixtieth Jubilee, he addressed the Club again, this time on 'Sixty Years of British Mycology', a history of the subject up to his own time in which he showed how, with the development of the microscope, interest had moved from the larger fungi to the thousands of microscopic species which had been newly discovered, and how, with better knowledge of the complex life cycles of fungi, it had been possible to reduce the number recognised as distinct species; but he regretted that microscopic characters could be used to split as well as to lump:

> There is no group of fungi in which the microscope has not been employed; even the Hymenomycetes, such as the *Agarics*, for instance have had to submit to investigation. Perhaps amongst these plants microscopical distinctions have been pushed to their extreme limits. It seems to be 'sharp practice' to separate two *Agarics*, so closely resembling each other as to be indistinguishable by the naked eye, simply because one of them has spores which are two or three micromillimetres larger or smaller than the other.

There is sarcasm here. 1 micromillimetre = 1/1000 micron! As to:

> all that concerns the inner life of the minute fungi ... we [the British] have freely taken advantage of all researches and disquisition in France, Germany, Italy, Austria, and United States, to increase our own store of knowledge, if we have not contributed much ourselves [6].

There is much truth in this remark. It was not until some way into the twentieth century that British mycologists made any notable contribution to experimental mycology.

It is not recorded that he spoke at the 1898 foray, but the:

> occasion was taken by those present at that large meeting, at the suggestion of Professor R. Meldola, F.R.S. and Mr. G. Massee, F.L.S., to pass a special vote of thanks to Dr. M. C. Cooke, for the active share he had taken in promoting the accurate study of fungi during Her Majesty's reign and in particular for the very valuable services Dr. Cooke had rendered to science and to popular education by the number and excellence of the works he had written on the Cryptogamia, and also for the assistance he had always been so ready to afford to beginners and earnest students.

Later, the Council endorses these sentiments and also:

> gratefully acknowledge the Club's indebtedness to Dr. Cooke for the aid he has rendered to the Society in the capacity of principal 'Conductor' and Recorder at the Fungus Meetings ... [hoping that] the Doctor may be spared to carry on his good work and to assist at these gatherings for many years to come [7].

Perhaps it was appropriate that Cooke's contribution to science should be recognised by a field club some years before he was honoured by the botanical establishment.

Despite his poor health and constant fear that he had reached the end of his useful life, Cooke must, in fact, have been working hard at his next publication, of which the first part appeared in 1897. Following on the success of his *Rambles* series for children, Nelson's

> commissioned me to prepare a series of Wall Diagrams, about a yard across on Structural Botany subjects, for the purpose of teaching Botany in schools, and I prepared forty three diagrams [actually 45], which were soon executed and afterwards adopted by the London School Board for teaching in Elementary Board schools (Aut).

After forty years his interest was again, for a short while, centred on the schoolroom. The 45 'diagrams'–to-day we should call them posters–are each 2'3" × 3' in size and are divided into six series of seven or eight sheets each. Series 1 and 2 cover the structure, germination, fertilisation and seed dispersal of flowering plants, and Series 3 introduces the Cryptograms. Series 4–6 were published a year later and are concerned with 'Useful Plants', that is, their designer's speciality of Economic Botany. A number of these sheets include not only the plants themselves, but illustrations of the preparation and uses of their products. Most of the 'diagrams', which are very large, clear, beautifully executed and attractively coloured, are not in fact diagrammatic at all, but fully modelled illustrations. Working on them must have reminded him of the much admired exhibitions he used to put on for the Metropolitan Schoolmasters' Association. Each of the six series is accompanied by a booklet of teaching notes entitled *Object Lesson Handbooks* beginning: 'the following lessons deal with the ordinary phenomena of common life, and with objects familiar to the children'.

Among Cooke's papers in the archives of the Tolson Museum, Huddersfield, are some fragmentary notes headed 'Chats at Breakfast with my Children' by 'Uncle Matt', the pseudonym he had used for his *Rambles* series. Like the wall diagrams, the notes are concerned with economic botany, and until 1983 there was no suggestion that he had ever developed them further. In that year, however, the City Museum, Leeds, discovered in its archives the undated, unfinished manuscript, in Cooke's handwriting, of a book for children entitled 'Chats about useful Plants', also by 'Uncle Matt', datable to approximately 1899, and obviously the expanded version of the Tolson Museum notes. In this manuscript, instead of Uncle Matt chatting to his niece Cissy, as in the *Rambles*, Papa talks to his young family, Tommy, Kitty, Flossie and others over the breakfast table. He discusses in turn each of a huge range of British and exotic plants of economic importance, under the main headings, The Staff of Life, Edible Fruits, Beverages, Condiments, Narcotics and Medicine, Clothing, Arts and Manufactures, and Building and Furniture, the sections vary in length from 61 pages for the Staff of Life and 53 for Beverages, to a mere 15 for Building and Furniture. The whole manuscript runs to 271 pages.

At the introductory breakfast the eldest child complains to Papa that history is all about dates and wars, but never about how people lived: once again, as many years ago at Lambeth, Cooke is critical of current educational methods. He sets about remedying the matter by making Papa hinge his first talk on 'What did Julius Caesar have for breakfast?' and dealing with the diet of the Ancient Britons, Anglo-Saxons, Normans and others. The following discussion of wheat occupies 11 pages and includes a history of its development from ancient times, a description of the different types, the world wheat trade, the history of milling, of flour and of bread making, and even a discussion of the Luton and Dunstable straw hat trade. There is a comment that 'all harvesting is done by machinery now-a-days'. Under 'Coffee', Cooke describes the destruction of the plantations in Ceylon by 'a fungus parasite', which he says was controlled by growing an African variety of the coffee bush resistant to the fungus. The presentation is so closely related to the *Rambles* series that on one occasion (in the section on Pepper) the children round the breakfast table are exhorted to emulate their 'Cousin Cissy' and take an interest in wild flowers.

At Junction Road, life for Leila Cooke (Fig. 15.2) in these years, as the only child among all the grown-ups, would have been sad and lonely but for the fact that her Uncle Ebenezer's large family was within easy reach. Leila saw a lot of the children and was especially friendly with her youngest cousin, Winnie, almost her own age, who 'used to sort of drop in, and I used to drop in there'. Another favourite was Winnie's elder brother, Gilbert, who 'always used to come in with some comic stories'. Cousin Edith was less popular:

> Edith married and the marriage wasn't a success . . . Her husband had no money. When she was having a baby I remember her coming in to tell my father how hard up she was. And he always had half a guinea or something to give them . . . All the Ebenezer Cooke family used to come and cadge money off him. He always seemed to be giving this one half a sovereign, somebody else a sovereign. Pater was very generous.

Her father's generosity is a trait constantly reiterated by Leila, and is in strong contrast to the reputation for tight-fistedness he has gained among mycologists; maybe it was only by penny-pinching outside the family that he could finance his natural generosity inside it.

Financially, if for no other reason, it probably came as a relief when, on August 5th, 1897, poor shadowy Sophia died, aged 74. She was attended by the family doctor, Doctor J. F. Haines (he had also seen Leila through her attack of diphtheria), who certified that the cause of death was 'epilepsy, exhaustion'. That she was not a life-long epileptic we may be sure, or she would never have been a school-mistress, so it is difficult to guess at the actual condition from which she died or how long she had been ill beforehand. It was Annie, not Mordecai, who registered the death. She did so truthfully, giving her own name and relationship to the deceased as 'Annie Cubitt, daughter', and entering Sophia as Mordecai's wife. She herself had moved once again and was living at Monnery Road, another of the turnings off Junction Road–she had not far to go for her weekly allowance. That Mordecai failed to sign his wife's death certificate cannot, I think, be taken as proof that the two were estranged, for never once during his life had he witnessed an official document concerning his family. Sophia was buried in Finchley Cemetery in the same grave as her eldest grandson, Harry Linnaeus, but whether this was recorded on the original gravestone is not known. Certainly the fact was omitted from the new stone which Leila erected after her father's death, and the only record now remaining of the last resting place of his unhappy wife is in the registers of the Cemetery Superintendent.

With Sophia's death there was no possible excuse for further postponing a move to a smaller house, which Mordecai had first intended to do soon after retiring and which, with his recent financial losses, must now have become urgent. Willie was married, and Ada and Caleb, with the three little daughters who had already been born to them, had moved to Southsea; only he and Leila were left, but the move must have been a terrible wrench for both of them. Mordecai had to part with his library, which he sold almost in its entirety, including a complete run of *Grevillea*, copies of all his own books, and reprints of most of his papers, to his favourite booksellers, Messrs. Wheldon's specialists in biological books. As for Leila, she hated leaving the garden:

> In my time there were lovely old trees, a white and purple lilac, a red hawthorn, pear and plum trees, a laburnum with the longest flower sprays I have ever seen, 16 to 18 inches: gooseberries, white and red currants and two willow trees at the bottom of the garden. In my time everything was much neglected but very beautiful.

They moved towards London, to 53, Castle Road, just south of Kentish Town West station and a far less salubrious district than Junction Road; Leila describes it as 'a very poor house and a very poor place', and a grand-daughter remembers its twisted stairs and tiny garden. The Castle public house still stands at the corner of the road, but No. 53 was one of many houses swept away after the second World War to make way for a high block of council flats.

For Leila, one of the many disadvantages of the move was that she was now further from her cousins in St. John's Road, but worse was to come for her. Ebenezer's reputation as a progressive teacher was by now considerable. He not only taught in private schools and had his own private pupils, but took part in experimental teaching projects in State schools, sat on committees, published articles on his method, and had written the preface to the 1894 English translation of Pestalozzi's *How Gertrude Teaches her Children*, the handbook of the great teacher's method. Recently his fame had spread to America and he had been invited to lecture there but, his wife Ellen having been gravely ill for some time, it was not till after her death that he accepted the invitation, travelling there in the summer of 1898. In doing so he had achieved what his brother, despite his many years of contact with that country's mycologists, never did succeed in doing. When Ebenezer returned from the States he too moved house, but in the opposite direction from Mordecai, settling in 1899 in South Hill Park, Hampstead, too far from Castle Road for casual 'dropping in'. For Leila and Winnie the parting was even more final for, to Leila's lasting regret, her cousin was sent to live with an elder sister in Manchester, and from then on the two families, despite having been so close, seem to have drifted apart.

Chapter Sixteen

'Nature Never Did Betray the Heart that Loved Her'. (W. Wordsworth)
1899–1914

Not long after the move to Castle Road, Cooke was prevailed upon by the Committee of the Quekett Microscopical Club to return, after his long absence, and give members a talk. It was ten years since he had last attended a meeting, and at that time the Club was still using the Library of University College. But in 1899 it had been asked to vacate these premises because, the authorities said, the fumes from the gas lights members used for their microscopes were damaging the books, so meetings were now held in the rooms of the Royal Medical and Chirurgical Society at 20, Hanover Square. Cooke called his talk, given on June 16th, 1899, 'Early Memories of the Quekett Microscopical Club', and it has supplied much of the material for my account of the setting up of the organisation.

In May [1] he attended another meeting to hear a talk given by a young mycologist, Mr. E. S. Salmon (1871–1959), on the life history of the Erysiphaceae (the powdery mildews) of which he later wrote the classical account. Asked afterwards by the Chairman to 'give them some of his ideas on the subject', Cooke replied:

> that in the first place he was not so pugnaceous as he used to be half a century ago; in the next place he believed that with old age he was getting exceedingly indolent, and in the

last place his hearing was not as good as it was, and therefore it was only with the greatest difficulty that he could follow the paper–not being able to hear all.

Then, oblivious of the subject of the paper, as so often happens when distinguished elders are recalled from a long retirement, he mounted his favourite hobby-horse and roundly condemned the ascendency in modern times of the 'splitters' over the 'lumpers', a sentiment with which the Chairman expressed his heartfelt agreement. Turning to the overwintering of conidia, Cooke then spoke briefly of the phenomenon in a species totally unrelated to the Erysiphaceae and continued that:

> he had not read all the abstruse investigations or imaginations which had been put forward on these subjects, and could therefore say little about them.

Not surprisingly, poor Mr. Salmon seems to have found himself unable to contribute to a 'discussion' so unrelated to his paper. Cooke then expressed his pleasure at attending the meeting, his hope that he would be able to take part in some of the outings that summer, and his regret that, of faces so familiar in the old days, 'so many had ... gone over to meet that great majority to which he himself should also go before long'. This seems to have been the last meeting that he ever attended of the Club of which he is now known as the 'Father'.

However, he remained active in the Essex Field Club for some years, helping to conduct its forays, usually in company with George Massee, until at least 1904, when he was 79. A fellow member of the Club, Miss Gulielma Lister (1860–1949)*, remembered Cooke thus:

> Dr. Cooke was ... a familiar figure on the occasion of our Club's annual Fungus Forays, when he gave his valuable services as referee, and when, in the evenings, he would sum up the results of the day's work, or give some pithy address on the wider aspects of fungus life. We well remember his short lean figure, his quiet energy, and the quaint humour with which he would enliven the most technical discourse [2].

A slightly more revealing, but still fond, description of the old man at this time is given by J. F. Rayner of the Hampshire Field Club, at the forays of which Cooke still sometimes acted as referee. For Rayner, Cooke 'may well be entitled the Nestor of British mycology' [3], Nestor being the aged hero of the Trojan War who is renowned for his experience, wisdom and garrulity, but whose advice, ever freely given, tends to the platitudinous.

G. G. Desmond [4] a correspondent for the *Daily News*, was inspired by an exceptionally memorable foray, which may have been Cooke's last with the Essex Field Club, to a most lyrical piece of journalism:

---

*Niece of the great Joseph Lister, pioneer of antiseptics, and herself an authority on slime moulds.

When one looks for fungi at this time of year they are everywhere. Among the ferns are velvet-topped boleti, bearing underneath a red, straw-coloured, or saffron yellow sponge, according to kind. One termed edulis is dainty eating; another, satanus, is noxiously worthy of its name; a third, piperatus, scorches the tongue like fire. From the trees other fungi look down till they rot and fall under foot; the 'beef-steak' juts out like a succulent ox tongue from high up an oak; quaint shapes peep from holes in the pollarded hornbeams. Under foot bright yellow candles gem the long grass, or rosy-topped russula push their way through the mould to have their beauty spoiled the first night by slugs. A bank down which the water is ever oozing seems thickly strewn with dead, wet leaves, but they all turn out to be fungi rejoicing in the moisture they love. Every fallen log and sawn-off stump is gemmed more or less with the fungus whose duty it is to destroy it and adorn it in its death. . . .

Once I came upon a tree stretching out a hollow branch, into the tube of which I looked. The ceiling was beautifully sculptured with the gills of a few large red-brown fungi, which had scattered their spores of brightest chloride of gold throughout the little grotto. First a spider had spun its cobweb diagonally from floor to ceiling. The red-gold dust had converted the gossamer into the richest scarf that fairy ever wore. On the floor a curled beech leaf rested daintily on its keel. We know how the autumn beech leaf shines, as if with an inner light. This one shone three times more brilliantly, for the red dust upon it and upon the whole grotto looked as though lighted up by red limelight.

The Club's little Museum at Chingford, though a considerable success, was hardly large enough even to cater for the needs of Epping Forest, let alone for those of the rest of the county as well, so at the turn of the century a new Museum, the Passmore Edwards, was built at Stratford, and Cooke was asked 'to prepare a series of coloured wall diagrams' for it. He 'duly executed forty-nine botanical diagrams, partly of the Cryptogamia', and though they are no longer in existence it seems likely that they were modelled on those he had prepared for Nelson's some years earlier, and used the principles he had put forward for the Chingford Museum display. He completed the diagrams in 1903 and contributed his last article to the Club's journal in 1905.

As long as he was physically able, and undeterred by the worst of weather, Cooke continued to attend the fortnightly meetings of the Science Committee of the Royal Horticultural Society. One of his fellow members, H. T. Gussow, a Canadian plant pathologist who worked for a while in England and became intimately acquainted with him at this time, describes Cooke as 'indefatigable', and 'a modest, kind old gentleman with a pronounced sense of humour' [5] (Fig. 16.4).

We have seen that from the start of his interest in mycology he had been drawn to the pathogenic fungi, and had, indeed, published a number of useful papers on them in addition to his early popular book, *Rust, Smut, Mildew and Mould*; but they had inevitably taken second place to the huge volume of taxonomic work he undertook. Now, as his ties with taxonomic mycology dropped away, he had time to spend on the parasites, and could take an interest not only in the organisms themselves, but in the diseases they cause. In 1902 his work for the Scientific Committee on the fungal diseases of plants, together with his eminence as a mycologist, were recognised by the Society when it presented him with its Victoria Medal of Honour, an award first instituted in 1897, the year of the Queen's Jubilee. No record survives of the actual presentation, but at a meeting of the Scientific Committee on November 18th, 1902, its Chairman:

> rose to offer in the name of the Committee their hearty congratulations and goodwill to Dr. M. C. Cooke having been presented with the Society's Victoria Medal of Honour. Dr. Cooke in reply observed, in thanking the committee, that he had no anticipation of the honour, as it was quite unexpected, since whatever he had done was always *con amore* [6].

George Massee, Cooke's protégé and successor at Kew, was honoured at the same time.

The presentation of the Victoria Medal was followed by an even more prestigious and richly deserved honour, the Linnean Society's Gold Medal (Fig. 16.1). At the Society's Annual General Meeting on May 25th, 1903, the President, Professor S. H. Vynes of Oxford, said in his Address:

> I take this opportunity of formally announcing that the Linnean Medal this year has been awarded by the Council to Dr. M. C. Cooke, who has been an Associate of the Society for over 25 years, and is so well known as a high authority in the department of Mycology [7]

Fig. 16.1   The Linnean Society's Gold Medal, presented to M. C. Cooke.

Later, in his presentation speech, Professor Vynes pointed out that only one previous medal winner had been a mycologist, and that not the Rev. M. J. Berkeley, as the Medal had not been instituted in his day. But since Berkeley's name

> does not adorn our list, nothing can be more appropriate than that it should include one who was his collaborator and has proved himself to be his legitimate successor ... For more than forty years you have unceasingly engaged in describing, depicting, naming and classifying the enormous mass of material that has been submitted to you from all parts of the world; and it is not too much to say that few have contributed so materially as yourself to the reduction of the mycological chaos ...

Cooke bequeathed his Linnean Medal to Leila and his Victoria Medal to Frank, the executrix and executor of his will, with instructions that each should be passed on to the recipient's children, if any, or if not to others of his grandchildren. In the event, neither had a family. Leila believes that, on Frank's early death, the Victoria Medal went to Willie who, as the eldest son, was in any case jealous that his father had not bequeathed it to him. Willie probably sold it when he fell on hard times. Long afterwards, in 1952, she herself presented the Linnean Medal to the Royal Botanic Gardens, Kew, where it can still be seen.

In view of Cooke's strong feelings on the admission of women to the Quekett Club, it is perhaps interesting to note that he would not for long be able to escape from them in the Linnean, to whose rooms, his daughter informed me, he often repaired to meet his mycological friends. For in the same Presidential Address in which he welcomed Cooke as a medalist Professor Vynes, whose feelings on the matter of women scientists were apparently not dissimilar from Cooke's, commented on a recent vote by members which was overwhelmingly in favour of their admission, and continued:

> I have little doubt that ... it will fall to my lot to admit the first Lady-fellow ... regarding the matter, as I am bound to do, from the point of view of the welfare of the Society, I must confess that I am not altogether free from apprehension as to the future. We are making a somewhat heroic experiment, with no precedent, no working hypothesis, to suggest to us what the results are likely to be. If purity of motive can deserve success, then it should certainly be ours ... However, we must not shut our eyes to the fact that the Society is passing through a serious crisis ...

The last mycological enterprise of any note which Cooke had undertaken before receiving the Victoria Medal had been the publication of the *Introduction to the Study of Fungi* some seven years earlier. Since then the inexorable inner drive which had for so long forced him into a ceaseless round of mycological activity, had seemed to be dying away, to be replaced by the slow realisation that he must now resign himself to the frustrating limitations of old age.

But resignation was not in his nature, and his post-foray talk to the Essex Field Club the previous year had pressaged the new direction his work would take. There, no doubt influenced by the problems brought to the Scientific Committee of the Royal Horticultural Society, he had pointed out that nowadays, rather than toadstools, 'the strongest interest is in the parasites which infect and destroy our crops, and thus compel attention to the invaders not from choice but from self-interest'. He mentioned particularly the current research on the silver-leaf disease of fruit trees and 'the causes affecting club root' (of cruciferous vegetables), and ended with a strong appeal for more work on crop diseases. Perhaps it was the award of the Victoria Medal which provided the stimulus for his new venture, but in 1903, when he was 78, it became clear to all that the old man was far from finished yet, for there began to flow from his pen, in the *Journal of the Royal Horticultural Society*, a major series of articles which continued until 1905 under the over all title of 'Fungoid Diseases of Cultivated Plants' (Figs 16.2 & 16.3). There were six articles in all, concerning pests of the orchard and of fruit (on which he had already written in the Woolhope *Journal* in 1892), of the vinery and stove (or hothouse), the flower garden, garden vegetables, ornamental shrubberies and forest trees. Each article was illustrated in colour with drawings by his own hand, many of which he had used before but including some which were specially prepared for the series.

His working method for these articles, as doubtless for all the many other compilations he had produced during his long life, can be exactly followed, for he eventually sold part of his preparatory material to the Royal Horticultural Society, in whose archives it remains. It is contained in eleven box files, three each for the pests of the orchard, shrubbery, and vinery and stove, and two for an unpublished article on the diseases of wild flowers. The whole collection is neat, tidy and methodical. There is a folder for each host plant with the diseases to be included listed on the front, and inside a further folder for each disease. Each of these is headed with the name of the disease, the host, the fungus, its country of origin, and its reference in Saccardo's *Sylloge*; inside is the material from which the text will be written, consisting either of manuscript notes or of clippings and reprints from his own writings and those of others. The illustrations are carefully arranged on separate sheets, usually having been clipped from a printed text and pasted in, but sometimes drawn directly on to the sheet; only a few of them were selected for the final article. In that article are included the description of each disease, general information about it, instructions for its prevention or treatment, and references.

The articles were so well received that, in Cooke's own words:

> ... they were revised, somewhat enlarged–and the Society published them in a separate volume neatly bound in calf at half a guinea–but afterwards, in order to induce the purchase by gardeners, and others of small income, they reduced the price one half, so that the book could be purchased for five shillings. (Aut)

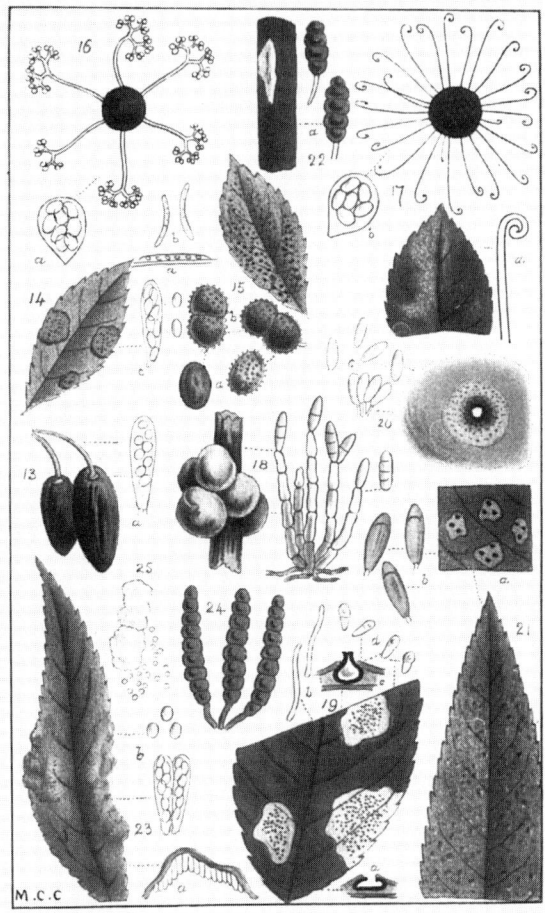

Fig. 16.2 Plate from *Fungoid Pests of Cultivated Plants* showing diseases of plum and peach. Original in colour.

The additions included a short Introduction giving 'the main facts in the life history of some of the parasites to be recorded', a new section on the pests of field crops, and a list of fungicides in use at the time. The book appeared in 1906, to a general welcome from mycologists and horticulturalists alike.

The section on fungicides includes two groups–'Fluids', which may be applied 'by sprinkling, spraying or sponging the foliage', and 'Powders', applied 'by dredging from a flour dredger, or pepper pot, or enclosed in a canvas bag'. Methods of spraying are not specified, but the 3-gallon knapsack sprayer was already in use, having been developed in the French vineyards in about 1888, and on a larger scale, horse-drawn cart sprayers appeared soon afterwards in

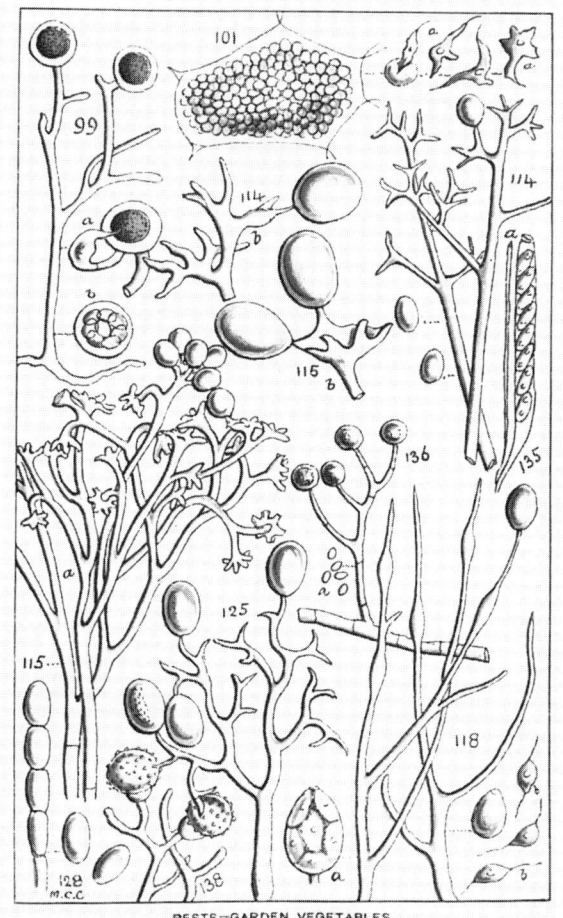

Fig. 16.3 Illustration from *Fungoid Pests of Cultivated Plants* showing fungi causing disease in vegetables.

the blighted potato fields of Britain and the U.S.A.

Compared with the complex organic fungicides in general use to-day, those listed by Cooke are few and simple. Sulphur, in the form of lime sulphur, had first been used in England in 1802 against mildew of fruit trees, and was further developed and widely used in this country, the U.S.A. and France during the century. Copper, on the other hand, did not make its appearance until 1882, when P. M. A. Millardet saved the French vineyards from the depredations of the downy mildew, newly introduced from N. America, with his Bordeaux Mixture, a preparation of copper sulphate and lime. This was found to be equally effective against potato blight and diseases of other crops and quickly rivalled sulphur as the leading fungicide, as is reflected in the list

Fig. 16.4 M. C. Cooke aged 78.

given by Cooke; for out of 17, six are copper based, three sulphur based, and one is a mixture of the two. Others, such as Jeye's fluid, paraffin and permanganate of potash are recommended for limited and specific purposes only.

In 1908 the manufacture of proprietory fungicides had not yet begun, for the ingredients were simple and easily obtainable and the preparations could still be made up by the user. Cooke, therefore, gives recipes for all. Yet the time and labour involved in handling many gallons of liquid and slaking many pounds of lime on large holdings and farms was enormous, and by 1911 the first factory-made lime sulphur preparation had come on to the market in Great Britain and the U.S.A., so starting what would become a major and highly lucrative branch of the chemical industry.

One paragraph of the Introduction to the book is worthy of special comment here as being another example of Cooke's obstinate conservatism in science as well as in politics. In discussing the five spore forms produced by most rust fungi he says:

> Besides these cases in which aecidiospores, uredospores, and teleutospores are produced on the same species of host-plant, there is another group which those who have implicit

faith in heteroecism contend produce the aecidiospores with spermogonia on one species of plant, and the uredospores and teleutospores on another and quite different species of host-plant. Let each be persuaded to his own mind, as it will serve no good purpose to enter upon discussion here.

Heteroecism, the need of certain parasitic organisms for two distinct host species in order to complete their life cycle, had first been established for fungi in 1865 by Anton de Bary, working on that destructive pathogen, the wheat rust fungus, *Puccinia graminis*. It had for centuries been observed by farmers that an attack of wheat rust often started in those fields which were in close proximity to barbery bushes, only spreading later to crops which were further off. De Bary, was able to show that there was a perfectly good scientific explanation for this phenomenon (Fig. 16.5). The first spores to appear on the wheat plant in the spring are the rust-red uredospores; these are the spores which spread the disease far and wide over the wheat fields. They give rise to the mycelium producing the teleutospores, the stripes of which disfigure the wheat leaves in summer and which, with their thick, protective walls, enable the fungus to survive the winter. In the spring the teleutospores germinate to produce, not mycelium, but a third type of spore, the minute, almost naked, sporidia,* and it was these last which caused so much confusion, for no-one had been able to coax them to reinfect wheat under experimental conditions. Where, then, did the uredospores on the new young spring wheat come from? De Bary had disentangled the complex life cycle of *Puccinia graminis* and shown what excellent naturalists the farmers were by demonstrating that the sporidia could easily be made to germinate, not on wheat but on barberry leaves, there to produce two more spore forms, the spermogonia and the aecidiospores, the last of which completed the cycle in the spring by infecting young wheat in adjacent fields. This was heteroecism, but though the novel concept was accepted on the Continent, British mycologists for long regarded it with scepticism, and Cooke was the greatest sceptic of all. One of his two contributions to the weekly magazine of science, *Nature*, then only three years old, was an article written in 1872, seven years after de Bary's discovery, on 'Alternation of Generations in Fungi', in which the writer is only prevented by de Bary's eminence from denying outright the possibility of heteroecism. He goes so far as to state that he considers the evidence insufficient to prove it and concludes: 'I claim the privilege of doubting where I dare not deny' [8]. The correspondence columns of the same issue include an almost equally doubting letter from Berkeley who, however, allows that he would support the hypothesis if the experiments were successfully repeated.

Opinions in England changed little for about ten years, but then the matter was raised again by, among others, C. B. Plowright, the respected mycologist

---

*The word is used here in its modern sense. Rust sporidia have no connection with the sexually produced arcospores for which the word was coined in the nineteenth century.

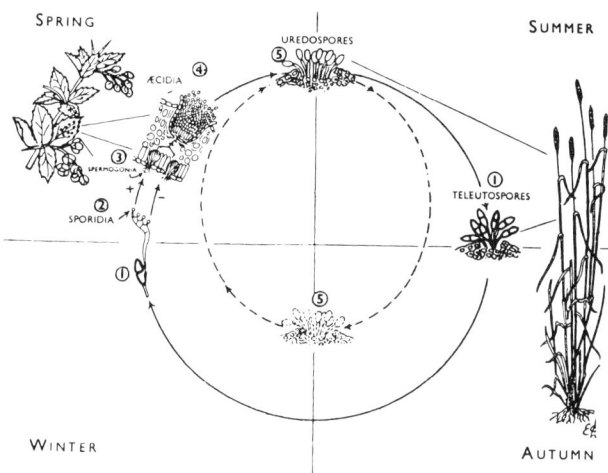

Fig. 16.5 The life cycle of the black stem rust fungus of wheat. *Puccinia graminis*. (From E. C. Large, *The Advance of the Fungi*).

from King's Lynn, who began to publish articles in *Grevillea* in support of the hypothesis. Cooke was quick to reply, surprisingly not in his own, or any other, scientific journal, but in a lengthy talk to the Hackney Field Society, reprinted in its *Report*. The talk was given in March 1883, and was probably of more topical interest to his lay audience than we would expect, for 1882 had been a bad year for wheat rust:

> some of us have heard, or read, during the past twelve months, dreadful prognostics of the doom of the wheat plant. The barberry and the Mildew [rust] have been stalking the land like ghosts.

His attitude to what he still persisted in regarding as merely a hypothesis is made clear in the first few scathing paragraphs of the talk:

> During the past year we have heard much of a theory, which was at its greatest on the continent ten years ago, but in this island we were spared the infliction until of late, when for each other's amusement, some enthusiastic members of the advanced school, have revived the theory that the wheat mildew is produced from Barberry bushes.

His strategy for the attack is to try to show that the hypothesis is unnecessary and unproven, not that it is impossible; that the problem of the overwintering of the fungus can be explained in simpler ways. His counterproposal is to the effect that *P. graminis* is seed-borne; that it survives the winter as mycelium within the seed, to proliferate within the young seedling in the spring. Concerning the experimental evidence that the barbery stage, when artificially inoculated into wheat, produces the spores of the wheat stage, he is of the opinion that:

> it is more reasonable to suppose that by setting up a local irritation, by means of a foreign parasite, a local eruption of the latent *Aecidium* should be caused, than that the sowing of the seeds of one plant, should result in the production of another, not at all like its parent. 'Do men gather grapes of thorns, or figs of thistles?'

Not content with a verbal reply, it seems that he set about carrying out some experimental work for himself, for in a letter to Farlow on Sept. 3rd that year he wrote:

> Every day and every investigation strengthens me in the conviction that I am right in strongly doubting the theory of *heteroecism*–and I am accumulating evidence. Next year–if I find time I shall have a large number of cultivations to prove the extent of *hereditary* transmission of *Puccinia* from generation to generation–through seeds and buds, ... I have no time to enter fully on the question here–but–is it not a fact–that in these experimental cultivations we have conditions and results which are never obtained in a state of nature.

After calling to his support the cases of some other rust species, which we now know to have incomplete life cycles, he continues, apparently oblivious of the fact that his remarks are double edged:

> Like all other theorists the supporters of *this* can see only their own views. It is a great tax on one's time to have to waste it in opposing errors of this kind.

A threat to publish his views '*in extenso*' seems never to have come to pass, but as we have seen, he was still unconvinced about heteroecism in 1906, by which time it had been fully accepted even by other British mycologists. As with the 'dual lichen hypothesis', Cooke remained to the end of his life, totally unable to open his mind to a fundamentally new concept.

After describing in his autobiographical notes the publication of *Fungoid Pests of Cultivated Plants*, Cooke continued:

> This was unfortunately destined to be my last book, for within twelve months after its publication my eyesight had failed to the extent that my microscope was locked away and I could not read a line of ordinary printed matter, and had to relinquish all my botanical pursuits.

The cataracts which had been menacing his sight for some years were now having a disastrous effect on his activities, though he was still able to continue with limited 'botanical pursuits' for a few years more and it was not until 1910 that:

> I offered my collection of drawings and notes of diseases of plants to the Royal Horticultural Society, as I could not hope to augment them further, or make any use of them. There were 1650 sheets or drawings which the Council accepted and presented me with a cheque for one hundred pounds in acknowledgement.

This is the collection contained in the eleven box files just described, together with his own copy of *Fungoid Pests*, interleaved with blank pages, many of which he has used to annotate, update or correct the text. Perhaps he gave the Society his poem to 'Flamen Pomonalis' at the same time.

His last taxonomic paper appeared in the *Transactions of the British Mycological Society* in 1908–a short paper on an agaric species–but even now he had two larger projects in hand, for one of which Alfred Clarke was giving him much needed assistance. His increasing blindness was forcing him to rely more and more on others and it was often Clarke to whom he turned and to whom he was perpetually appologising for his helplesness:

> 'It seems that when I get into a difficulty I think of you to help me out of it'.
> 'I am afraid that I have become a nuisance to my friends. . . . I hope you will forgive me for imposing such a task upon you'.
> 'I am sorry that I am so useless–but I cannot help it'.
> 'I know that I ought to be ashamed of myself for troubling you.' (Tol)

However, it seems that Clarke's assistance to the older man was ungrudging and generous.

Cooke had concieved the idea of publishing a *Catalogue and Field-Book of British Basidiomycetes*, largely in protest against the way in which authorities for species were now being cited, but also intended as a check-list for use in the field, though its shape, $10\frac{1}{2}'' \times 4''$, and its soft cover, would make it somewhat unsuitable for carrying in the pocket. Because of his failing sight he had to rely on Carleton Rea to see the volume through the press, but unfortunately, over the years, his dislike of Rea had grown in intensity, as is obvious from a letter to Clarke written at this time. Apparently Clarke's opinion of Rea was little higher than Cooke's, for on April 26th, 1909, the latter wrote, still in excellent hand-writing despite his deteriorating sight:

> Thank you so much for your letter, and I fully sympathise with you in all you say in regard to the Secretary' [Rea, Secretary of the British Mycological Society] 'I heartily wish I was able to see the Catalogue through the press–and am almost afraid I shall have to abandon it at the eleventh hour–because I do not like leaving it to the tender mercies of C.R. [Rea]. In fact I do not think now that *he* will wish it, since I have declined to accept some of his suggestions–and ventured to tell him that my experience was a little longer than his–His latest response was to designate me as retrograde to which I have only replied that if it cannot be printed as I have left it I shall not object to put it in the kitchen fire. . . .
> The worthy Secretary seems to have got a swelled head now that he is President,

Secretary and Treasurer as well as Editor for the B.M.S. . . .

I must confess that I am getting tired of Rea–and want to have as little to do with him as possible–and I should not have hinted at his having anything to do with getting out the *Catalogue*, if I had felt able to do it myself but at just 85 years of age one does not feel the pluck of 'auld lang syne'. . . .

I am inclined myself to think him [Rea] a big puppy and this opinion is strengthened by every additional experience, and am not displeased to find another body of a similar opinion.

If I were twenty years younger I should be pleased to enter into a fair fight with C.R. but I am not surprised to read between the lines, that he is now aiming at getting himself recognised as the British authority on Mycology–and putting Massee in the shadow–but he will never do it with people of sense. (Tol)

In fact, though both men did useful work and Massee's, as would be expected from a professional, was much more extensive than Rea's, Rea was the more reliable of the two.

A month later Cooke wrote again to Clarke:

I wrote a long letter to a friend explaining my views on Genera and Species, which I should like to become more widely known–and I have consent to have it published in Yorkshire Naturalist–if you think it advisable–and I will abide by your decision if you will read it, and send it on to Crossland to hear if he approves. If you two are of my opinion will you forward it to the 'Naturalist'* for me.

It is sad to find Cooke reduced to seeking anxiously the approval of others before daring to publish. The original letter to which he refers is lost and we do not know to whom it was written, but a copy of it's substance made by Clarke still exists. It contains a restatement and enlargement of its writer's beliefs on the unbroken chain of nature, first enunciated to the Society of Amateur Botanists more than forty years before.

I recognise *Genera* and *Species* as purely artificial groups, constituted by human authorities–for the purposes of classification, but that really there are no such things in Nature–and in justification of this view I affirm that it may be discovered that in all genera there are abnormal forms, or intermediate links which do not belong strictly to any genus, but are intermediate, or 'missing links' which join one genus to another, and combine the whole into one continuous harmonious whole, gliding the one into the other without a gap between–and thus the whole scheme of vegetation is a unity . . .

And he applies the same reasoning to orders and species. Strangely, considering that he first put forward his creed so long ago, he continues:

---

*The journal of the Yorkshire Naturalists' Union.

> I am often deeply grieved, that these views did not present themselves to me many years ago, when I had the facilities for discovering 'missing links' all over the world, which could not occur within the limits of a local flora'.

One can only suppose that he means that he regrets using his ideas solely to justify himself as a 'lumper', and that it had not occurred to him deliberately to collect intermediate forms in order to support these ideas. 'Missing links' in the chain of evolution, especially of man, were being eagerly sought at the time, and Cooke, as he points out, was simply applying this concept to fungi.

> ... it is no new theory that one form, or so-called species merges into another by links, which are now sometimes obscure or missing, but which unite, not only all the members of one Natural Order with another, but also the larger groups, such as the Fungi with the Algae and other of the Cryptogamia, although we may have to travel back thousands of years to realize the association.

A taxonomist who holds such views will of necessity be a 'lumper' and will ignore small differences between infinitely variable individuals, confining his definitions of species to a few outstanding characters to which numerous individuals can conform. Calling Berkeley to his support, Cooke quotes him as saying: 'Any good description can be contained in three lines, and exceeding that, it is but the picture of an individual'.

Cooke was 84 when he wrote this, his last mycological testament. It epitomised his whole philosophy of the world of nature to which he had devoted his long life, and he wished it to go down to posterity; but there were difficulties. The letter concludes:

> I think the views which I have attempted to explain should be more widely known, but now that 'Grevillea' is dead, I have no means of communicating with students. Some day in the future, if these notes are in existence, it might be of service to allow them to be printed, but that is not for me to decide.

Now, in his old age, the self-assurance with which he had previously been accustomed to attack colleagues whose views differed from his own had been replaced by a diffidence which passed over to others the decision as to whether or not to publish; and when the decision was taken, by whom is not known, publication was in the magazine of a local natural history society rather than in a national scientific journal. The *Transactions of the British Mycological Society* had more than replaced *Grevillea*, so the consideration which prevented Cooke from apparently even considering publication in its pages may well have been his loathing of Rea and his experiences with him over the *Catalogue*. However, perhaps it was fitting that the last, deeply felt testament of a man who had done so much to bring his subject to the attention of ordinary people, should be submitted to their judgement in the columns of the

Naturalist of 1909, under the title 'Genera and Species of Fungi', rather than to that of the now growing class of professional biologists who might pass it by with little more attention than a shrug.

While Cooke's phenomenal energy, the vast numbers of fungi he had examined and reported on over the years, the world-wide reputation he had gained, and publications such as the *Handbook*, the *Illustrations* and *Grevillea*, were undoubtedly responsible for making British mycology after Berkeley a force to be reckoned with, he had done little to steer the subject away from its nineteenth century image and on into the experimentalist world of the twentieth century where, as he himself had pointed out thirteen years earlier, Continental workers had pushed far ahead. Partly because of the overwhelming volume of taxonomic work thrust upon him by mycologists all over the world, but partly too because, the circumstances of his life having conspired to make it so, he lacked both the breadth of vision and the self-confidence to pioneer a new direction in his chosen field, he could only consolidate and build on, albeit in a massive way, the foundations that the great Berkeley had laid. Again, never at any time in his career had he been allowed to express his opinions on matters of administration or policy making: indeed both at the India Museum and at Kew he was relegated to the position of a skilled technician whose business was practical science and nothing more; even his contribution to the Science Committee of the Royal Horticultural Society was only at a technical level. It is true that before he reached the age of sixty his outlook had become incurably reactionary, and that he was by then quite incapable of helping to guide mycology into a new era. But one cannot help remembering the pioneering schoolmaster of the 1850s and wondering whether it was not the treatment meted out to him by a university-educated establishment, determined to keep an outsider in his place, that was not ultimately responsible for sapping his confidence, and thus stultifying his vision and closing his mind to new ideas. If, for instance, the botanists amongst whom he worked had been, like his fellow teachers, drawn from similar backgrounds to his own; and if the gentlemen of the botanical hierarchy had welcomed his efforts to join their circle instead of snubbing him; might Cooke have felt sufficiently sure of himself to be able to divert some of his phenomenal drive from simply naming and classifying ever more fungi, to putting forward ideas and initiating projects which would be the basis for the mycology of the future?

The inhibitions which limited his contribution to mycology in no way affected his other notable service to the science of his day, namely the awakening of the lay public to a greater understanding and enjoyment of nature; for he remained throughout his life one of her most enthusiastic and energetic propagandists. Starting with the elementary schoolchildren of Lambeth and the teachers of London, he went on to spread a love of natural history first among working men and later among those from rather more affluent, though very varied, walks of life. When he himself became almost entirely preoccupied by fungi his missionary efforts did not cease, but turned from general biology to expounding mycological problems at meetings of field clubs in many parts of the country, while his popular writings spread his gospel even further. It is for his missionary work for natural history in general and for mycology in particular, as much as for his contribution to mycology as a science, that so great a debt is owed to him.

More and more now, Cooke was dependant on his daughter Leila, who still lived with him and attended him devotedly. She was in her late twenties and was employed by the Post Office as a 'governmental telegraph teacher' at the Central Telegraph Office, by her own account a very dull job. Either she worked a shift system, or her hours must have been very long, for her father told Alfred Clarke that she seldom arrived home before about nine o'clock. She was a lively and adventurous young woman who longed to lead her own life and, much as she loved her father, she found her situation trying indeed. In addition to the frustrations of his dependence upon her, her relations with her mother were strained, as she described many years later in a letter to one of Ada's daughters:

> She [Annie] used to come and see him [Mordecai] about once every week ... When I was growing up she tried hard to be friendly with me. As I grew older I realised I had been left as a toddler and I resented it and just wouldn't be friendly. I think now I was wrong. Now I am old and know more of the world I feel sorry for her ... I think I was wrong to have been so bitter and unfriendly to her, but I was young.

On arriving home from work it was Leila's first duty to read aloud to her father from the *Daily Telegraph*, the only daily or weekly paper they ever took. The one hint we have of Mordecais's political outlook at the time is a brief comment he made to Clarke in 1912:

> Sorry to hear of the distress in the North. Infatuated coal miners only punishing and starving their own class at the pleasure of a few bloated agitators–ought to be shot.

While still retaining some sympathy for the working classes, he clearly felt that it was their proper duty to accept with resignation the fate to which destiny consigned them. The *Telegraph* at this time carried a weekly article on popular science by Sir Ray Lankester (1847–1929), the son of Mordecai's former benefactor Edwin Lankester and erstwhile member of the Society of Amateur Botanists, and Mordecai looked forward eagerly to these. He remembered Sir Ray as 'a boy in a short Eton jacket frequenting Hardwicke's shop in company with his father' (J.R.), but this brilliant lad had grown up to a stormy career. In 1898 he had been appointed to the Directorship of the Natural History Museum, but had fallen out with the Trustees, and in 1907 had been asked to resign. The beautifully written articles that he now contributed to the *Telegraph* and other publications were a means of augmenting his limited pension.

As well as the daily paper, Leila would also read to her father from his favourite essayists and from the journals of learned and natural history societies which still arrived regularly. When on his own he would sometimes while away the hours by sorting through and giving away the still voluminous collections which had survived the move from Junction Road.

He was still very worried about the state of his finances and 'offered to the authorities at Kew my collection of drawings of fungi generally, numbering upwards of 6000 which they at once accepted, and I was left barren and incapable'. He asked £120 for them for, 'my end must now be getting near and I had to leave my daughter the means to put me away decent and out of sight', as he wrote to Clarke.

More and more, now that he could no longer move far from his home, he depended on correspondence to keep in touch, but fewer and fewer of his old colleagues found time to write to him. Alfred Clarke was very faithful, and Miss Georgina Decima Graham, a keen naturalist from Carlisle, was both a great support to him and a close friend of Leila's. When or how Mordecai first made Decima's acquaintance is not known, but it seems unlikely that it was through his friend Dr. Carlyle, for she would only have been about 20 when the Doctor died in 1892. However, the Carlisle Field Naturalists' Club sometimes held its field meetings at Edmund Castle, the home of the Graham family outside the city, and Mordecai may have attended some of them before Dr. Carlyle's death. Be this as it may, when Mordecai mentioned Decima in a letter to Clarke in 1909 he undoubtedly knew her well. At some time she and several of her sisters moved south, living in Kensington for part of the year and visiting the Cooke's often. Sometimes, too, Leila would travel north to stay with Decima at Edmund Castle. Decima's letters were undoubtedly a great solace to Mordecai in his frustrated old age, and he sent her books and other objects from his collections:

> I was so glad you received Vol. III all right–as I packed it off all in a hurry–because I got afraid lest I should be so laid up that I might fail to keep my promise–and I was determined that you should complete the 8 vols of the Illustrations.

He sent her, too, a copy of the *Myxomycetes*, with an account of how he came to translate it from the Polish.

Cooke's main comfort, and his 'sole prevailing vice' at this time was his pipe (Fig. 16.6). To Clarke he wrote:

> Many thanks for the 'weed'. I will sample it as soon as my stock is exhausted.
>
> I did not know that you were a votary. I fancied it was too plebeian an occupation for you–but I should have known better.
>
> In my time I have known fungus hunters who could not smoke and I never found them good for anything–they were too delicate to venture out of the beaten paths–and to turn over dead leaves offended their sense of smell–I have a vivid recollection of a Wool-

Fig. 16.6 M. C. Cooke in old age. Photograph by his daughter, Leila Cooke.

hope hunter who was an enemy of 'baccy' and would not fraternise with anyone who carried a pipe–and–oh my–wasn't he a cure.

One of the complaints Smith made to Britten about Cooke after the latter's death was that he reeked of tobacco, but as the two men met at many other forays besides the Woolhope, Smith was not necessarily referring to Cooke here.

As Cooke grew older he bothered less and less about his personal appearance. There is good evidence that in his younger days he had been smart enough, but as early as 1880 even Berkeley, the most charitable of men, commented to Hooker on Cooke's dirty fingernails, and by the time he reached old age both his descendants and his fellow mycologists agree that he let himself go completely and his clothes became shabby to a degree.

For many years he had suffered from severe and chronic 'indigestion', sometimes being forced to rest on his couch for two or three days on end, a waste of time which had worried him considerably while he was still able to work. According to Dr. Ramsbottom 'his fondness for patent medicines probably made matters worse'. This dyspepsia added to the misery of his old age but at first at least, he had a family doctor, Dr. Haines, who took a gratifying personal interest both in the patient and his work. However, Dr. Haines' early death deprived Cooke of a sympathetic friend, and to his new young doctor he was just another elderly charge.

Enforced idleness afforded one great pleasure to Mordecai; he was able to enjoy his grandchildren in a way that he had never had time to enjoy his own children, and the pleasure was reciprocal. Only Ada's daughters remember their grandfather; Willie's only child died in infancy, and Frank and Bertie did not marry until they were in their thirties, at the end of the their father's life. I have met three of Ada's four daughters and all agree that Mordecai was a wonderful grandfather. One of them said:

> my memories of him couldn't be any better, at Christmas he never forgot us. We had moved from London to Portsmouth, I was about $2\frac{1}{2}$ yrs I imagine at that time. Every Christmas a large box came with all kinds of dried fruit and nuts to make puddings. Also a small box of chocolates for each of us, we thought that wonderful, as in those days children didn't get any delicacies. Every birthday also he would remember in some way. I think he loved children. Every summer holiday we would go to London and stay with my grandfather and my aunt Miss Cooke. My grandfather would take us to the store to buy groceries for the house [Leila confirmed that he did most of the household shopping at this time] then he would buy clothes for us. I remember I used to like to sew, I must have been about 10 yrs or younger. He took me to a store and bought me a whole lot of remnants of material so I could sew all I wanted. He would take us to the Zoo. My Aunt and Uncle Frank would take us to all places of interest . . . I have only memories of a lovely kind Grandfather and also a wonderful Aunt. My grandmother I never really knew.

It must have been in about 1909 or 1910 that the childhood companionship between Mordecai and his younger brother William was resumed after many years of estrangement. William had done well at the iron-works at Chichester, soon becoming the Manager, and being proud of the part he played in the manufacture of the new gates for the Cathedral. But he had retired somewhat precipitately:

> to the evident chagrin of the Principal whose sons just released from school were admitted to the firm—without any business training—resulting in friction which caused W.C.C. to retire.

He moved to London, to Mottingham, Maida Vale, where he suffered 'severe pecuniary losses and the more bitter pains of family bereavement' (four of his six children died young). Now he began to pay weekly visits to Castle Road, and Leila remembers with amusement the two old gentlemen gossiping together for hours on end about their Norfolk childhood, the shooting, the fishing, and the 'little jokes (that) were made between them about the girls they'd had'. Leila sometimes returned these visits, riding to Maida Vale on the outside of the bus. It would have been these conversations which led William to write the history of the family, 'The Cookes of Horning' (Chap. 1), which has proved to be of such importance in the writing of this biography. William, who was two years younger than his brother, outlived him by two years.

In the same year in which William wrote 'The Cookes of Horning', Mordecai completed his autobiographical notes (Fig. 16.7). There is evidence from the small, beautifully neat handwriting of the first ten pages that he had started them some years earlier when his sight was still good, then left them for a while, to pick them up again after the cataracts had begun to take their toll; though always legible and quite remarkably tidy, from page 11 onwards the previously neatly formed letters grow larger and less precise. In his last, helpless years the books he had written assumed a tremendous importance for him:

> He would often write on slips of paper the dates and titles of his more important works, the number of illustrations, etc.–a craving for the feeling of having once done something, 'to look back with proud satisfaction at all of which we have individually been able to do in aid of the onward progress of this department of Natural Science'. (J.R.)

This craving is reflected in his notes, which to start with give a concise account of the events in his life, but towards the end become little more than an annotated list of his books. The final, tragic paragraph reads:

> And now in 1912 I can neither read nor write except very occasionally in the bright light of day, still my writing capacity is very small, and I suppose I must accept the dictum of the experts that my eyes are left behind in the micrsocope, and will never return.

One small ray of light lit up this dark period of his life when, on July 13th, 1912, in celebration of his 87th birthday the previous day, the *Morning Post* published a long interview with him*:

> ... the oldest mycologist in Europe, having been continuously writing and working on the subject for nearly half a century.... Indeed, his whole career has been one of strenuous endeavour and successful achievement'.

After summarising the highlights of that career the article continues:

---

*This article contains much inaccurate information, whether given by Cooke himself, or a distortion of his statements by the reporter cannot be known. The remark that 'his father could not read or write' was angrily repudiated in a letter to the editor by Mordecai's brother William, a copy of which is included in the 'Liber de Cookeis Horningi. (p. 3).

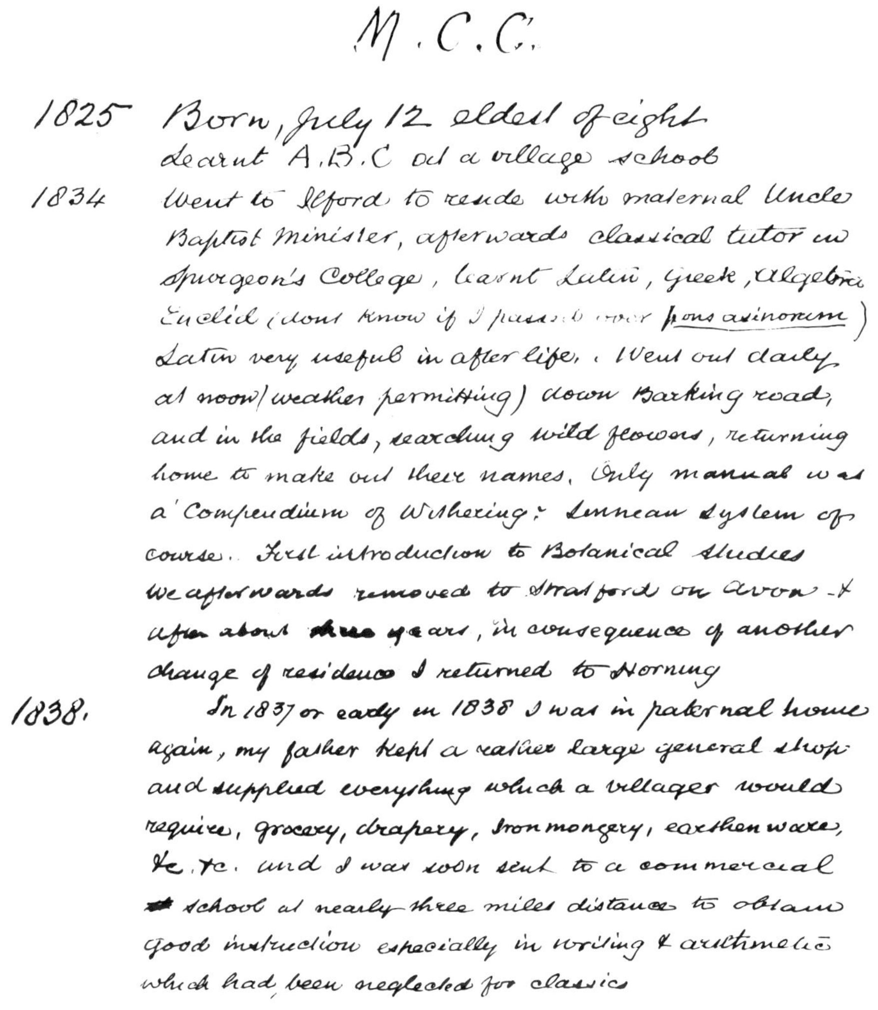

'NATURE NEVER DID BETRAY THE HEART THAT LOVED HER'

M.C.C.

1825 — Born, July 12 eldest of eight, learnt A.B.C at a village school

1834 — Went to Ilford to reside with maternal Uncle Baptist Minister, afterwards classical tutor in Spurgeon's College, learnt Latin, Greek, Algebra Euclid (don't know if I passed over *pons asinorum*) Latin very useful in after life. Went out daily at noon (weather permitting) down Barking road, and in the fields, searching wild flowers, returning home to make out their names. Only manual was a 'Compendium of Withering': Linnean system of course. First introduction to Botanical Studies. We afterwards removed to Stratford on Avon & after about three years, in consequence of another change of residence I returned to Horning

1838. — In 1837 or early in 1838 I was in paternal home again, my father kept a rather large general shop and supplied everything which a villager would require, grocery, drapery, ironmongery, earthenware, &c. &c. and I was soon sent to a commercial school at nearly three miles distance to obtain good instruction especially in writing & arithmetic which had been neglected for classics

Fig. 16.7 Part of the first page of Cooke's autobiographical notes, now at the Royal Botanic Gardens, Kew.

Those are the simple biographical facts of a simple career, but they do little more than suggest the unceasing struggle of a poor, self-taught son of the people to achieve some measure of distinction in the world of work.... 'I have never had any occupation or interest', said Mr. Cooke, 'that was not afterwards of practical value to me in my life's work. From the time I toiled at Latin in my uncle's house, or gathered flowers on the Barking-road, or taught a botanical class at Lambeth, all my interests have been useful, and all my work has been delightful.

A few months later all the pleasure he had derived from this interview turned to anger and distress, for someone,* reading the article quickly, took it to be an obituary and alerted the authorities at Kew. On October 26th a notice appeared in the *Kew Bulletin* announcing the death of Dr. M. C. Cooke, to be followed by obituaries in *the Times, The Globe* and elsewhere. In his younger days Cooke would have taken this as a good joke but now, depressed, half blind, and with death in any case in the forefront of his mind, he was shaken. On October 30th he wrote to an unknown correspondent [9], still with a flash of humour:

> My dear Sir, Permit me to warn you against falling into the error of the Kew Bulletin, The Times, D.T. and some others of announcing my *death*, which has not yet taken place–nor have I taken to my bed except at the orthodox sleeping hours 10.30 p.m. to 8.30 a.m. daily. My doctor visits me occasionally, inspects my tongue, feels my pulse, and nods his head, but has not yet declared me to be dead.
>
> This absurd report has caused very great annoyance to
>   Yours Truly

Perhaps to make amends for the pain it had inadvertently caused, the *Morning Post* published a shortened version of the same article on Cooke's 88th birthday.

One unfortunate result of the premature announcement of his decease was that the societies, both British and foreign, whose journals Cooke still received, stopped sending them. Leila, who used to read them aloud to him, told me that this upset him greatly, making him feel even more cut off, and that 'although we wrote and said he wasn't dead, it was hard to start the machinery going again'.

By 1913 Cooke had dispersed almost all his most treasured possessions–his herbarium, his original drawings and plates, his library, even his storage boxes–only one thing he still clung to, the microscope with which he had carried out his life's work. Now he parted with that too. He wrote to the Royal Horticultural Society:

> As my working microscope has been cast aside as useless to me for the past six years I have resolved to offer it for sale and I have hoped that you would see your way to recommend your worthy secretary to advise the Council to sanction its purchase for the use of Laboratory students as it was constructed for use and not for ornament, of the simplest form and with the best glasses which could be procured.

The Society paid him ten pounds for the instrument, which is still kept at its Wisley Gardens. Pasted to the stand is a sheet of paper inscribed:

> With this microscope in use for 30 years upwards of 15,000 drawings were made and

---

*On his own confession many years later, the perpetrator was Dr. J. Ramsbottom.

published, mostly from Cryptogamia, and many thousands of examinations made of specimens sent from various contries of the world, most of the results recorded in Grevillea. M. C. Cooke.

Now came a family loss which Mordecai must have felt keenly. Ebenezer was at this time 'only' 76. His pioneering days in art education were over and in recognition of his services he had just been awarded a Civil List pension. He still, however, had some private pupils, and it was while travelling between two teaching appointments that he suffered a heart attack at Southfields Station and died instantly. Sadly, the biography of this remarkable man, projected by his son Arthur, was never written, and few educationalists to-day fully appreciate that it was Ebenezer Cooke who laid the foundations of modern methods of teaching art to children.

Only a short time before this Mordecai had written to Alfred Clarke:

> I get no news of anything now a days for I hear from no-one except my own family and I can only smoke and watch the pigeons in the street. You can hardly realise what a loss is the proper use of the eyes until they fail you.

Gradually now he was becoming aware that the sort of existence he was forced by his helplessness to impose on his daughter was not fair on a young woman with her life before her, and in any case, while she was at work his days were bitterly lonely with only the maid, Florie (Florence Dunn, who witnessed his will), for company. So No. 53 Castle Road was sold, Leila moved into a flat and celebrated her independence by embarking on a course in Botany at Birkbeck College, in which she did extremely well though never completing her Degree; while Mordecai went down to Southsea to live with Ada, Caleb and their family, thinking that the sea air and the company of his four grand-daughters might remedy his ill-health and loneliness. His physical health, apart from his blindness, was remarkably good for his age, and he remained sound in mind to the end, but his loneliness was for something his grand-daughters were in no position to replace. It was not dispelled by the move. He wrote to Decima Graham in pathetically scrawled writing:

> Dear Decima, You must excuse me not writing but my sight is so bad I cannot see what I am doing–or read at all–I hope you are well as I have not heard for many months–and no one remembers me now that I am unable to remind them–I often think of you and wonder where you are and what is the hobby.
> This is only a short note but it is the best I can do.*

To be forgotten in his old age by fellow mycologists, after the long and wearying struggle which he had been through to achieve recognition, must have

---

*Original in the possession of the author.

been a particularly bitter experience for Cooke.

For a while after the move he rallied, but in August the First World War began, he could no longer sit in the sun on the sea front, his strength began to fail, and he died peacefully at the age of 89 on November 12th, 1914. As he grew weaker he had asked to see Leila, and Ada had sent for her sister by telegram, but to Leila's lasting grief, she arrived too late. To use her own expression, she had '*adored*' her father, and despite enjoying her freedom after he moved to Southsea, she had missed him badly and was desolated at not seeing him before he died.

They brought his body back to London and buried him with his wife, Sophia, and young son, Harry, in Finchley Cemetery.* Of the funeral Leila

Fig. 16.8   M. C. Cooke's gravestone at Finchley (now Islington) Cemetery.

*Now Islington Cemetery. Grave No. 7620, Section U, Block 2.

Fig. 16.9   The grave in 1974.

wrote only: 'Nobody came. The War was on'. I believe that by 'nobody' she meant no mycologists, no-one from Kew. Mordecai Cooke had lived too long: he was lonely to the end.

It was Leila who designed his tombstone, insisting on such an unusual embellishment that she had to obtain special permission from the Cemetery Company to erect it. The title pages of the two volumes of the *Handbook of British Fungi* are decorated with one of her father's drawings of a clump of *Coprinus* toadstools (Fig. 10.3), and Leila had this beautifully carved in relief on the rounded head of the stone. Underneath is Mordecai's name, 'Late of the Royal Botanic Gardens, Kew', followed by his favourite line of poetry: 'Nature never did betray the heart that loved her' (Fig 16.8). After this is inscribed the dedication to his two young sons which she presumably copied from the original stone. That is all; there is, as has been noted, no mention of Sophia, though whether Leila deliberately omitted her grandmother's name, or whether she had been kept in ignorance of a burial place which had never been marked, will never be known.

When I visited this unique and most personal grave in November, 1974, in company with a fellow mycologist, that part of the cemetery where it lies was neglected, overgrown with long grass and self-sown bushes, and must have been teeming with small fauna and flora. Nature had run wild (Fig. 16.9). I think this would have given great pleasure to the man who, some fifty years before, had railed against the taming of the countryside to make way for neat parks with well-kept lawns and flower-beds.

Postscript

In May, 1909, Mordecai Cooke had made his will, a brief and uncomplicated document in which he left his modest estate (with the exception, as we have seen, of his two medals) to be equally divided between his five surviving children. No mention is made of Annie, which would be understandable except that her grand-children say that she was still a frequent visitor at Castle Road (under the name of Mrs. Cubitt) and Mordecai certainly continued to pay her an allowance for many years after the separation. However, he may have felt that she was sufficiently well provided for, for her grandson, Herbert Stone, was by then 17 years old and could very soon be expected to support his grandmother, and Frank and Bertie would certainly see that she did not want.

By 1914 Herbert was 21 and was still living with Annie in Islington, earning his living as a barman. But he was a very sick man, suffering from that scourge of the age, tuberculosis, and on September 19th, less than two months before Mordecai's death, he followed his parents to an early grave: all three members of the little family had died while still in their twenties. Annie, now a lonely, but by no means defeated, old lady of 70 with only her sons Frank and Bertie to depend on, signed her grandson's death certificate as Annie Cooke, though ever since the marriage of her daughter Mabel in 1891 she had been calling herself by her rightful name, Annie Cubitt. She buried Herbert with his mother in Highgate Cemetery, placing his name on the gravestone below those of his parents, but leaving room for another at the bottom.

When she left Mordecai for the second time, the rift between Annie and her eldest daughter Ada must have been complete, for Ada never thereafter took her children to see their grandmother when they came to London to visit Mordecai. They remember meeting her on only one occasion, which left an indelible impression on their minds. While they were spending a holiday with their grandfather a 'frosty-faced female' came in, looked them up and down, said 'You must be Ada's girls', and walked out of the room: it was left to the maid, Florie, to tell them that this was their grandmother, Mrs. Cubitt. Though she had cut herself off from Ada's children, and Herbert Stone was dead, Annie still had one more chance of enjoying the pleasures of grandchildren, for in 1911 Bertie had married a local girl, Laura, whose two married sisters lived in Junction Road. Unfortunately, however, Annie disapproved of the union and did her best to make trouble between husband and wife. This so angered Laura that for a time she refused even to see her mother-in-law, though after the birth of her first child she relented sufficiently to allow Annie to see the infant. It seems that Annie used to cause considerable embarrassment by knitting garments for the children which were not at all to their mother's taste. As for Frank, he too married, but not until the end of the First World War, when he settled a mile or so west of Junction Road. He had no family.

No more is heard of Annie until 1920, when she was living at 37, Elthorne Road, Islington. She was now 76, and to judge by her granddaughters' impressions of her, a difficult old lady. Soon after New Year she had an accident, falling from a chair and breaking her ribs, a serious matter at her age. The inevitable complications set in, she was moved to the Islington Infirmary* with pneumonia and pleurisy, and died there three days later, her death being registered under the totally fictitious name and status of 'Annie Cubitt Cooke, widow of Mordecai Cubitt Cooke'. Presumably these were the particulars that she, or whoever accompanied her to the Hospital, gave to the authorities on her admission, so of all the names she had used during her long life, this must have been the one she finally decided to adopt. But even this was not the last, for when she was buried with the Stone family in the grave which she herself had bought, her mourners, presumably Frank and Bertie, added in the vacant space on the gravestone the name 'Anne Elizabeth Cubitt Cooke'. Below this the lettering is so eroded as to be virtually illegible, leaving one curious to know what text or quotation had apparently been inscribed at the bottom. How much Frank and Bertie knew of the facts of their mother's extraordinary life when they commemorated her under this name can never now be discovered, for even Bertie's children were ignorant of them when we met.

To complete this history, the fate of Mordecai Cooke's five surviving children and their families can be traced briefly. All Ada's daughters married, two remaining in this country and two emigrating to Canada. All have families, but none of the children have shown a special aptitude for any branch of biology. Willie continued to work as an illustrator and artist, but as photography came into ever greater commercial use he fell on hard times and had a struggle to make a living. On the outbreak of the Second World War he left London to join an artists' colony in Stroud. His only child, a daughter, died young.

Bertie and Laura had four children, a son who died in infancy, and three daughters, all of whom married, two emigrating with their husbands to Australia and one to New Zealand, after the Second World War. All three have families, but again, none of the children has become a biologist. Bertie himself died of tuberculosis in 1927, a sister-in-law from Junction Road being with him when he died, but Laura lived on and eventually joined her daughters in Australia. Within a few months of Bertie's death Frank, the only one of Mor-

---

*Now part of the Whittington Hospital.

decai's children to have kept in touch with both sides of the divided family, also died of tuberculosis.

The chances of a man who had had five sons, as Mordecai Cooke had, failing to perpetuate his family name, might seem very small, but Mordecai had flouted family superstition by failing to pass on his hated Christian name to any of his boys. So no-one in his family, or in William's or Josiah's families either, was surprised when two of his sons were killed in childhood, one was childless, one had no surviving children, only the daughters of the fifth grew to adulthood, and thus the name of Cooke died out in one generation. They already knew that it was bound to happen.

Mordecai's youngest child, Leila, had an innate initiative and longing for adventure far greater than those of her elder brothers or sister, but her ambitions had been held in check, first by her devotion to her father and then, after his death, by the limitations on travel imposed by war. As soon as the First World War was over, however, she wasted no time in indulging them. First, she left her stultifying job in the Post Office and moved up to Bolton as personnel manager in a blouse factory. Her many friends at this time included Mr. Bulley of Liverpool, an ardent botanist and alpinist whom she had met through her father; according to one of her nieces, she once walked from Bolton to Liverpool to visit him and his wife and daughter. It was probably during this period too that she travelled to Russia and back by sea with Decima Graham, whose father owned a line of steamers trading with that country.

While she was between jobs Mr. Bulley, who had business interests in Kenya, suggested that she might like to join his firm there, selling trucks, an opportunity which she accepted with alacrity. But with increasing age Mr. Bulley lost interest in the project and Leila, left to fend for herself as best she could (though Miss Bulley watched over her welfare to the end) had to find a way of earning her living, never easy for a single woman in a colonial country. She settled in the small community of Eldoret and there opened a shop selling ladies' clothing and underwear to the wives of European settlers, later taking up an agency for books as well. Through this successful venture she became well known in the rapidly growing local community, began to explore Kenya, and became a correspondent, especially on popular science, for one of the local newspapers. In the early years of the shop the Kakameyo gold-rush took place about 70 miles from Eldoret, an event of which Leila took full advantage, transporting her wares across the sparsely populated countryside in her own van to meet the demands of the sudden influx of Europeans.

In 1951, when she was nearly 70, she sold everything and returned to England, planning to find a job and remain here for good. Needless to say, her hopes were dashed; most of her relatives had died or emigrated, and without a job she could not afford to stay. While she was here, however, she visited Norwich and Kew Gardens, and made contact with the Quekett Microscopical Club. At Kew she gave a lecture on her father, and afterwards the staff there and at the neighbouring Commonwealth Mycological Institute gave her immense pleasure by presenting her with a copy of his *Plain and Easy Account of British Fungi* which they had all signed, a treasure which she left in the care of her friend Dr. Cotton (see below) who, after Leila's death, kindly gave it to me.

Having returned to Kenya, the indomitable old lady still had to keep herself, so she went back to Eldoret and opened a travel agency, the Bestway, which was at first a considerable success, but which began to fail as the political situation caused increasing numbers of Europeans to leave the country. Until she was in her mid-eighties, while the agency was still profitable, she and a friend travelled widely through the country by rough roads and tracks in their Land-Rover. Leila was in Kenya throughout the Mau Mau emergency and the fight for independence, and though she did not believe that 'the natives' were capable of governing themselves, she was in no doubt that they had just cause for complaint against British rule, recognising that Mau Mau was, in fact, a liberation movement following the encroachment of Europeans on African reserves. In 1954 she wrote to a niece:

> The Government solution is to force the native into industrialism, but from a pastoral age to a modern factory is too much. . . . Britain wants cotton mills etc. in the vicinity of the Jinja Dam. She hopes for an unlimited supply of labour cheaper than in Lancashire.

Although Leila herself held no religious beliefs, one of her closest friends at this time was Dr. Diana Cotton, a medical missionary, who has told me, among many other things, that even then, Leila always kept a photograph of her father above her bed. But Dr. Cotton, like most of Leila's friends, eventually left Kenya, and the old lady, now almost alone, began to hate Africa and long, once again to return to her native land. Sadly, it was not to be, for she was now nearly 90, there were no relatives here to look after her, and she was not sufficiently well off to settle here on her own. So she returned to Nairobi, living for a while in a home before being moved to hospital, where she stayed for three or four years, dying on September 2nd, 1977, at the age of 95. It is a pleasure to know, from her letters to Miss Bulley and Dr. Cotton, that I was able to cheer her a little during my visit the previous year, with an assurance that a biography of her father was planned, and that he was known and recognised throughout the mycological world.

Post-Postscript

I first consulted 'The Cookes of Horning in 1974, very early on in my researches into the family. In his book William Cooke states that Mordecai married Sophia Elizabeth, née Biggs, in 1846, and that she died in 1897 at the age of 75: a check on her marriage and death certificates confirmed this. The book also names Mordecai's seven children, giving the dates of birth of all except Frank Harry, and it took no great mental effort to realise that, if these dates were correct, Sophia would have been 60 when she gave birth to her youngest child, Leila. This needed investigation.

The birth certificates of all the children except Frank Harry were easily found, and it was from these that the astonishing fact emerged that none of them were Sophia's children–the mother of all was one Annie Elizabeth Thornton Biggs, who changed her surname to Cooke at the birth of her third child. Who was Annie? was she Sophia's sister? Sophia's cousin? was it a chance coincidence of a fairly common surname? There was no means of telling. It was impossible to obtain Sophia's birth certificate, as she would have been born before 1836 when the central registration of births, deaths and marriages began; and although Annie must have been born at least ten years after that date, I was unable to discover her birth certificate despite hours of searching. It was surprising in the circumstances to find, from Sophia's death certificate, that she had died in Mordecai's home. Had she been living in the house for 35 years while some relative of her's was co-habiting with Mordecai and bearing his children? The mystery was only deepened by the name of the witness to Sophia's death, Annie Cubitt, who gave an address a few hundred yards from where the Cookes lived. Surely she must be a relative on the Cooke side? but though every possibility I could think of was investigated, I could find no trace of her. Also, I wondered where Sophia was buried, for I had already visited Mordecai's grave and found no mention of her on his tombstone. It was then that I learned from the authorities that despite this, she was indeed buried with him.

For several years the whole mystery of Sophia, Annie, Frank Harry and Annie Cubitt lay dormant. I believe that Leila Cooke knew the whole story and deliberately kept me in the dark, when I visited her in Nairobi in 1976, but it was out of the question to press her in any way. However, from 1978 onwards I began to get in touch with Mordecai's grand-daughters in Canada,

Australia and England, all of whom knew Annie as their grandmother. Some certainly knew absolutely nothing of their complex family history; others may have known part or all of the story, but did not divulge it to me. However, it was one of them who unwittingly provided the next clue. It was a Canadian granddaughter who showed me a letter from her aunt, Leila Cooke, written some years previously in answer to a query about the family history. This letter mentioned two children not listed in 'The Cookes of Horning'–Mabel, who was said to have died at 16, and Priscilla, born after Leila, who died as a baby. I had always remarked on the fact that Annie's first four children had been born at two-yearly intervals from 1862 to 1868, and that then there was an eleven year gap until 1879 when Herbert Stephen was born, to be followed in 1882 by Leila. I had assumed that Frank Harry, whose birth certificate was missing, had been born during that gap and had somehow never been registered, and that perhaps there had been some stillbirths during this period. Now I searched for Mabel's birth certificate during the gap, but failed to find it, neither could I trace it to any other possible date. Nevertheless, it was Mabel who would eventually lead me to the solution of the mystery. I could not find then, and have never found since, any trace of Priscilla. Once again my investigations came to a standstill.

Leila Cooke told me, during my visit to Nairobi in 1976, that a few years after her own birth Annie had left Mordecai and, while always keeping in touch with him, had never gone back to him permanently. She had outlived Mordecai, and because of the estrangement was not buried with him. On my return from Kenya, no amount of searching for Annie's death certificate around the dates suggested by Leila yielded success. But in 1978 one of Mordecai's Australian grand-daughters provided the information that at last led to the unravelling of the whole complicated story.

This grand-daughter told me that Annie had died shortly after the First World War and had been buried in Highgate Cemetery. With the field of search greatly reduced I soon found a death certificate in the name of 'Annie Cubitt Cooke'. The change of name was surprising, but there was no doubt of the identity of the deceased, for she was registered as the 'widow' of M. C. Cooke. Now that the date of her death and the name under which she had been buried were at last known, perhaps it would be possible to find Annie's grave. The authorities at Highgate Cemetery were able to confirm that she was buried there, and they also told me that she shared her grave with George Henry and Mabel Katherine Stone, who were much younger than her, and their son, Herbert George, names that were completely new to me. This surprising news seemed quite inexplicable. Even if Annie had been living with some young people who had grown fond of her, surely she had not become so much a part of the family as to be buried with them? Then the name 'Mabel' struck me. Could Mabel Stone be Mabel, the daughter of Mordecai and Annie, mentioned by Leila Cooke, but not by her Uncle William in his history of the family? Had Annie gone to live with a married daughter? Obviously

Mabel's marriage certificate must be found, as that would tell me her maiden name. She had died so young that the date of her marriage could be placed within very narrow limits, and the search was not a long one. Mabel's name had not been Cooke, but Cubitt, her father being one John Quincey Cubitt; and the marriage to George Stone had been witnessed by M. C. Cooke and Annie Cubitt. Clearly my guess was along the right lines in that it was all in the family, and here, presumably, was the Annie Cubitt whom I had already met as a witness to Sophia's death. But I still did not know who she was. Mabel's birth certificate supplied the answer. Her mother–John Cubitt's wife–was none other than Annie Elizabeth Thornton, née Biggs, who was already the mother, at the time of her marriage, of four of Mordecai's children. Mabel's birth filled in part of the eleven year gap between the births of Mordecai's children, Ernest Frederick and Herbert Stephen.

The discovery of the exact date of Annie's death, and of her age at the time, meant that at last there was a real chance of finding her birth certificate, and hence of discovering who she really was. The narrow time span which now had to be dealt with meant that the search could be much more thorough than on previous occasions, and eventually the birth was found–not under Annie Elizabeth Thornton Biggs, but with the first two names reversed, a fatal difference when one is working on the assumption of alphabetical order. It was this certificate that yielded the most astonishing information of all that was to be unearthed in the whole of this extraordinary family saga. Annie was Sophia's illegitimate daughter by a clerk, David Thornton, and was two years old at the time of Sophia's marriage to Mordecai. Mordecai's children were his wife's grandchildren!

Only one problem now remained–Frank's missing birth certificate. Could he, like Mabel, be John Cubitt's child rather than Mordecai's? I searched the registers under the name of Cubitt in the years between the births of Mabel and of Herbert Stephen, and duly found Frank Harry. But I was in for one last surprise. No father was named on the certificate: Frank was illegitimate, and so took his mother's surname. Only later events suggested that Mordecai was his father.

Appendix A

Chronological Table of the Life and Work of Mordecai Cubitt Cooke.

1825. Born at Horning, Norfolk, on 12th July.
1844. Left Norfolk for London.
1846. Married to Sophia Elizabeth Biggs.
1848. (about) Interest in fungi first aroused by Richard Ward.
1851–9. Teacher at Holy Trinity School, Lambeth.
1861. Cataloguing Indian and New Zealand exhibits for 1862 International Exhibition.

Family and personal matters.	Activities, books and articles	
	Mycology	Natural History
1862	Assistant at the India Museum, Whitehall.	
1862. First child, Harry Linnaeus, born to Annie Biggs.	Start of correspondence with M. J. Berkeley (till 1889). 'Plain and Easy' *British Fungi*.	Founds Soc. of Amateur Botanists (1862–64). Introduced to Robert Hardwicke. *Manual of Structural Botany*.
1863.	First articles for *J. Botany* (to 1872). *Index Fungorum Brittannicorum*.	Activities as above.
1864. Ada Gordon born. Move to 6, Springfield Ter.	Articles for *Popular Science Review* (1864–9).	Activities as above.
1865.	Non-resident Fellow of Royal. Soc. Edinburgh. Publication of *Exsiccati* (1865–72). *Rust, Smut, Mildew and Mould*.	Founds Quekett Microscopical Club. Co-founds *Science Gossip*; editor and contributor 1865–1871. *British Hepaticae*. *Our Reptiles*
1866. Willie Cubitt born	Activities as above.	QMC excusions; Fibre Committee. Articles for *J.Q.M.C*. Other activities as above.

APPENDIX A

Family and personal matters.	Activities, books and articles	
	Mycology	Natural History
1867. Promotion to Reporter on Indian Products.		To Paris for International Exhibition.
1867.	Activities as above.	Portfolios on mites and spiders. *Fern Book for Everybody.* Other activities as above.
1868. Ernest Frederick born. Move to 2, Junction Villas.	First American fungi received. Other activities as above.	Corr. Member Lyceum of Natural History, New York and Portland Nat. Hist. Soc.. Vice President Q.M.C.. *Science Gossip.*
1869.	Collating material for *Handbook*. First paper on American fungi. *Exsiccati* etc.	Lectures for Q.M.C. *Science Gossip.* *1000 Objects for the Microscope.*
1870. Move to 146, Junction Rd.	First attends Woolhope forays. Writing *Handbook*. Arranging Edinburgh herbarium.	Activities as above.
1871. Annie marries John Cubitt. Their child Mabel born.	*Handbook of British Fungi.*	Quarrel with Hardwicke. Ceases to edit *Science Gossip*. Editor J.Q.M.C..
1872.	Articles for *Pharmaceutical Journal*.	
1872.	Start of the 'maze of correspondence', world wide. Founds and edits *Grevillea*.	Ceases to edit *J.Q.M.C.* but still active in Q.M.C.
1873.	To Vienna for International Exhibition.	
1873 Harry killed.	M.A. (Yale). Arranging Nat. Hist. Museum herbarium. 12 papers for *Grevillea*. Correspondence (for 10 yrs).	Q.M.C.

Family and personal matters.	Activities, books and articles	
	Mycology	Natural History
1874.	Promotion to Assistant Curator. *Report on Gums etc.*	
1874.	Hon. Member Woolhope Club. Nat. Hist. Museum herbarium. Catalogue of Discomycetes for Nat. Hist. Museum. Correspondence. *Grevillea.*	Hon. Member Norfolk and Norwich Naturalists. Q.M.C.
1875. Frank Harry born. Annie returns bringing Mabel and Frank.	Corr. Member Acad. Nat. Sciences, Pa. Hon. Member Cryptogamic Soc. Scotland. Massive correspondence with U.S. Start of correspondence with Saccardo. *Grevillea. Exsiccati.* *Mycographia.* *Fungi, Their Nature, Influence and Uses.*	Q.M.C.
1876.	*Report on Oil Seeds etc.*	
1878.	R. Horticultural Soc. Scientific Cttee. Correspondence. *Grevillea.* *Mycographia. Exsiccati.*	Q.M.C.
1877.	Associate Linean Soc. RHS Science Cttee. Correspondence. *Grevillea.* *Mycographia. British Mycetozoa.*	Q.M.C.
1878.	Corr. Member Soc. Crittogamologica Italiana. Paris foray. Ravenel's *Exsiccati.* *Clavis Hymenomycetum.* Other activities as above.	Q.M.C. *Songs for Queketters.*
1879. Herbert Stephen born.	Activities as above, but end of *Mycographia.*	Q.M.C. *The Woodlands.*
1880.	Secondment to Kew 3 days a week	

APPENDIX A

Family and personal matters.	Activities, books and articles	
	Mycology	Natural History
1880 .	Activities as above. Start of Farlow correspondence. *Report on Salmon Disease.*	Hon. Member Hackney Nat. Hist. Soc.. Vice-President Q.M.C.. *Ponds and Ditches.*
1881. Paralytic illness.	Activities as above. Start of *Illustrations of British Fungi.*	Hon. Member Essex Field Club. Vice-President Q.M.C. President Hackney Nat. Hist. Soc. *Freaks and Marvels . . .*
1882. Weak from illness. Leila Annie born.	Activities as above, including *Illustrations.* Start of Australian Correspondence. End of Ravenel's *Exsiccati.*	Hon. Member Herts. Nat. Hist. Soc.. Vice-Pres. Q.M.C.. President Hackney Nat. Hist. Soc.. *British Fresh-Water Algae.*
1883.	Activities as above. *Fungi Australiani.*	President Q.M.C.. President Hackney N.H. Soc.. *British Fresh-Water Algae.*
1884.	Appointment at Kew renewed. Activites as above.	As above. 'Flament Pomonalis'
1885. Ernest killed. Primrose League Medal and Star.	Herbarium sold to Kew. Activities as above.	Vice-Pres. Q.M.C.. Articles for *J.Q.M.C..*
1886. Primrose League Bar.	Activites as above, plus forays.	Pres. Hackney Nat. Hist. Soc.. Resignation from Q.M.C. Committee.
1887. Primrose League Bar.	Activities as above, plus forays. 11 papers in *Grevillea*	Pres. Hackney Nat. Hist. Soc.. *British Desmids.*
1888. Primrose League Bar.	Activities as above, plus forays. 15 papers in *Grevillea.*	Pres. Hackney Nat. Hist. Soc.
1889. Approx. date at which Annie left M.C.C. permanently.	Activities as above, plus forays. 21 papers in *Grevillea.*	Pres. Hackney Nat. Hist. Soc.. *Toilers in the Sea.*

Family and personal matters.	Activities, books and articles	
	Mycology	Natural History
1890.	Activities as above, plus forays. 17 papers in *Grevillea*.	Hon. Member Glasgow Nat. Hist. Soc..
1891.	Activities as above. *British Edible Fungi*. *Handbook of Australian Fungi*. *Synopsis Pyrenomycetum*.	
1892.	Activities as above. End of *Illustrations*. Sale of *Grevillea*. *Vegetable Wasps and Plant Worms*.	
	Retirement.	
1893.		*Romance of Low Life*. *1000 Objects* (new edition).
1894.	*Edible and Poisonous Mushrooms*.	*British Hepaticae*.
1895.	*Introduction to the Study of Fungi*.	*British Hepaticae*. *Rambles* childrens' series. Hon. Member Worcs. Nat. Hist. Soc..
1896.	Founder Member British Mycological Society.	Wall diagrams for schools.
1897. Death of Sophia.		Object Lesson Handbooks.
1898. Move to 53, Castle Rd. with daughter Leila.	Sale of Library.	
1902. 1903. 1906. 1909. 1911. 1913. 1914.	Victoria Medal of Honour. Linnean Society Gold Medal. *Fungoid Diseases of Cultivated Plants*. *Catalogue of Basidiomycetes*. Sale of fungal drawings to Kew. Sale of microscope to R. Hort. Soc. Died at Southsea, Hants. on 12th November.	

Appendix B

List of Works by M. C. Cooke in Chronological order of their Appearance

1860.
The Seven Sisters of Sleep. pp. xv. 371. James Blackwood.

1861.
A Manual of Structural Botany. pp. iv. 123. Robert Hardwicke. New editions and reprints, other publishers: 1876, 1883, 1884, 1889, 1893, 1908.

1862.
A Manual of Botanic Terms. pp. iv. 90. Robert Hardwicke: New editions and reprints: 1865, 1873, 1875. Other publishers: 1884, 1885.
A Plain and Easy Account of British Fungi. pp. vi. 148. Robert Hardwicke. Further editions: Hardwicke and Bogue, 1876, 1878: W. H. Allen, 1884, 1898.

1863.
Index Fungorum Britannicorum. pp. 58. Robert Hardwicke.

1865.
Our Reptiles. pp. vii. 199. Robert Hardwicke. Another edition: *Our Reptiles and Batrachians*, 1893, W. H. Allen.
Rust, Smut, Mildew and Mould. pp. vii. 238. Robert Hardwicke. Further editions: 1870, 1872. Revised and enlarged edition (pp. 262) Hardwicke and Bogue, 1878. Further editions W. H. Allen; 1886, 1898. Reprinted 1902.

1867.
A Fern Book for Everybody. pp. 124. Frederick Warne. New editions, 1881, 1889.

1869.
One Thousand Objects for the Microscope. pp. iv. 123. Frederick Warne. Enlarged edition (pp. x. 179): 1900.

1871.
Handbook of British Fungi. 2 vol. pp. 981. Macmillan. Second and revised edition (not completed) pp. 398, 1883–91.

1874.
Report on the Gums, Resins, Oleo-Resins and Resinous Products in the India Museum. pp. iv. 152. India Office–India Museum.

1875.
Fungi: Their Nature, Influence and Uses. Edited by M. J. Berkeley. pp. xii. 209. Kegan Paul and Trench. (International Scientific Series, vol. 14). Further editions: 1881, 1888, 1894. French edition, 1875. pp. 274. Further editions: 1878, 1882, 1892.

1876.
Report . . . on the Oil Seeds and Oils in the India Museum. India Office–India Museum.

1877.
The Myxomycetes of Great Britain. Translated from the Polish (J. T. Rostafinski). pp. iv. 96. Williams and Norgate.

1878.
Clavis Synoptica Hymenomycetum Europaeorum. With L. Quelet. pp. 240. Hardwicke and Bogue. Reprinted 1887.

1879. (1875–1879)
Mycographia, seu Icones Fungorum. Vol. I, Discomycetes, part i. pp. 267. Williams and Norgate. No more published.
The Woodlands. (Natural History Rambles). pp. 288. Society for Promoting Christian Knowledge. Another edition: 1885.

1880.
Ponds and Ditches. (Natural History Rambles). pp. 254. S.P.C.K. New editions: 1892, 1897.

1881.
Freaks and Marvels of Plant Life. pp. viii. 463. S.P.C.K.

1881-1891.
Illustrations of British Fungi. 8 vols. pp. 20. pl. 1198. pp. 20. Williams and Norgate.

1882–1884.
British Fresh-Water Algae. 2 vols. pp. viii. 329. Williams and Norgate.

1883.
Fungi Australiani. pp. 72. Melbourne. (Reprinted from *Grevillea*).

1887 (1886–1887).
British Desmids. pp. xiv. 205. Williams and Norgate.

1889.
Toilers in the Sea. pp. viii. 373. S.P.C.K.

1890.
Introduction to Fresh-Water Algae. pp. vi. 334. Kegan Paul, Trench and Trübner. (International Scientific Series, Vol. 69).

1891.
British Edible Fungi. pp. 237. Kegan Paul, Tench, Trübner.

1892.
Handbook of Australian Fungi. pp. 458. Williams and Norgate. *Vegetable Vegetable Wasps and Plant Worms*. pp. v. 364. S.P.C.K.

1893.
Handbook of British Hepaticae. pp. vii. 310. W. H. Allen. Reprinted 1907, Edinburgh: John Grant.
Romance of Low Life amongst Plants. pp. vii. 320. S.P.C.K.

1894.
Edible and Poisonous Mushrooms. pp. 126. S.P.C.K.

1895.
Across the Common after Wild Flowers. By 'Uncle Matt'. pp. 98. T. Nelson and Sons.
Around a Cornfield in a Ramble after Wilf Flowers. By 'Uncle Matt'. pp. 98. T. Nelson and Sons.
Down the Lane and Back in Search of Wild Flowers. By 'Uncle Matt'. pp. 114. T. Nelson and Sons.
Through the Copse. By 'Uncle Matt'. pp. 106. T. Nelson and Sons.
A Stroll on a Marsh in Search of Wild Flowers. By 'Uncle Matt'. pp. 94. T. Nelson and Sons.
Introduction to the Study of Fungi. pp. x. 360. Adam and Charles Black.

1897, 1898.
Object Lesson Handbooks; to accompany the Royal Portfolio of Pictures and Diagrams. Plant Life. 5 parts. T. Nelson and Sons.

1906.
Fungoid Pests of Cultivated Plants. pp. xv. 278. Spottiswoode and Co. (Reprinted from the Journal of the Royal Horticultural Society).

1909.
Catalogue and Field-Book of British Basidiomycetes up to and Inclusive of the year 1908. pp. 98. London.

Periodical publications edited by M. C. Cooke.

1863–1871.
Hardwicke's Science Gossip, vols. 1–8. Robert Hardwicke.

1869–1873
Journal of the Quekett Microscopical Club, vols. 2, 3. Robert Hardwicke.

1872–1890.
Grevillea, vols. 1–20. Williams and Norgate.

Journals in Which M. C. Cooke Published Articles and Communications.

Below is a list of the periodicals, journals and transactions in which M. C. Cooke published, together with the date span over which his articles appeared in each. They are arranged in chronological order of the appearance of the first article. Those in which few or none of his contributions are concerned with fungi are marked*. The list is as complete as the author has been able to make it, but does not claim to be exhaustive.

The School and the Teacher*, vols. 1–7, 1854–1860.
Journal of the Royal Society of Arts*, vols. 6–8, 1858–1860.
The Technologist*, vols. 2–3, 1861–1862.
Journal of Botany, vols. 1–4, 1863–1866; vols. 8–10, 1870–1872; vol. 21, 1883; vol. 36, 1898.
Popular Science Review, vols. 3–8, 1864–1869; vols. 14–15, 1875–1876.
Hardwicke's Science Gossip*, vols. 1–7, 1865–1871.
Journal of the Quekett Microscopical Club*, Ser. 1, vols. 1–6, Ser. 2, vols. 1–2, 1865–1886; Ser. 2, vol. 11, 1898–1900.
Proceedings of the Portland Society for Natural History, vol. 1, 1869.
Pharmaceutical Journal*, Ser. 3, vols. 1–3, 1870–1873.
Land and Water, vol. 12, 1871.
Nature, vol. 5, 1871.
Grevillea, vols. 1–22, 1872–1894.
Bulletin of the Society of Natural History, Buffalo, vol. 2, 1875.
Hedwigia, vols. 14–17, 1875–1878.
Transactions and Proceedings of the Royal Botanical Society of Edinburgh, vol. 12, 1875.
Transactions of the Woolhope Naturalists Field Club, 1876–1892.
Annals of the Lyceum of Natural History, New York, vols. 1–2, 1877–1878.
Bulletin du Société Botanique de France, vol. 24, 1877.
Field and Forest, vol. 2, 1877.
Journal of the Linnean Society, vols. 14–20, 1877–1884.
Proceedings of the Natural History Society of Philadelphia, vol. II, 1877.
Gardener's Chronicle, vol. 10, 1878; Ser. 3, vols. 9–10, 1891–1892.
Journal of the Royal Horticultural Society, New Ser., vols. 5–6, 1878–1879; vols. 26–34, 1902–1910.
Nuovo Giornale Botanico Italiano, vol. 10, 1878.
Transactions of the Botanical Society of Edinburgh, vol. 8, 1878.
Journal of the Royal Microscopical Society, vol. 3, 1880.
Reports of the Hackney Microscopical and Natural History Society*, Reports 5–11, 1881–1888.
Transactions of the Epping Forest and County of Essex Field Club, vol. 2, 1881.
Proceedings of the Royal Society of Edinburgh, vol. II, 1882.

Bulletin of the Californian Acadamy of Sciences, vol. I, 1884.
Journal of Mycology, vol. I, 1885.
Essex Naturalist, vols. 1–14, 1887–1905; vol. 18, 1909.
Victorian Naturalist, vol. 9, 1893.
Transactions of the British Mycological Society, vol. 3, 1908.
*The Naturalist**, 1909.

References and Bibliography

Certain sources which have been drawn on very frequently during the preparation of this biography are listed below, together with the abbreviations used for them both in lists of references and in the text. Sources supplying the general background information for each chapter are grouped together at the head of the list of specific references for that chapter.

Main sources and abbreviations.

Aut.   Cooke's autobiographical notes. Manuscript at the Royal Botanic Gardens, Kew.

CEB   Correspondence of C. E. Broome. In the Botany Library, British Museum (Natural History).

CH   'The Cookes of Horning' (1913). Manuscript volume by W. C. Cooke of which there is a photocopy in Norwich Central Library.

LAC   Miss Leila Cooke's memories of her father. Typescript notes deposited at the Royal Botanic Gardens, Kew, the Quekett Microscopical Club, and with members of the family. Also a transcript of the author's conversation with Miss Cooke in Nairobi in 1976.

MJB   Correspondence of the Rev. M. J. Berkeley, in the Botany Library, British Museum (Natural History).

JQMC   *Journal of the Quekett Microscopical Club.*

JR   Ramsbottom, J. (1916). Mordecai Cubitt Cooke. 1825–1914. *Transactions of the British Mycological Society*, 5, 169.

PAS   Cooke's letters to P. A. Saccardo, at the Institute of Botany, University of Padua, Italy.

SAB   'Proceedings of the Society of Amateur Botanists', 1863 and 1864. Manuscript in the Botany Library, British Museum (Natural History).

SG   *Hardwicke's Science Gossip.*

Tol.   Collection of Cooke's letters and manuscripts at the Tolson Museum, Huddersfield.

WGF   Cooke's letters to W. G. Farlow and others, 1876–1905, at the Farlow Reference Library, Harvard University, U.S.A.

REFERENCES AND BIBLIOGRAPHY

Chapter 2. Cookes and Cubitts.

General sources: Aut; CH; JR; LAC.

1. Neatishead Chapel Register.
2. Poll books for the Eastern Division or Norfolk, 1835 and 1841.
3. Cooke, M. C. (1865a). *Our Reptiles*, preface.
4. Cooke, M. C. (1865b). *SG, I*, 26.
5. Cooke, M. C. (1879). *Natural History Rambles. The Woodlands*, 177.
6. Cooke, M. C. (1866). Puff-balls. *SG, 2*, 270.
7. Cooke, M. C. (1893). *Romance of Low Life amongst Plants*, 274.
8. Anon. (1912). Mr. Mordecai C. Cooke. *Morning Post*, July 13.
9. Cooke, M. C. (1880). *Natural History Rambles. Ponds and Ditches*, 122.
10. M. J. C. (1864). Memoir of the Rev. James Cubitt. *Baptist Messenger*, 23.
11. Morris, F. O. (1855). *A History of British Birds*. London: Groombridge, Vol. 4.

Chapter 3.   Holy Trinity School, Lambeth.

General sources: Aut., JR.
Sturt, Mary. *The Education of the People*. (1967). Routledge and Kegan Paul.
Silver, P. and Silver, H. (1974). *The Education of the Poor. The History of a National School, 1824–1974*. London: Routledge & Kegan Paul.
Butterworth, H. (1968). *The Science and Arts Department, 1853–1890*. Ph.D. Thesis. Sheffield University.
The School and the Teacher (1854–1860).

1. Longmate, M. (1966). *King Cholera, the Biography of a Disease*. London. p. 171.
2. Gillman, A. W. (1895). *The Gillmans of Highgate*. London.
3. *Minutes of Council on Education*. Inspectors' Reports, 1853–4, II.
4. *Sixth and Seventh Reports of the Science and Arts Department*, 1859 and 1860.

Chapter 4.   Out of School-Hours.

General sources: Aut.
Richardson, A. E. (1971). *Ebenezer Cooke, 1873–1913: A Pioneer of Art Education*. Thesis. University of Newcastle-upon-Tyne, Institute of Education.
The School and the Teacher, (1854–1860).
Muspratt, E. Personal Communication. Information on the Working Tailors' Association.

1. Pestalozzi, J. H. (1894). *How Gertrude Teaches her Children*. Translation, A. E. Holland and F. C. Turner. Edited Ebenezer Cooke. Introduction by Ebenezer Cooke.
2. Cooke, Ebenezer (1905). In 'Collegiana'. *Working Men's College Journal, 9*, 133
3. Anon. (1851). *Christian Socialist, 2*, 300.
4. Cook, E. T. (1905). *Life of John Ruskin*, Vol. 1. Chap. 19.

5. Anon. (1895). *Men and Women of the Time*. Simmond-Lund, Peter.
6. Anon. (1897). *Atheneum*, ii, 493.
7. Cooke, M. C. (1911). Letter to A. Clarke dated 8th July. Tol.

Chapter 5.   Interregnum.

General sources: Aut., JR.
   Anon. (1892). Edwin Lankester (1814–1874). *Dictionary of National Biography*, 578.
   Anon. (1874). Edwin Lankester, M. D., M.R.C.P., F.R.S. *Lancet*, November 7, 676.
   English, M. P. (1985). Robert Hardwicke, publisher of medical and scientific books. *Archives of Natural History, 13,* 25.
   Ramsbottom, J. (1917). Alfred Grugeon (1826–1913). *Journal of Botany, 55,* 193.

1. Anon. (1895). A pioneer of popular education. *Richmond and Twickenham Times*, February 23.
2. Anon. (1860). Twickenham Economic Museum. *The School and the Teacher, 7,* 266.
3. Kapp, Y. (1972). *Eleanor Marx*. Vol. 1. London.
4. Anon. (1863). *Popular Science Review, 2,* 258.
5. Cooke, M. C. (1869). *One Thousand Objects for the Microscope*, I.
6. Anon. (1861). (Report of General Meeting). *Working Men's College Magazine, 3,* 173.
7. Ramsbottom, J. (1928–1932). The Society of Amateur Botanists and the Quekett Microscopical Club. *J.Q.M.C.*, Ser. 2, *16,* 215.
8. Anon. (1913). Alfred Grugeon. *Working Men's College Journal, 8,* 49.
9. SAB.

Chapter 6.   The India Museum. Family Matters.

General sources: Aut; JR; LAC;
   Archival material at the India Office Library and Records, 197, Blackfriars Road, London.
   Desmond, R. (1982). *The India Museum, 1801–1897.* H.M.S.O.

1. Anon. (1858). A Visit to the India Museum. *Leisure Hour*, 469.
2. Anon. (1861). *Illustrated London News*, August 3.
3. Anon. (1871–1880). Museums. *Kew Reports*, 56.
4. Cooke, M. C. (1870–1871). Jew's Ear. *Pharmaceutical Journal*, Ser. 3, *1,* 681.
5. Cooke, M. C. (1911). Letter to A. Clarke, July 8. Tol.
6. Cooke, M. C. (1867). A voice from Paris. *SG, 3,* 97, 121.
7. Anon. (1821). *Norwich Mercury*, November 3.
8. Smith, W. G. (1915). Letter to J. Britten dated January 14. Box 4, General Archives, Botany Library, British Museum (Natural History).
9. Watson, F. J. (1873). *Vienna Universal Exhibition, 1873. A Classified & Descriptive Catalogue of the India Department*. London: W. H. Allen. Preface.
10. Cooke, M. C. and Berkeley, M. J. (1874). *Fungi, Their Nature, Influence and Uses*, 84.
11. Cooke, M. C. (1879). *Natural History Rambles. The Woodlands,* 66.

REFERENCES AND BIBLIOGRAPHY

Chapter 7. Lovers of Nature.

General Sources: Aut.; SAB.
    Allen, D. E. (1975), *The Naturalist in Britain. A Social History.* London: Allen Lane.
    Cooke, M. C. (1898–1900). Early Memories of the Q.M.C. *J.Q.M.C.*, Ser. 2, 7, 229.
    Dyer, J. (1978). *Worthington Smith and other Studies.* Bedfordshire Historical Records Society, Vol. 57.
    Ramsbottom, J. (1928–1932). The Society of Amateur Botanists and the Quekett Microscopical Club. *J.Q.M.C.*, Ser. 2, 16, 215.

1. SAB (1863).
2. Cooke, M. C. (1898–1900). As above.
3. Anon (1864). *Journal of Botany*, 2, 287.
4. Smith, W. G. (1914). The late Dr. M. C. Cooke. *Gardener's Chronicle*, November 28, 356.
5. Smith, W. G. (1915). Letter to J. Britten (see Chap. 6, ref. 8).
6. Cooke, M. C. (1911). Letter, July 22. Tol.
7. Large, E. C. (1940). *The Advance of the Fungi.* London: Cape, 162.
8. Smith, W. G. (1867). Toads and Frogs. *S.G.*, 3, 234.
9. Cooke, M. C. (1865). Toadstools. *S.G.*, 1, 225.
10. Editorial (M. C. Cooke) (1866). Science-Gossip. *S.G.*, 2, 1.
11. English, M. P. (1977). William Tilbury Fox and dermatological Mycology. *British Journal of Dermatology*, 97, 573.
12. Seeman, B. (undated). Letter to M. C. Cooke. Kew Archives, Letters to M. C. Cooke, COO, no. 28.
13. Cooke, M. C. (1895). *Handbook of British Hepaticae*, v.
14. Maddox, J. Quoted in Roos, D. A. (see below).
15. Roos, D. A. (1981). Aims and intentions of *Nature. Annals of the New York Acadamy of Sciences*, 360, 159.
16. Turner, G. L'E. (1978). Dr. M. C. Cooke's microscope designed by W. Moginie. *Microscopy*, 33, 339.

Chapter 8. The Quekett Microscopical Club.

General sources: *J.Q.M.C.*, vols. 1–6 (1866–1880) contain full reports of business and scientific meetings of the Club.
    Minutes of the Committee of the Quekett Microscopical Club, August 4, 1865–June 2, 1880. In the archives of the Club, c/o British Museum (Natural History).
    Ramsbottom, J. (1928–1932). See Chapter 7.

1. Allen, D. E. (1967). *The Victorian Fern Craze.* London: Hutchinson.
2. Cooke, M. C. (1879). *Natural History Rambles. The Woodlands*, 186.
3. English, M. P. (1977). See Chap. 7, ref. 11.
4. Cooke, M. C. (1866–1869). On the hairs of Indian Bats. *J.Q.M.C.*, I, 33.
5. Cooke, M. C. (1878). *Songs written for the Excursionists' Annual Dinners*, Q.M.C. Pamphlet, London: Keating.
6. Burgess, G.H.O. (1967). *The Curious World of Frank Buckland.* London: Barber.

7. Smith, W. G. (1914). See Chap. 7, ref. 4.
8. Cooke, M. C. (1886–1889). Address. *Transactions of the Woolhope Naturalists' Field Club*, 62.
9. Cooke, M. C. (undated). Old Foragers. Tol.
10. Dyer, J. (1978). See Chapter 7, General sources.
11. Anon. (1875). *Publisher's Circular*, March 16, 202.
12. Huxley, T. H. (1870). On the relations of *Penicillium, Torula* and *Bacterium. Quarterly Journal of Microscopical Science*, New ser., *10*, 355.
13. Anon. (1883–1892). Saturday September 29th, 1883. Annual Cryptogamic and Botanical Meeting. *Journal of the Proceedings of the Essex Field Club*, *4*, xliii.
14. Cooke, M. C. (1880). Observations on the salmon disease. *Salmon Disease*, Inspector's Report, H.M.S.O. [c. 2660] XIV, London. Appendix 2, 103.
15. Large, E. C. (1940). *The Advance of the Fungi*. London: Cape. 175.

Chapter 9. The Making of a Mycologist.

General Source: Aut.

1. G.M. (1889). The Rev. M. J. Berkeley, M.A., F.R.S. *Journal of Botany*, *27*, 305.
2. Cooke, M.C. (1889–1890). The Rev. M.J. Berkeley. *Grevillea*, *18*, 17.
3. Darwin, C. (1859). *The Origin of Species*. Penguin edition, 1978, 101.
4. Cooke, M.C. (1864). S.A.B.
5. Editorial (1865). Splitters and Lumpers. *S.G.*, *1*, 73.
6. Broome, C. E. (1866). Letter dated May 8. MJB.
7. Large, E. C. (1940). As Chap. 7, ref. 7, p. 285.
8. Cooke, M. C. (1865). *Rust, Smut, Mildew and Mould*, 182.
9. Cooke, M. C. (1892). Fungi, past, present and future. *Transactions of the Woolhope Naturalists' Field Club*, 360.
10. Hottes, C. F. (1940). Personal recollections of Thomas J. Burrill and his work. *Illinois Alumni News*, 6.
11. Cooke, M. C. (1978). Letter 7, WGF.
12 British Museum, Bloomsbury. Minutes of Committee Book 30, p. 10, 494.
13. Turner, G. L'E. (1978). See Chapter 7, ref. 16.
14. Cooke, M. C. (1884–1886). On some remarkable moulds. *J.Q.M.C.*, Ser. 2, *2*, 138.

Chapter 10. A Maze of Correspondence.

General sources: Aut; CEB.
Fletcher, H.R. (1969). *The Story of the Royal Horticultural Society*. O.U.P.
Information on American mycologists was obtained mainly from the following:
Haines, J. H. (1978). Charles Horton Peck. *McIlvainea*, *3*, 3.
Hesler, L. R. (1975). *Biographical Sketches of Deceased North American Mycologists*. University of Tennessee, unpublished.
Lloyd, C. G. (1913–1916). *Mycological Writings*, *4*. Cincinnati.
Rogers, D. P. (1977). *A Brief History of Mycology in North America*. Publication of the Second International Mycological Congress.
Stevenson, J. A. (1971). An Account of fungus exsiccati coming from the Americas. *Nova Hedwigia*, *36*, 1.
All Cooke's seventy papers on American fungi, up to and including 1886, are listed in:
Farlow, W. G. and Trelease, W. (1887). List of works on North American Fungi. *Harvard University Bulletin*, *4*, 449.

1. Ravenel, H. W. (1867). Letter in the Archives of Clemson University, South Carolina, U.S.A.
2. Anon. (1866–1869). June 26th, 1868–Ordinary Meeting. *J.Q.M.C.*, *1*, 113.
3. Cooke, M. C. (1874–1877). On black moulds. *J.Q.M.C.*, *4*, 246.
4. Cooke, M. C. (1881–1884). Circumnutation in fungi. *J.Q.M.C.*, Ser. 2, *1*, 309.
5. Broome, C. E. (1881). Letter dated November 3. MJB.
6. Cooke, M. C. (1897). Sixty years of British Mycology. *Essex Naturalist*, *10*, 216.
7. Britten. J. (1922). William Carruthers. *Journal of Botany*, *60*, 249.
8. This exchange is taken from the Curruthers Correspondence, held in the Botany Library, British Museum (Natural History).
9. Cooke, M. C. (1876). Letter 2, WGF.
10. Cooke, M. C. (1878–1879). Observations on Peziza. *Grevillea*, *8*, 129.
11. Anon. (1875). Mushrooms: their anatomy and uses. *Popular Science Review*, *14*, 186.
12. Smith, W. G. (1875). Fungi. A review, a gossip and a confession. *Gardeners' Chronicle*, New Series, *3*, 398.
13. Cooke, M. C. (1877–1878). 'Session Mycologique' of Paris. *Grevillea*, *6, 61*.
14. Stevenson, J. A., (1971). As above, p. 307.

Chapter 11.   Royal Botanic Gardens, Kew.

General sources: Aut., LAC, WGF.
Kew Report, 1880.
Archives of the Royal Botanic Gardens, Kew; especially the Correspondence of J. D. Hooker, 1860–1879, Herbarium Staff, Letters 282–374.

1. Anon. (1880). Pharmaceutical Meeting. *Pharmaceutical Journal*, *2*, 635.
2. Lloyd, C. G. (1923). Notes on the Herbarium at Kew. *Lloyd. Mycological Writings*, *7*, 1182.
3. Cooke, M. C. (1881). *Illustrations of British Fungi*, Vol. 1, Preface.
4. Smith, W. G. (1915). Letter to J. Britten. Botany Library, British Museum (Natural History), General Archives, Box 4.
5. Cooke, M. C. (1882–1883). The Perisporaceae of Saccardo's Sylloge Fungorum. *Grevillea*, *II*, 35.
6. Cooke, M. C. (1879–1880). On Sphaeria quercuum Schwz. *Grevillea*, *8*, 35.

Chapter 12.   Years of Disaster.

General sources: Archives and Library of the Society for the Propagation of Christian Knowledge, Holy Trinity Church, Marylebone Road, London.
Kew Archives; LAC;
Richardson, A. E. (1971), as Chap. 4, General sources.

1. Cooke, E. (1878). *Scholastic World*, November, p. 9.
2. Nuth, M. (1948). *The Story of Camden House*. London: Putnam.
3. Cooke, M. C. (1884). The President's Address. *J.Q.M.C.*, Ser. 2, *2*, 64.
4. Cooke, M. C. (1878–1892). Abstract of the President's Address. Tenth Annual Report, 1886–1887. *Reports of the Hackney Microscopical and Field Club*, 1–15, p. 6.
5. Cooke, M. C. (1886–1887). British Desmids. *Grevillea*, *15*, 30.

6. Lloyd, C. G. (1913–1916). Letter 57. *Index of Mycological Writings*, *4*.
7. G.M. (1893). Romance of Low Life amongst Plants. *Journal of Botany*, *30*, 379.
8. Ramsbottom, J. (1917). George Edward Massee. *Journal of Botany*, *55*, 225.
9. Cooke, M. C. (1893). Letter pasted on to the fly-leaf of Vol. 1 of Cooke's *Illustrations of British Fungi*, Library of Dept. of Botany, Oxford University. My attention was drawn to it by Dr. S. T. Buczacki.

Chapter 13. Forays Amongst the Funguses.

General sources: JQMC; Tol.
Mays, R. H. (1978). *Henry Doubleday, the Epping Naturalist*. London: Precision Press.
Reports of the Hackney Microscopical and Natural History Society, (1878–1892), 1–15.
Thompson, P. (1930). *A Short History of the Essex Field Club, (1880–1930)*, London: Essex Field Club.

1. Smith, W. G. (1880). Epping Forest Club–Fungus Foray. *Gardeners' Chronicle*, October, 459.
2. Cooke, M. C. (1878–1892). The Student of Nature. *Reports of the Hackney Microscopical and Natural History Society*, 1–15, 12.
3. Cooke, M. C. (1884). The President's Address. *J.Q.M.C.*, Ser. 2, *2*, 64.
4. Cooke, M. C. (1878). *Songs Written for the Excursionists' Annual Dinners. Q.M.C.*. London: Keating.
5. Cooke, M. C. (undated, a). 'Old Foragers'. Tol.
6. Cooke, M. C. (1879). The Woolhope Club Forays, *Gardeners' Chronicle*, October, 458.
7. Smith, W. G. (1874). The Woolhope Club Fungus Foray. *Gardeners' Chronicle*, October, 455.
8. Cooke, M. C. (1886–1889). Address. *Transactions of the Woolhope Naturalists' Field Club*, 62.
9. Anon. (1874–1876). Ibid., 241.
10. Cooke, M. C. (1884, approximately). 'Flamen Pomonalis, or Pomona Unbound'. Manuscript in the archives of the Royal Horticultural Society, Vincent Square, London, SW1.
11. Cooke, M. C. (undated, b). 'Old Foragers'. Tol.
12. Cooke, M. C. (1889–1890). The Rev. M. J. Berkeley, M.A., F.R.S. *Grevillea*, *18*, 17.
13. Cooke, M. C. (1890–1892). Fungi, past, present and future. *Transactions of the Woolhope Naturalists' Field Club*, 360.

Chapter 14. Missionary for Mycology.

General Sources: Cooke, M. C. 'Confessions of a Mycophagist'. Tol.
Jones, M. M. (1980); *The Lookers-Out of Worcestershire*. Worcestershire Naturalists' Club.

1. Rayner, J. F. (1904). *A List of Fungi of the New Forest*. Pamphlet.
2. Cooke, M. C. (1890–1891). Confessions of a mycophagist. *Grevillea*, *19*, 67.
3. Cooke, M. C. (1862). *British Fungi*. First Edition.
4. Anon. (1847–1896). Ockeridge and Shrawley Woods. *Transactions of the Worcestershire Naturalists' Club*, *I*, 388.
5. Ramsbottom, J. (1947). Carleton Rea. *Transactions of the British Mycological Society*, *30*, 180.

Chapter 15. Troubled Retirement.

General sources: LAC, Tol.

1. Cooke, M. C. (1893). See Chap. 12, ref. 9.
2. Cooke, M. C. (1895). Letter dated 19 November. Tol.
3. Ramsbottom, J. (1948). Presidential Address. *Transactions of the British Mycological Society*, 30, 1.
4. Blackwell, E. M. (1961). Links with past Yorkshire mycologists. *Naturalist*, 887, 53.
5. Cooke, M. C. (1897–1898a). The Essex Fungi and how they should be represented in the Epping Forest Museum. *Essex Naturalist*, 10, 1.
6. Cooke, M. C. (1897–1898b). Sixty Years of British mycology. *Essex Naturalist*, 10, 216.
7. Cooke, M. C. (1897–1898c). Annual Report. Memoirs contributed. *Essex Naturalist*, 10, 250.

Chapter 16. 'Nature Never did Betray the Heart that Loved Her'

General sources: LAC; Tol.

1. Anon. (1898–1900). Proceedings of the Ordinary Meeting, May 18th, 1900. *J.Q.M.C.*, Ser. 2, 11, 431.
2. Lister, G. (1918). *Essex Field Club Special Memoirs*, 6.
3. Rayner, J. F. (1904). *List of Fungi of the New Forest*. Pamphlet.
4. Desmond, G. G. (1904). A fungus foray. *Daily News*, October 20.
5. Gussow, H. T. (1916). Mordecai Cubitt Cooke (1825–1914). *Phytopathology*, 6, 1.
6. Anon. (1903a). Scientific Committee, November 18, 1902. *Journal of the Royal Horticultural Society*, 27, 201.
7. Anon. (1903b). *Proceedings of the Linnean Society*, October, 1903, 115th Session, 17, 25.
8. Cooke, M. C. (1872). Alternation of generations in fungi. *Nature*, 5, 108.
9. Cooke, M. C. Letter in the archives of the Royal Botanic Gardens, Kew.

Index

Agarics, 43, 63, 159, 188, 215–8, 230, 261, 292–3, 303
Agaricus (*Armillaria*) *melleus* (see *Armillaria mellea*)
Agaricus arvensis, 82, 275
Algae, 227–8. 241–2
Allen, W. H., publishers, 64, 106, 161, 285
Amanita muscaria, 50
Amanita phalloides, 286, 288
Amanita virosa, 43–4
Armillaria mellea, 63, 87
Ascomycetes, 157, 185, 188, 192, 201, 217, 221
Aspergillus, 167–8
Astley Cooper, Violet, 278

Bacteria, 141–2, 148, 256
Baptists, 9, 11, 16, 20
Beck, Richard, 115, 117
Berkeley, Revd. Miles Joseph, 43–4, 125, 136, 150–3, 155, 159, 171, 175, 184, 186, 193–7, 200, 214–5, 221, 226, 263–7, 302, 307, 312
Berkeley, M. J. *Gardeners Chronicle*, 43–4
Berkeley, M. J. *Gardeners Chronicle*, 43–4
   Vegetable Pathology, 43, 152
Biggs, Annie (see Cooke (Cubitt), Annie Elizabeth Thornton)
Biggs, Sophia (see Cooke, Sophia Elizabeth)
Birds, 14, 19, 34
Bogue, David, 140

Bolles, Revd. Edwin, C., 125, 172–4
Bombazine, 10
Botanic Gardens, Edinburgh, 178–9
Boudier, J. L. E., 190
Braithwaite, Robert, 120, 178, 200, 226
Brefeld, Oscar, 157
British Schools, 23
Britten, James, 97, 118, 316
Broome, Christopher Edmund, 152–3, 155, 164, 167, 171, 175, 178, 188, 215, 267
Buckland, Frank, 132–3, 147
Bull, Dr. Henry Graves, 136–8, 215, 221, 261–6, 289
Bull, Dr. Henry, *Herefordshire Pomona*, 263–6
Burdett Coutts, Angela, 56, 69, 274
Bywater, Witham M., 97, 113, 116, 118, 131

Capron, Dr. Edward, 164, 175
Carlyle, Dr. David, 217–8, 315
Carruthers, William, 183–6, 190, 243, 279
Cholera, 26, 58
Christian Socialists, 40
Clarke, Alfred, 288–292, 310, 314–5, 321
Cooke (Cubitt), Annie Elizabeth Thornton, née Biggs, 28–9, 66, 70, 78, 80, 83, 89, 179, 199, 220–1, 244–5, 280–3, 296, 314, 325–6, 329–31
Cooke, Ada Gordon, 75, 78, 233, 280, 291, 297, 314, 317, 321–2, 325–6

INDEX 351

Cooke, Bertie (see Cooke, Herbert)
Cooke, Clement, 10
Cooke, Ebenezer, 13–4, 39–42, 48, 52, 63, 69, 78, 105–7, 133, 236–8, 296, 321
Cooke, Ellen (née Lane), 52, 63, 69, 72, 78, 237
Cooke, Ernest Frederick, 2, 80, 233–5
Cooke, Frank Harry, 89, 199, 282, 302, 317, 325–6, 329–31
Cooke, Harry Linnaeus, 70, 87–89, 182, 296, 322
Cooke, Herbert Stephen, 89, 244, 282–3, 317, 325–6, 330–1
Cooke, Josiah Henry, 13, 80, 327
Cooke, Leila Annie, 2, 14, 81, 85, 89, 91–2, 102, 104, 221, 233–5, 239, 241, 259, 274, 280–1, 283–4, 291, 296–7, 302, 314–5, 320–2, 327–30
Cooke, Mary (née Cubitt), 10, 12–4, 82, 236
Cooke, Mordecai, 11, 80
Cooke, Mordecai Cubitt
  events: (see also Appendix A)
    Norwich apprenticeship, 18
    Marriage, 20
    Introduction to economic botany, 45
    Introduction to Hardwicke, 58 et seq.
    Appointment to India Museum, 73
    Purchase of No. 146, Junction Road, 80
    Honorary degrees, 174, 182
    Associate of Linnean Society, 198
    Appointment to Kew, 205, 207 et seq.
    Paralytic illness, 220, 253
    Eyesight, 285, 309–10
    Royal Horticultural Society Victoria Medal of Honour, 301
    Linnean Society Gold Medal, 301–2
  personal:
    appearance, habits, 86, 91, 100, 127, 252, 299, 300, 316
    as artist, 14, 34–5, 63, 101, 216–7, 287
    ″ classicist, 17, 107–8, 263
    ″ conservationist, 100, 146, 251
    ″ entomologist, 123–4
    ″ musician, 18, 93
    ″ teacher, 28 et seq., 33 et seq., 46, 313
    ″ versifyer, 21 et seq., 127 et seq., 263 et seq.
    childhood, 14 et seq.
    conservatism, 177, 238, 257, 306, 314
    family relationships, 71, 83, 85, 204, 244, 317
    finances, 6, 105, 163, 186 et seq., 284, 296
    isolation, 181, 315, 321
    library, 86, 281, 297, 320
    on museums, 293
    origin of name, 13
    religious opinions, 28, 33–6, 154
    social relations, 114, 138, 313
  mycology:
    as experimentalist, 142, 177–8, 309
    ″ microscopist, 64, 114, 164–70
    Exsiccati, 135, 161–3, 192, 202
    first new species described, 67
    herbarium, 159, 214
    importance of field work, 94
    introduction to, 43
    on amateur naturalists, 104, 253, 256, 270
    ″ American fungi, 171, 179, 190, 223, 280
    ″ Australian fungi, 230, 279
    ″ classification, 158, 190, 201, 221, 294
    ″ common names, 273
    ″ genera and species, 154 et seq., 311–2
    ″ herbarium methods, 158–9

Cooke, Mordecai Cubitt
mycology: *contd*
  on Latin diagnoses, 179
    " lichens, 143–4
    " metric measurements, 169
  working practice, 146, 181, 216, 229, 303
  manuscripts:
    Autobiographical Notes, 3, 149–50, 288, 318–9
    'British Fungi arranged on the basis of Fries', 217
    'Catalogue of Economic Products', 77
    'Chats about useful Plants'. 295
    'Chats at Breakfast with my Children', 295–6
    'Confessions of a Mycophagist'. 270
    'Flamen Pomonalis', 263 et seq., 310
    'Illustrations of Oeconomic Botany', 49
    'Indian Plants', 77
    'Monographia Acarinae', 123
  publications:
    British Desmids, 241–3
    British Edible Fungi, 274–6
    British Fresh-Water Algae, 227–8, 230, 241
    Catalogue and Field-Book of British Basidiomycetes, 310–2
    Clavis Synoptica Hymenomycetum Europaeorum, 200
    Easy Guide to the Study of British Hepaticae, 112, 285
    Edible and Poisonous Mushrooms, 286
    Freaks and Marvels of Plant Life, 219, 226
    'Fungi Australiani', 231, 245
    Fungi, their Nature, Influence and Uses, 193, 195, 243, 287
    'Fungoid Diseases of Cultivated Plants', 303–4
    Fungoid Pests of Cultivated Plants, 304–5, 309–10
    Genera and Species of Fungi, 313
    Handbook of Australian Fungi, 246
    Handbook of British Fungi, 134, 155, 175–8, 181, 273, 313
    Handbook of British Hepaticae, 285
    Illustrations of British Fungi, 43, 214–8, 230, 240, 245, 248, 267, 284–5, 292, 313, 315
    Index Fungorum Britanicorum, 158
    Introduction to Fresh-Water Algae, 242–3
    Introduction to the Study of Fungi, 287–8, 302
    Manual of Botanic Terms, 62, 257
    Manual of Structural Botany, 61–2, 159
    'Microscopic Fungi Parasitic on Living Plants', 159
    Microscopic Fungi, (see *Rust, Smut, Mildew & Mould*)
    'Monographia Polyporum', 245
    Mycographia, 188–90
    Myxomycetes of Great Britain, 199, 315
    Object Lesson Handbooks, 295
    'Observations on the Salmon Disease', 147
    On the Gums, Resins, Oleo-Resins & Resinous Products in the India Museum or Produced in India, 76
    On the Oil Seeds and Oils in the India Museum, 76
    One Thousand Objects for the Microscope, 129, 164, 168
    Our Reptiles, 100, 105–6, 132
    Plain and Easy Account of British Fungi, 62–4, 159, 328
    Ponds and Ditches, 145–7, 219
    Rambles Among the Wild Flowers, 286–7, 295–6
      A Stroll on the Marsh, 286
      Across the Common, 286
      Around a Cornfield, 287

Down the Lane and Back, 286
Through the Copse, 239, 286
Romance of Low Life, 247
Rust, Smut, Mildew & Mould, 98, 160–1, 163, 300
Seven Sisters of Sleep, 50–1
'Synopsis Pyrenomycetum', 245
Toilers in the Sea, 235, 246
Vegetable Wasps and Plant Worms, 246
Wall Diagrams, 294
Woodlands, 145–6, 219, 251

Cooke, Sophia Elizabeth (née Biggs), 20–1, 28, 31, 66, 71–2, 88, 90–1, 220, 280, 296, 322, 324, 329, 331
Cooke, William Cubitt, 3, 13, 18, 52, 164, 317–8, 327, 329–30
Cooke, W. C., *The Cookes of Horning*, 3, 318, 329–30
Cooke, Willie Cubitt, 78, 244, 280, 297, 317, 326
Cooke, Zenas, 13, 163
Cookella microscopica, 201
Cookia, 201
Coprinus, 323
Coprinus atramentarius, 99
Coprinus comatus, 215, 261
Cortinarius, 277–8
Crossland, Charles, 289–90
Cubitt, Annie (see Cooke (Cubitt) Annie)
Cubitt, Benjamin, 7
Cubitt, David, 5
Cubitt, George Eaton Stannard, 5, 206
Cubitt, James, 7, 10, 83
Cubitt, James (Uncle James), 16–7, 107
Cubitt, John Quincey, 83, 89, 245, 280
Cubitt, Mabel, 85, 89, 178, 199, 233, 244, 282–3, 325, 330–1
Cubitt, Mary (see Cooke, Mary)
Cubitt, Naomi, 10, 24
Cubitt, Thomas, 53
Cubitt, William II, 7, 82

Cubitt, William III, 7, 9
Cubitt, William IV, 10, 20
Cubitt, William Quincey, 7, 83
Curtis, M. A., 171

Darwin, Charles, 42, 59, 151, 154, 184, 206, 218–9
Origin of Species, 35
de Bary, Anton, 157, 198–9, 223, 307
Department of Science & Art, South Kensington, 35–8, 57, 159
Teachers' Examinations, 36–8, 61, 66
Diatrype sordida, 155
Discomycetes, 175, 185–8
Dual-Lichen Hypothesis, 144, 243, 256, 309
Dyer, William Thiselton, 97, 118, 198, 206–9, 214, 241, 243, 249

Economic Botany, 44, 77, 287, 295
Economic products, 44–46, 53, 79
Ellis, Job, B., 191–2, 223–5
English, James Lake, 251–2
Entomogenous fungi, 246–7
Epping Forest, 110, 251–2, 300
Euplectella aspergillum, 235
Exsiccati, preparation of, 161, 202–3
Farlow, W. G., 223–7, 279, 284, 309
Fern craze, 122
Ferns, 62, 122–3
Field clubs, 96
Fife House, Whitehall Yard, 68, 73, 74
Finchley Cemetery, 89, 296, 322
Foray (see Fungus foray)
Fox, Dr. W. Tilbury, 110, 120–1, 124–5, 257
Fries, Elias, 157, 175, 222
Fungicides, 304–6
Fungus forays, 136–9, 250, 253, 258–61, 269–72, 277–8, 289, 291–2, 299

Germ theory of disease, 151, 256
Gillman, Revd. James, 25–6, 28–30, 33, 40, 51

Ginger-beer plant, 142
Graham, Georgina Decima, 315, 321, 327
Green Dragon Hotel, Hereford, 136, 138, 261–3, 266
Grugeon, Alfred, 65–6, 95, 98, 103, 118, 149, 163

Hampstead Heath, 120, 163, 237
Hardwicke, Robert, 57–62, 96–7, 100, 107–110, 114–117, 120, 133–4, 139–40, 156, 159–61, 314
Harkness, H. W., 192
Herbaria, preparation of, 158
Herbaria:
  Berkeley, M. J., 152, 214, 240
  Botanic Gardens, Edinburgh, 178, 183
  British Museum, Botany Department, 183–8
  Chingford Museum, 293
  Cooke, M. C., 214, 239–40, 320
  Farlow, W. G., 223
  Royal Botanic Gardens, Kew, 212, 226, 241, 280
High Road Church, Ilford, 16, 17
Highgate Cemetery, 283, 325, 330
Highley, Samuel, 114, 117
Hirneola auricula-judae, 67, 77
Holly Lodge, Hampstead, 55
Hooker, Sir Joseph Dalton, 43–4, 75, 143, 184, 196, 205–11, 219–20, 231–2, 241, 280
Hooker, Sir William, 59, 150, 206
  English Flora, 43
Horning Post Office, 11–4, 39
Huxley, Prof. Thomas Henry, 32, 34–6, 38, 57, 59, 74, 140–2, 144–5, 219

India Office, 73–4, 205, 207, 209–11, 220, 232, 243, 249
Insects, 123–4
International Exhibition, Dublin, 1865, 79
International Exhibition, London, 1862, 65, 67–8
International Exhibition, Paris, 1867, 79
International Exhibition, Vienna, 1873, 87, 135, 182, 185

Jew's ear fungus (see *Hirneola auricula-judae*)
Journals:
  Botanists' Chronicle, 103–4
  Gardener's Chronicle, 152, 194, 197, 202, 252
  Grevillea, 134, 177–81, 189–90, 192–3, 197, 200, 221, 224–226 230–1, 243, 245, 248, 270, 289, 297, 308, 313
  Hardwicke's Science Gossip, 105–10, 113–4, 121–3, 125, 134, 145, 178, 251
  Journal of Botany, 60, 102, 104, 110, 155, 159, 192, 241
  Journal of Royal Horticultural Society, 303
  Journal of the Quekett Microscopical Club, 121, 130, 133–5, 139, 257
  Journal of the Society of Arts, 49–50
  Lancet, 110
  Land and Water, 132, 251
  Naturalist, 311, 313
  Nature, 112, 307
  Pharmaceutical Journal, 76, 124, 208
  Popular Science Review, 60, 97, 159, 186, 193
  Quarterly Journal of Microscopial Science, 58
  Report of the Hackney Microscopical & Field Society, 308
  School and the Teacher, 33–36, 48, 55, 103
  Science Gossip (see *Hardwicke's Science Gossip*)
  Technologist, 53
  Transactions of the British Mycological Society, 3, 289, 310, 312

INDEX

Transactions of the Woolhope Club, 135, 303
Junction Road, Upper Holloway, 78–82, 88, 182, 199, 237, 282, 296–7, 315, 325–6

Kalchbrenner, Revd. C., 230–1
Kentish Town, 52, 56, 69–70, 101, 133, 163, 236, 283, 297
Ketchup, mushroom, 275
Ketteringham, Thomas, 97, 103, 113, 116, 118
Kew, Kew Gardens (see Royal Botanic Gardens, Kew)
Knaggs, Henry Guard, 220

Lambeth, Holy Trinity Church, 25–6, 28, 51
Lambeth, Holy Trinity School, 25–30, 37, 47, 57, 204, 238
Lane, John, 40
Lankester, Dr. Edwin, 38, 55, 57–8, 60, 74, 88, 116, 119–20, 124, 140, 159, 314
Lankester, Phebe, 58, 62, 122, 159
Lankester, Sir Ray, 97, 112, 314
Lankesteria, 58
Large, E. C., 101, 148, 156
Le Cornu, M., 200, 252, 258–9
Le Neve Foster, Peter, 117, 120, 131
Leatherhead, 125–6, 255
Lepiota procera, 273–5
Lichens, 143–4, 288
Lindley, John, 58
Lithography, 40
Liverworts, 112, 122, 285
Lycoperdon giganteum, 15, 274–5
Lycoperdon pusillus, 72

Marasmius oreades, 133
Marx, Karl, 49, 58
Massee, George, 248–9, 272, 289–90, 294, 299, 301, 311
Maurice, F. D., 40, 42

Microscopy, 79, 109, 113, 129, 164, 169, 174
Moginie, William, 165, 167
Monitorial system, 23–4
Morchella, 187
Murray, George, 243, 247, 279
Museums, educational, 46 et seq., 293
  British Museum (Nat. Hist.) see Natural History Museum
  British Museum, Botany Dept., 163, 172, 182–4, 188, 196, 226, 240
  Chingford Museum, 293, 300
  India Museum, 68, 73–7, 87, 94, 119, 149, 164, 186, 204, 208, 210, 313
  Natural History Museum, 36, 67, 152–3, 183, 217, 243, 248, 255, 279, 314
  Passmore Edwards Museum, 300
  Scholastic Museum, 46–7, 49, 52, 67
  Twickenham Economic Museum, 54–6, 67
Myxomycetes, 199

National Schools, 23, 30
Natural History, definition of, 32
Neatishead, 7, 18, 72, 80, 163, 220, 235
Neatishead Baptist Chapel, 9, 28, 80, 236
Neatishead Post Office, 39
Negretti, Henry, 80
Norfolk Villa, 20, 81
Norfolk, 7, 9, 15, 19, 34, 42–3, 71, 80, 82, 135, 163, 236, 238, 318

Opium, 19, 51, 209

Pasteur, Louis, 141, 194
Peck, Charles H. 190, 223, 225–7
Penicillium, 141, 167
Pestalozzi, J. H., 24, 39, 133, 237, 287
  How Gertrude Teaches Her Children, 297
Peziza, 188–189
Phytophthora infestans, 151, 196–7
Piccadilly, 58–9
Pimlico, 53–4, 56, 70

Plant pathogenic fungi, 43, 149, 151–2, 156, 159–61, 196–8, 203–9
Plowright, Dr. C. B., 135, 252, 289, 307
Potato blight, 151, 196–7, 305
Potato murrain (see potato blight)
Primrose League, 238–40, 257
Priority of scientific names, 226
Prudential Insurance Company, 26
Puccinia graminis, 307–9
Puff-ball, giant (see *Lycoperdon giganteum*)
Pure culture techniques, 141–2, 194

Quelet, Lucien, 200, 258

Ramsbottom, Dr. J., 255, 276, 291, 320
Ravenel, Henry W., 171-3, 175, 192, 202, 223
Rea, Carleton, 276–8, 289–90, 310–12
Reptiles, 14, 105
Rose, Amy, 277–8, 291
Rostafinski, Joseph, 199, 223
Royal Botanic Gardens, Kew, 44, 47, 75, 90, 108, 184, 196, 205–13, 220, 227–9, 231, 233, 243, 249–50, 302, 313, 315, 320, 323, 327
Royal College of Surgeons, 59–60
Ruffle, W. M., 61, 77, 97, 109, 118, 125
Ruskin, John 42

S.P.C.K. (see Society for the Promotion of Christian Knowledge)
Saccardo, P. A., 190–1, 200–1, 221–5, 279, 285, 292
Sylloge fungorum, 190, 200, 221, 225, 245–6, 292, 303
Salmon disease, 147–8
Salmon, E. S., 298–9
Saprolegnia ferax, 147–8
Savage, Sophia, 21
School of Mines, 36, 57–8, 74
Science teaching in schools, 31 et seq.
Sexual reproduction of fungi, 157
Simmonds, Peter Lund, 45, 47–9, 52–3, 66, 159

Smith, Worthington George, 86, 97–101, 105–8, 112, 118, 134, 136, 142, 175, 188, 194, 215–6, 241, 252–3, 268, 316
Societies, Associations and Clubs:
Academy of Natural Sciences, Philadelphia, 192
British Mycological Society, 289–291, 311
Carlisle Field Naturalists' Club, 315
Cryptogamic Society of Scotland, 193, 197
Edinburgh Botanical Society, 178
Essex Field Club, 142, 250–2, 272, 293–4, 299–300, 303
Hackney Microscopical and Natural History Society, 104, 148, 253, 255, 308
Hampshire Field Club, 269, 299
Hertfordshire Natural History Society, 269
Linnean Society, 58, 181, 196, 198, 206, 226, 276, 301–2
Metropolitan Church Schoolmasters' Association, 33, 36, 48, 52, 295
Microscopical Society, 58, 114–6, 196
National Society, 23–4, 30
New York Academy of Sciences, 174
Norfolk and Norwich Naturalists, 135
Pharmaceutical Society, 208–10
Portland Society for Natural History, Maine, 125, 173
Quekett Microscopical Club, 115, 117, 120–32, 139–40, 143–4, 148, 167, 170, 196, 227, 250, 255–8, 298, 302, 327
Royal Horticultural Society, 68, 136, 164–7, 184, 195–8, 210, 252, 280, 300–3, 310, 313, 320
Royal Society, 58
Società Crittogamologica Italiana, 200, 258

INDEX   357

Société Botanique de France, 200, 258
Society for the Promotion of Christian Knowledge, 145–7, 219, 246–7, 286
Society of Amateur Botanists, 95–100, 102–3, 106, 111, 113, 115, 117–9, 154, 254, 311, 314
Society of Arts, 49, 55–7, 117, 120
Woolhope Naturalists' Field Club, 135–8, 193, 214, 238, 250, 252, 258–68, 276–7, 289
Worcestershire Naturalists' Club, 276–8
Working Tailor's Association, 40–1, 238
Yorkshire Naturalists' Union, 289–92
Southsea, 297
Sowerby, John Edward, 159–60
Sphaerella inaequalis, 156
Sphaeriales, 224
Splitters and Lumpers, 110, 153, 298
Spontaneous generation, 130, 194
St. Alphege's Church, Greenwich, 20, 28
St. Lawrence University, Canton, New York, 173, 182
St. Mary's Church, Lambeth, 26
Stock, Daniel, 43, 97
Stone, George Henry, 245, 282–3, 330–1
Stone, Herbert George, 282, 325, 330

Thornton, David, 20
Toadstools (see Agarics)
Touchwood, 15
Tricholoma gambosum, 275
Truffles, 267
Twining, Thomas, 54–6
Type specimens, 158

Uncle Matt, 286, 295
University College, London, 57–8, 117, 122, 298

Valsa, 155–6
Venus flower basket (see *Euplectella aspergillum*)
Vinegar plant, 141–2
von Mueller, Baron Ferdinand, 220, 230–1, 245

Ward, Richard, 43, 62
Wardian cases, 122
Watson, Dr. J. Forbes, 68–9, 73–4
Wharton, Dr. Henry, 259, 281
Wheldon's, biological booksellers, 121, 281–2, 297
Wherries, 10
Williams and Norgate, publishers, 134, 227
Withering, W. *Arrangement of British Plants*, 43
   British Botany, 17
Working Men's College, 41, 56, 64–66, 94–5, 133

Yale University, 48, 182
Yeasts, 142, 287

Zymotic theory, 256